Cannabinoids for Chronic Pain

From Research to Clinical Applications

Cannabinoids for Chronic Pain

From Research to Clinical Applications

Edited by

Alaa Abd-Elsayed

Departments of Anesthesiology and Chronic Pain Physician, University of Wisconsin-Madison, Madison, WI, United States

ELSEVIER

ACADEMIC PRESS

An imprint of Elsevier

Academic Press is an imprint of Elsevier
125 London Wall, London EC2Y 5AS, United Kingdom
50 Hampshire Street, 5th Floor, Cambridge, MA 02139, United States

ISBN: 978-0-443-36773-1

For information on all Academic Press publications visit our website at https://www.elsevier.com/books-and-journals

Publisher: Megan Ball
Acquisitions Editor: Sangeeta Yadav
Editorial Project Manager: Nayanika Saikia
Production Project Manager: Fahmida Sultana
Cover Designer: Mark Rogers

Typeset by Integra Software Services

To my two beautiful children, Maro and George

Contents

CHAPTER 6 Other phytocannabinoids

Roshni Patel

CHAPTER 7 Dronabinol

Ahmed Abdalla, Preethash Sundar, and Jamal Hasoon

Contributors

Alaa Abd-Elsayed
Departments of Anesthesiology and Chronic Pain Physician, University of Wisconsin-Madison, Madison, WI, United States

Ahmed Abdalla
Department of Anesthesiology, Critical Care, and Pain Medicine, The University of Texas Health Science Center at Houston, Houston, TX, United States

Alexander Bautista
Department of Anesthesiology and Perioperative Medicine, University of Louisville, Louisville, KY, United States

Kenneth D. Candido
University of Illinois College of Medicine; Pain Management Services, Thorek Memorial Hospital, Chicago, IL, United States

Matthew Chung
Department of Pain Medicine, Division of Anesthesia, Critical Care and Pain Medicine, The University of Texas, MD Anderson Cancer Center, Houston, TX, United States

Royce Copland
Department of Pain Medicine, Division of Anesthesia, Critical Care and Pain Medicine, The University of Texas, MD Anderson Cancer Center, Houston, TX, United States

Trent Emerick
Division of Chronic Pain, Department of Anesthesiology and Perioperative Medicine, University of Pittsburgh Medical Center, Pittsburgh, PA, United States

Orlando Florete Jr.
Orles Pain Management and Regenerative Medicine Group, Jacksonville, FL, United States

Robert Wayde Gilmore
Department of Anesthesiology and Perioperative Medicine, University of Louisville, Louisville, KY, United States

Jennifer M. Hah
Department of Anesthesiology, Perioperative and Pain Medicine, Division of Pain Medicine, Stanford University, Stanford, CA, United States

Jamal Hasoon
Department of Anesthesiology, Critical Care, and Pain Medicine, The University of Texas Health Science Center at Houston, Houston, TX, United States

Hemant Kalia
C.R.I.S.P. Center for Research & Innovation in Spine & Pain, Rochester, NY, United States

Anusha Kallurkar
Department of Anesthesiology, LSU School of Medicine, Shreveport, LA, United States

Alan D. Kaye
Departments of Anesthesiology and Pharmacology, Toxicology, and Neurosciences, Louisiana State University Health Sciences Center Shreveport; Department of Anesthesiology, LSU School of Medicine, Shreveport, LA, United States

Danielle Kohr
Department of Pain Medicine, Fachklinik Löwenstein, SLK Kliniken Heilbronn GmbH, Löwenstein, Germany

Michael Kugler
Department of Pain Medicine, Fachklinik Löwenstein, SLK Kliniken Heilbronn GmbH, Löwenstein, Germany

Jon B. Lee
Department of Anesthesiology, Perioperative and Pain Medicine, Division of Pain Medicine, Stanford University, Stanford, CA, United States

Robert Moghim
Interventional Pain & Spine, Colorado Pain Care, Denver, CO, United States

Roshni Patel
RNP Regenerative Medicine, Center of Excellence in Pain and Regenerative Medicine, Quantum Brain Health, Farmington, CT, United States

David Rivetti
Department of Physical Medicine and Rehabilitation, University of Pittsburgh Medical Center, Pittsburgh, PA, United States

Sajan H. Shah
Department of Anesthesiology, Perioperative and Pain Medicine, Division of Pain Medicine, Stanford University, Stanford, CA, United States

Kimberly L. Skidmore
Department of Anesthesiology, LSU School of Medicine, Shreveport, LA, United States

Preethash Sundar
Department of Anesthesiology, Critical Care, and Pain Medicine, The University of Texas Health Science Center at Houston, Houston, TX, United States

Preface

The use of cannabinoids for treating chronic pain remains a topic of debate, not only from the scientific standpoint but also from the legal and regulatory standpoints.

This book was authored by experts who provided the most up-to-date knowledge about this topic to guide readers on using cannabinoids for treating chronic pain.

Readers need to comply with the regulations in their organizations and states, given the broad range of regulations among different states and countries.

The knowledge related to this topic is evolving and in its infancy; I hope more research and clinical trials will be conducted to provide strong evidence about the safety and efficacy of using cannabinoids in treating chronic pain conditions.

I would like to thank all authors for the countless hours they spent searching literature and carefully writing the content of this critical and controversial topic to provide readers with the best knowledge available to date.

I encourage readers to read this book and associated references to gain in-depth knowledge about the use of cannabinoids in treating chronic pain to provide the safest practice with appropriate patient selection with understanding the adverse events that can be associated with different cannabinoids.

Alaa Abd-Elsayed, MD, MBA, MPH, CPE, FASA, FASIPP
University of Wisconsin

CHAPTER

The history of medical marijuana

1

Alexander Bautista[a], Alan D. Kaye[b], and Alaa Abd-Elsayed[c]

[a]*Department of Anesthesiology and Perioperative Medicine, University of Louisville, Louisville, KY, United States*

[b]*Departments of Anesthesiology and Pharmacology, Toxicology, and Neurosciences, Louisiana State University Health Sciences Center Shreveport, Shreveport, LA, United States*

[c]*Departments of Anesthesiology and Chronic Pain Physician, University of Wisconsin-Madison, Madison, WI, United States*

Origins and ancient history

Cannabis is a genus of annual flowering plants in the Cannabaceae family, encompassing several taxonomically debated species, primarily Cannabis sativa and Cannabis indica. Cannabis sativa is characterized by its tall growth, narrow leaves, and longer flowering cycles, often cultivated for its strong bast fibers used in textiles, ropes, and paper.

Varieties with low Δ9-tetrahydrocannabinol (THC), referred to as hemp, were historically used in ancient agriculture and trade. In contrast, Cannabis indica is shorter and bushier, with higher concentrations of THC, the primary psychoactive compound. Traditional Chinese medicine documents cannabis use as early as 2700 BCE under Emperor Shen Nung, who recommended it for conditions like rheumatism and gout.

Ancient Indian texts, such as the Atharva Veda, describe cannabis as one of the five sacred plants, used in religious ceremonies and to treat various ailments, including pain, insomnia, and gastrointestinal disorders. In ancient Egypt, cannabis pollen has been discovered on the mummified remains of Ramses II, and the Ebers Papyrus (c. 1550 BCE) refers to cannabis for antiinflammatory purposes. In this regard, Greek and Roman physicians like Dioscorides and Galen also noted its use in treating earaches, edema, and pain relief. These early applications demonstrate cannabis's wide historical reach as a therapeutic agent (Li, 1973; Russo, 2007; Zuardi, 2006; Abel, 1980).

Cannabinoids for Chronic Pain. https://doi.org/10.1016/B978-0-443-36773-1.00004-0

Medieval and islamic medicine

During the medieval period, cannabis continued to be integrated into medical practice across various cultures, particularly within the Islamic Golden Age. Islamic scholars, such as Al-Razi (Rhazes) and Ibn Sina (Avicenna), documented the use of cannabis in their pharmacopoeias. In *The Canon of Medicine*, Avicenna mentioned cannabis as a treatment for gout, headaches, and ear pain.

Medical use of cannabis was also common in Persia, where it was applied to wounds and used as an antiinflammatory and antiepileptic (Touw, 1981). Despite some religious reservations, the medical use of hashish, a concentrated form of cannabis, spread throughout the Islamic world. In medieval Europe, cannabis was used as an herbal remedy, with Hildegard von Bingen and other herbalists noting its efficacy in treating pain, nausea, and poor appetite (Booth, 2005). By the 12th century, cannabis had become a common component in herbal treatments throughout the Middle East and Europe, demonstrating cross-cultural recognition of its medicinal potential.

Early modern period and western pharmacopoeias

The early modern period saw increased European exploration and colonialism, bringing renewed exposure to cannabis. In the 18th century, European physicians began formally documenting the therapeutic potential of cannabis brought from India and the Middle East. Notably, Irish physician William O'Shaughnessy introduced cannabis to Western medicine after observing its analgesic and antispasmodic properties in India. His 1839 publication detailed the successful use of cannabis extracts for rheumatism, tetanus, and convulsions, sparking widespread interest across Europe (O'Shaughnessy, 1843).

By the late 19th century, cannabis extracts appeared in the British and American pharmacopoeias, prescribed for various ailments, including migraines, epilepsy, neuralgia, and menstrual pain. Pharmaceutical companies like Parke-Davis and Eli Lilly marketed cannabis tinctures and pills. However, variability in plant composition and the rise of synthetic drugs began to challenge its popularity (Mikuriya, 1973).

20th century: Criminalization and stigmatization

Despite its therapeutic use, cannabis faced increasing criminalization in the early 20th century. The 1937 Marihuana Tax Act effectively banned its medical and recreational use in the United States by imposing prohibitive taxes and criminal penalties. The campaign by Harry Anslinger, Commissioner of the Federal Bureau of Narcotics, painted cannabis as a dangerous narcotic linked to crime and racial minorities (Musto, 1972).

Reefer Madness-style propaganda contributed to public fear, and cannabis was removed from the United States Pharmacopoeia in 1942. Internationally, the

1961 United Nations Single Convention on Narcotic Drugs classified cannabis as a Schedule I drug, further restricting medical research. This period of prohibition limited both scientific inquiry and clinical application, despite anecdotal reports of benefits among certain populations for a variety of conditions.

Re-emergence and legalization movements

Medical cannabis began re-entering public discourse in the latter half of the 20th century. In the 1970s and 1980s, advocacy efforts highlighted its efficacy in managing chemotherapy-induced nausea, HIV-related cachexia, and glaucoma. Landmark legal efforts included the Compassionate Investigational New Drug program in 1978, which allowed limited federal access to cannabis for medical purposes (Randall and O'Leary, 1998).

California's 1996 Proposition 215 legalized medical marijuana, making it the first U.S. state to do so. At present, 40 states have legalized cannabis for medical use, with 24 states legalizing it for recreational purposes. Internationally, countries like Canada, Israel, Germany, Uruguay, Albania, Argentina, Australia, Barbados, Brazil, Chile, Colombia, Costa Rica, Croatia, Cyprus, Czechia, Denmark, Ecuador, Finland, Georgia, Greece, Ireland, Italy, Jamaica, Lebanon, Luxembourg, Malawi, Malta, Mexico, and the Netherlands have adopted legalized medical use of cannabis programs. Seven countries have legalized the recreational use of cannabis, including the United States of America, Canada, Mexico, Uruguay, Georgia, South Africa, and Australia, and it is believed that nearly 200 million people around the world use marijuana on a regular basis. The legalization movement has been propelled by patient advocacy, changing public perception, and emerging scientific evidence supporting its therapeutic utility (Galli et al., 2011).

Contemporary challenges and future directions

Despite growing public support and broader legalization, cannabis remains unevenly integrated into modern medical practice. Many physicians lack formal education in endocannabinoid pharmacology or cannabis-based dosing protocols, contributing to hesitation in recommending its use (Ziemianski et al., 2015). In a Canadian survey, most physicians reported limited training on cannabis, despite patient interest and growing clinical applications. As a result, patients often rely on dispensary staff or online forums for dosing advice, which raises concerns about safety and consistency. In contrast, structured medical cannabis programs in countries like Israel and Canada incorporate standardized formulations and specialist oversight. These models demonstrate the feasibility of integrating cannabis into evidence-based care while balancing regulatory oversight (Hurd et al., 2015).

Pharmaceutical innovation has also accelerated. Beyond Food and Drug Administration -approved medications like dronabinol (synthetic THC) and cannabidiol

(Epidiolex), new products are in development to target cannabinoid receptors using non-psychoactive formulations. This includes inhalers, sublingual sprays, and transdermal patches designed to offer targeted relief with fewer systemic effects (Pertwee, 2006). Cannabinoid research is also expanding into psychiatric disorders, inflammatory bowel disease, neurodegeneration, and pain modulation. However, clinical trials remain hindered by regulatory barriers and cannabis's Schedule I classification in many jurisdictions (National Academies of Sciences, 2017).

Policy debates increasingly center around health equity and justice. Communities disproportionately impacted by cannabis prohibition often face challenges in accessing the legal cannabis industry due to startup costs and prior convictions. Equity-focused licensing programs aim to redress these imbalances, but implementation varies by jurisdiction. Furthermore, insurance coverage for medical cannabis remains rare, forcing patients to pay out-of-pocket, which limits access for economically disadvantaged populations (Helling, 2016). Policymakers face the challenge of aligning laws with emerging scientific evidence while ensuring equitable access and safe use.

Conclusion

The history of medical marijuana reflects a dynamic interplay between scientific discovery, cultural perception, and regulatory policy. From its early roots in ancient medicine through periods of widespread acceptance and prohibition, cannabis has persisted as a complex therapeutic agent. Modern science has begun to validate many traditional uses while also uncovering new applications rooted in our understanding of the endocannabinoid system. Moving forward, addressing barriers in physician education, regulatory frameworks, and equitable access will be critical to fully integrating cannabis into patient-centered care. As evidence continues to accumulate and societal perspectives evolve, medical cannabis may reclaim its role as a legitimate and valuable component of modern therapeutics.

References

Abel, E.L., 1980. Marijuana: The First Twelve Thousand Years, Springer Science & Business Media.

Booth, M., 2005. Cannabis: A History, St. Martin's Press.

Galli, J.A., Sawaya, R.A., Friedenberg, F.K., 2011. Cannabinoid hyperemesis syndrome. Curr. Drug Abuse Rev. 18744745. 4 (4), 241–249. Available from www.eurekaselect.com/671/journal/current-drug-abuse-reviews. doi:10.2174/1874473711104040241.

Helling, M.K., 2016. Legalizing marijuana: a shift in public policy. J. Pediatr. Pharmacol. Ther. 21 (1), 4–6.

Hurd, Y.L., Yoon, M., Manini, A.F., Hernandez, S., Olmedo, R., Ostman, M., Jutras-Aswad, D., 2015. Early Phase in the Development of Cannabidiol as a Treatment for Addiction: opioid Relapse Takes Initial Center Stage. Neurotherapeutics. 18787479. 12 (4), 807–815.

Available from: http://www.springerlink.com/content/1933-7213/. doi:10.1007/s13311-015-0373-7.
Li, H.L., 1973. An archaeological and historical account of cannabis in China. Econ. Bot. 18749364. 28 (4), 437–448. doi:10.1007/BF02862859.
Mikuriya, T.H., 1973. Marijuana: Medical Papers 1839–1972. Medi-Comp Press, Oakland.
Musto, D.F., 1972. The Marihuana Tax Act of 1937. Arch. Gen. Psychiatry. 15383636. 26 (2), 101–108. doi:10.1001/archpsyc.1972.01750200005002.
National Academies of Sciences, 2017. Engineering, and Medicine, the Health Effects of Cannabis and Cannabinoids: The Current State of Evidence and Recommendations for Research. National Academies Press, Washington, DC.
O'Shaughnessy, W.B., 1843. On the Preparations of the Indian Hemp, or Gunjah: cannabis Indica Their Effects on the Animal System in Health, and their Utility in the Treatment of Tetanus and other Convulsive Diseases. BMJ. 0959-8138. s1-5 (123), 363–369. doi:10.1136/bmj.s1-5.123.363.
Pertwee, R.G., 2006. Cannabinoid pharmacology: the first 66 years. Br. J. Pharmacol. 14765381. 147 (1), S163. doi:10.1038/sj.bjp.0706406.
Randall, R.C., O'Leary, A., 1998. Marijuana Rx: The Patients' Fight for Medical Pot, Thunder's Mouth Press.
Russo, E.B., 2007. History of cannabis and its preparations in saga, science, and sobriquet. Chem. Biodivers. 16121880. 4 (8), 1614–1648. doi:10.1002/cbdv.200790144.
Touw, M., 1981. The Religious and Medicinal Uses of Cannabis in China, India and Tibet. J. Psychoact. Drugs. 0279-1072. 13 (1), 23–34. doi:10.1080/02791072.1981.10471447.
Ziemianski, D., Capler, R., Tekanoff, R., Lacasse, A., Luconi, F., Ware, M.A., 2015. Cannabis in medicine: a national educational needs assessment among Canadian physicians. BMC Med. Educ. 1472-6920. 15 (1). doi:10.1186/s12909-015-0335-0.
Zuardi, A.W., 2006. History of cannabis as a medicine: a review. Rev. Bras. Psiquiatr. 15164446. 28 (2), 153–157. Available from: http://www.scielo.br/pdf/rbp/v28n2/29785.pdf. doi:10.1590/S1516-44462006000200015Brazil.

CHAPTER

2 The endocannabinoid system and its role in the treatment of neuropathic and chronic pain

Anusha Kallurkar, Kimberly L. Skidmore, and Kaye Alan D.
Department of Anesthesiology, LSU School of Medicine, Shreveport, LA, United States

Introduction

The earliest descriptions of cannabis date back over 12,000 years in China, eventually expanding to different regions of Asia. More than 100 phytocannabinoids have been isolated from cannabis plants, where "phyto" refers to plant origin. Cannabinol was one of the first phytocannabinoids isolated in the 1930s and chemically synthesized in 1940. Cannabidiol (CBD) was extracted from cannabis in 1940. Tetrahydrocannabinol (THC) was first extracted from cannabis in 1942 as a mixture of (-)-delta-8-THC and (-)-delta-9-THC. THC and CBD are present in cannabis mainly as decarboxylated acids when cannabis is heated (Pertwee, 2006).

Endocannabinoid system

The discovery of THC and CBD inspired searches for synthetic cannabinoid analogs that mimic their effects. Subsequently, researchers identified specific receptors for cannabinoids and endogenously produced ligands. Specific cannabinoid receptors (CB receptors), CB1 and CB2, were discovered in the human brain and periphery (Fig. 2.1). Two endogenous cannabinoids (endocannabinoids [eCBs])) were isolated: arachidonoyl ethanolamide, or anandamide (AEA), and 2-arachidonoylglycerol (2-AG). AEA and 2-AG are the main eCBs that bind to cannabinoid receptors. These eCBs, along with CB1 and CB2 receptors, constitute the endocannabinoid system (ECS) (Sideris et al., 2024).

Cannabinoids for Chronic Pain. https://doi.org/10.1016/B978-0-443-36773-1.00014-3

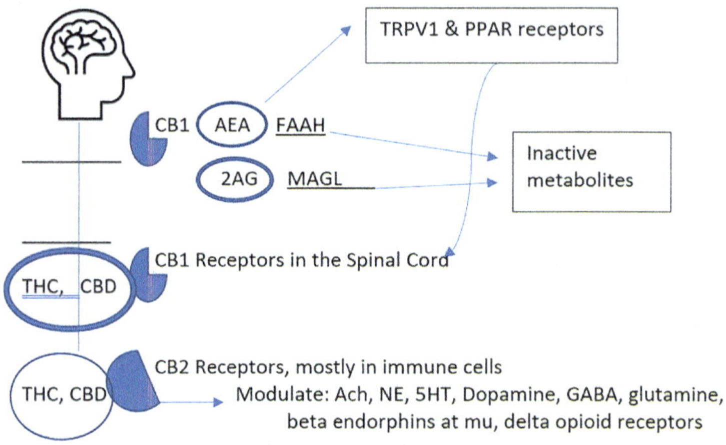

FIG. 2.1

CB1 and CB2 receptor sites and endocannabinoid binding in the body. *Image created in https://BioRender.com.*

Cannabinoid receptors

CB1 and CB2 receptors are eCB-binding sites and are primarily G-protein-coupled receptors. CB1 and CB2 receptors were cloned in the 1990s and differ in location and effects. While CB1 receptors reside in central and peripheral neurons, CB2 receptors are predominantly found in immune cells. In the brain, CB1 receptors are located in association areas, the amygdala, hippocampus, and basal ganglia, while in the peripheral nervous system, CB1 receptors are present on neurons of the dorsal root ganglia. CB2 receptors are induced in response to inflammation in microglial cells and peripheral cells such as macrophages, osteoblasts, spleen, and thymus. The location of CB1 receptors in the brain signifies their role in pain pathways, antiemetic effects, and memory formation (Howlett et al., 2002).

Endocannabinoids are synthesized in the postsynaptic neuron "on-demand" by cleavage of phospholipid precursors by enzymes, travel retrogradely, and bind to presynaptic CB1 receptors to inhibit neurotransmitter release through the inhibition of voltage-gated calcium channels or the activation of inwardly rectifying potassium channels (Fig. 2.2). These eCBs modulate the release of neurotransmitters, including acetylcholine, noradrenaline, dopamine, 5-hydroxytryptamine, gamma-aminobutyric acid, and glutamate (Pertwee, 2009; Lau and Schloss, 2008). eCBs and synthetic derivatives exhibit different affinities for CB1 and CB2 receptors. Of the eCBs, AEA has a higher affinity for CB1 but functions as a partial agonist. Conversely, 2-AG achieves lower affinity but is a full agonist at CB1 receptors. 2-AG is involved in use-dependent regulation of synaptic transmission, whereas AEA is tonically active (Sideris et al., 2024). AEA is metabolized by the enzyme fatty acid amide hydrolase (FAAH), an intracellularly bound serine hydrolase, while 2-AG is hydrolyzed by monoacylglycerol lipase (MAGL) (Fezza et al., 2014). Inhibitors of FAAH and MAGL have been developed to increase levels of AEA and 2-AG to treat pain.

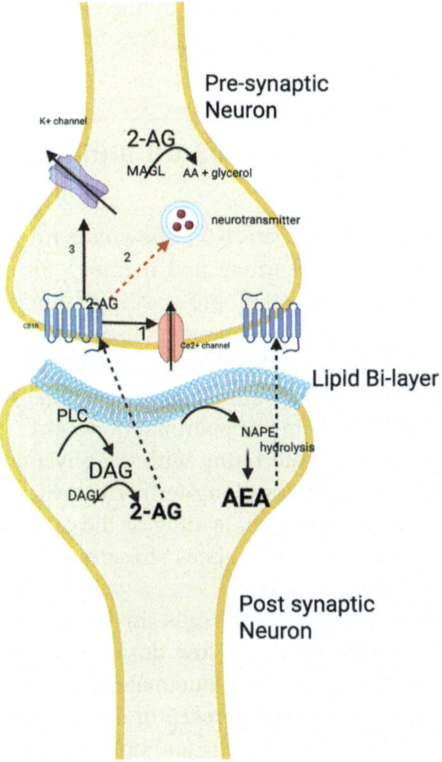

FIG. 2.2

Schematic representation of the synthesis and mechanism of action of endocannabinoids. *Image created in https://BioRender.com.*

The effects of phytocannabinoids provide insight into the clinical effects of related medications. Cannabinoids can function as positive or negative allosteric modulators at CB receptors. Similar to eCBs, THC can activate both CB1 and CB2 receptors. At CB1 receptors, THC has a similar affinity to AEA and behaves as a partial agonist but has less efficacy. The cannabinoid 2-AG has higher efficacy compared to THC at both CB1 and CB2 receptors. On the other hand, CBD exhibits lower efficacy at CB1 and CB2 receptors compared to THC (Malfitano et al., 2014).

The cannabinoid 2-AG is synthesized from diacyl glycerol (DAG) in the presence of diacyl glycerol lipase. DAG is derived from membrane phospholipids in the presence of phospholipase C. AEA is formed from the hydrolysis of N-arachidonoyl phosphatidyl ethanolamine) from the membrane phospholipids. Once synthesized, 2-AG and AEA travel retrogradely into presynaptic neurons and act on CB1 receptors, which are G-protein-coupled receptors. The activation of these receptors causes the following: (1) presynaptic inhibition of Ca^{2+} through voltage-gated calcium channels; (2) reduced neurotransmitter release; and (3) activation of presynaptic K^+ channels. Termination of the action of 2-AG is caused by its hydrolysis by the enzyme

MAGL, and the action of AEA is terminated by the enzyme FAAH (not included in the image). *Created in https://BioRender.com*

Clinical effects of cannabinoids on memory, anxiety, sleep, cognition, and motivation

Multiple studies of animals and humans have shown the role the ECS has in controlling bodily functions, including learning and memory formation, pain perception and modulation, temperature regulation, processing of emotions, sleep and appetite regulation, and inflammatory and immune responses. The ECS modulates synaptic plasticity in animals, a process that allows the brain to adapt to new information. The ECS modulates synaptic plasticity by affecting the strength of inter-neuronal connections and, ultimately, the functioning of neuronal networks. THC activates cannabinoid receptors in the brain, thus interfering with the physiological actions of eCBs. The ECS alters plasticity by inducing long-term changes in Gamma-aminobutyric acid release in areas like the amygdala, leading to the extinction of adverse memories. In the hippocampus, the ECS modulates stress response systems (Abate et al., 2021; Marsicano et al., 2002).

When escalating the dose of cannabinoids for the treatment of anxiety, the response is biphasic. Anxiolytic effects of low doses of cannabis occur when eCBs interact with the CB1 receptor on cortical glutamatergic terminals. In contrast, higher doses result in anxiogenesis via the CB1 receptor and GABA. The effect on anxiety is also mediated by cannabinoids at 5-HT_{1A} and opioid receptors. Cannabis preparations containing high amounts of THC are anxiogenic, especially in higher doses and in infrequent cannabis users, whereas preparations with a higher CBD:THC ratio are more anxiolytic and hence favored for the treatment of anxiety disorders.

The use of cannabinoids has increased with expectations of improving sleep, but research in humans remains inconclusive regarding their effects on maintaining sleep or wakefulness. Cannabinoids influence homeostatic and circadian sleep regulation via the suprachiasmatic nuclei at CB1 receptors through GABA effects (Acuna-Goycolea et al., 2010). THC aids in sleep consolidation, reduces sleep onset latency, increases total sleep time, and decreases wakefulness after sleep onset. However, chronic administration unfortunately decreases slow-wave sleep, suggesting the development of tolerance (Kaul et al., 2021). At low doses, CBD acts as a stimulant, but at high doses, it functions as a sedative, thereby showing promise in the treatment of insomnia (Babson, Sottile and Morabito, 2017).

Studies of adverse effects show that THC, the primary psychoactive constituent in cannabis plants, is responsible for cognitive disturbances during acute intoxication. In cannabis-naïve humans, acute use causes profound impairment of executive functions like attention, concentration, and decision-making, necessary, for example, to drive a vehicle. Though verbal memory remains intact, acute cannabis use impairs working memory, even in cannabis-tolerant humans (Crean et al., 2011). Heavy marijuana use is associated with decreased neurocognitive performance even after 28 days of abstinence (Bolla et al., 2002).

The ECS modulates reward pathways and motivation by influencing the mesocorticolimbic dopamine pathway, which arises from the midbrain ventral tegmental area. THC, through its partial agonism at CB1 receptors, increases limbic striatal dopamine release, which is believed to be responsible for its use in adolescents, elevating the risk for the development of schizophrenia (Arseneault et al., 2004). Although the effects of acute cannabis use on motivation are unclear, studies have shown that THC causes transient amotivation. This amotivation endures in those with cannabis dependence, manifested as impaired reward learning (Lawn et al., 2016).

Cannabinoids and their role in neuropathic pain

Research elucidates cannabinoid therapies for neuropathic pain in adults. A systematic review in 2014 found a prevalence of nearly 10% for neuropathic pain in the United States (Van Hecke et al., 2014). The International Association for the Study of Pain defines neuropathic pain as "pain arising as a direct consequence of a lesion or disease affecting the somatosensory system." Pain can cause spontaneous burning, tingling, or shooting sensations, amplified by noxious stimuli (pressure, heat, and cold), mechanical or thermal stimuli, and is associated with hyperalgesia or allodynia. Following injury or inflammation of a peripheral nerve, whether due to toxins, chemotherapeutics, hypoxic injury, direct compression, or trauma, the development of neuropathic pain is characterized by complex sensitization of central and peripheral neurons and immune cells.

At the level of peripheral nerve fibers, primary afferents are damaged, cytokines and growth factors are released, contributing to neuronal hyperexcitability and ectopic firing. Peripheral sensitization develops when it is related to repeated stimulation or an ongoing disease process. Chemical mediators like substance-P are released at the nerve terminal, which leads to vascular permeability, localized edema, and release of cytokines, compounding lower firing thresholds and ectopic discharges (Cohen and Mao, 2014).

At the level of the spinal cord, peripheral nerve injury induces hyperexcitability, thereby causing functional alterations in the neuronal network. For example, the loss of inhibitory neurons in the dorsal horn of the spinal cord and microglial activation cause the release of inflammatory mediators (Maldonado et al., 2016). The ECS modulates spinal cord immune responses during neuropathic pain (Racz et al., 2008). In the brain, there are changes at the level of the somatosensory cortex and thalamus in the form of a reduction in tonic noradrenergic inhibition (Wei et al., 2010) and plastic changes in areas of the cingulate cortex, amygdala, hippocampus, and nucleus accumbens, which may explain the emotional and cognitive alterations associated with neuropathic pain (Maldonado et al., 2016).

Lack of medications that effectively target neuropathic pain has led to exploring the role of the ECS. Animal studies have demonstrated key components of the ECS, such as the CB1 and CB2 receptors, and phytocannabinoids. Cannabinoids target pain by binding to multiple sites in the central nervous system and the periphery. In animal models, CB1 and CB2 receptors are upregulated in structures involved in pain processing following peripheral nerve damage, especially in the superficial laminae of the ipsilateral spinal cord dorsal horn (Lim et al., 2003).

In another animal study, peripheral nerve injury induced CB2 receptor expression on spinal microglial cells, reducing excitatory agent secretion (Zhang et al., 2003). Also, AEA acts as an endogenous ligand at transient receptor potential vanilloid type 1 (TRPV1) channel in the spinal cord and causes analgesia by channel activation and subsequent desensitization, attenuating cold allodynia and thermal hyperalgesia (Starowicz et al., 2012). Along with CB and TRPV1 receptors, eCBs also modulate pain through inhibition of 5-HT3A receptors and T-type voltage-gated calcium channels, and activation of peroxisome proliferator-activated receptor-γ (PPAR-γ) receptors (Schouten et al., 2024; Vučković et al., 2018).

Cannabinoids are also known to exhibit opioid-sparing effects. In animals, CBD is a positive allosteric modulator at mu and delta opioid receptors (Kathmann et al., 2006). Peripherally, CB2 receptor agonists lead to beta-endorphin release from keratinocytes, inhibiting nociception at opioid receptors on primary afferent neurons (Ibrahim et al., 2005). The eCB-degrading enzymes FAAH and MAGL are also potential targets for analgesia.

Adverse effects of cannabis use

Cannabis plants have been cultivated into hundreds of new strains, differentiated mainly by their relative content of THC or CBD, but also introducing multiple ill-defined chemicals, generally with energizing psychoactive or depressive effects, such as beta-myrcene (Surendran et al., 2021). Cannabis use disorder (CUD) may develop in up to a third of those using cannabis. After matching hundreds of people according to use—on average, four times weekly—there is up to a tenfold higher (upper limit of 95% confidence interval) risk of anxiety, depression, and psychosis, especially in established CUD (Lawn et al., 2022). Long-standing addiction during pregnancy was investigated at Kaiser Permanente. Nearly 300,000 patients were retrospectively screened over a decade, finding that 6.8% used cannabis and 0.2% had a CUD. Among pregnancies with cannabis use, bipolar disorder was 2.8 times more common, and stimulant use disorder was 21.9 times more common, especially if using daily rather than less than monthly (Young-Wolff et al., 2024). Cardiovascular and pulmonary adverse effects are ubiquitous. The addictive potential and concomitant permanent psychological harm translate into extreme caution when prescribing cannabinoids for pain.

Cannabis preparations available for consumption

Table 2.1 shows some of the cannabis preparations available for use in general. This table includes cannabinoid preparations used for both recreational and medicinal purposes (Shah et al., 2023).

Table 2.1 Human clinical trials and a summary of their results examining synthetic Cannabis preparations to treat neuropathic pain.

Research group/ Author	Types of pain	Patient demographics	Cannabis preparation studied	Outcome	Reference attached
Mathias Karst et al.	Treatment-resistant neuropathic pain with hyperalgesia and allodynia	8 women and 13 men, aged 29–65 years, in a crossover trial	Oral 1′,1′ Dimethylheptyl-Δ^8-tetrahydrocannabinol-11-oic acid(CT-3)	Mean VAS score at the 3-h period was lower in the CT-3 group compared to the placebo	(Karst et al., 2003)
Wilsey et al.	Central and peripheral neuropathic pain resistant to traditional treatment	39 patients in a randomized, double-blind, placebo-controlled, crossover design trial	Smoked cannabis—low dose (1.29% delta-9 THC) vs. medium dose (3.53% delta-9 THC) vs. placebo cannabis	Number needed to treat to achieve 30% pain reduction was 3.2 for placebo vs. low dose, 2.9 for placebo vs. medium dose, and 25 for medium vs. low dose	(Wilsey et al., 2013)
Turcotte et al.	Neuropathic pain secondary to relapsing-remitting multiple sclerosis	15 patients on oral gabapentin doses > 1800 mg/day with inadequate pain relief	Nabilone or placebo—titrated over 4 weeks (0.5 mg/week increase), followed by a 5-week maintenance of 1 mg oral nabilone twice daily	Decreased pain score in treatment group suggests nabilone could be an effective adjunct to gabapentin	(Turcotte et al., 2015)
Nurmikko et al.	Patients with unilateral peripheral neuropathic pain and allodynia	125 patients (63 in the Sativex group vs. 62 in the placebo group) participated in a 5-week, randomized, double-blind, placebo-controlled, parallel-design trial	Patients were allowed to self-titrate their dose of Sativex at home, with a maximum dose of 8 sprays per 3-h interval and a maximum of 48 sprays per 24 h	22% reduction in pain score at the end of treatment in the Sativex group, as compared to an 8% reduction in the placebo group	(Nurmikko et al., 2007)

Name/Preparation	Description/Examples	
Cannabis	Derivatives of the cannabis plant—from flowers, leaves, and seeds—contain the psychoactive component; common cannabis plant species include *Cannabis sativa*, *Cannabis indica*, and *Cannabis ruderalis*	 Image showing a single cannabis leaf. *Image created in https://BioRender.com.*
Marijuana	Terminology used interchangeably with cannabis in reference to plant strains containing high THC	
Endocannabinoids	Produced endogenously in the body; include AEA and 2-AG	
Phytocannabinoids	Cannabinoids derived from the cannabis plant commonly include Δ9-THC, CBD, cannabinol, and cannabigerol	
Pharmaceuticals or prescription cannabinoids	Nabilone (Cesamet), dronabinol (Marinol), cannabidiol (CBD; Epidiolex), and nabiximols (Sativex) (these are described below in detail)	 Image showing a bottle with cannabis label. *created at https://BioRender.com.*
Illicit/Recreational cannabis preparations	"K2" or "spice," AM2201, JWH-122, JWH-210, and XLR11 (Mills et al., 2015)	

Images created in https://BioRender.com.

Cannabinoid preparations used in clinical practice

Numerous cannabis preparations are in clinical use, as mentioned above, following encouraging results from human trials in neuropathic pain or treatment-resistant chronic pain (Haleem and Wright, 2020). The following is a brief overview of common synthetic cannabis preparations, drawn from some studies.

1. **Smoked and vaporized cannabis**: This route of administration is usually preferred by patients, especially in the presence of CUD, due to the rapid onset of action. There is no drug approval from the US Food and Drug Administration (FDA) for any form of smoked cannabis; however, it is now legalized in 36 states in the United States (Baratta et al., 2022). The drug bud or resin is inhaled either through rolled paper, a bong, or vaped. THC is the most active form in these preparations at varying concentrations. In a landmark study of 23 patients with postsurgical or posttraumatic neuropathic pain, a single inhalation of 25 mg of 9.4% tetrahydrocannabinol herbal cannabis three times daily for 5 days reduced the intensity of pain and improved their sleep (Ware et al., 2010). Compared to smoked cannabis, vaporized cannabis results in higher peak concentrations of THC in the blood and stronger effects (Spindle et al., 2018).
2. **Oral cannabis preparations**: The FDA has approved two synthetic cannabis-related products, Nabilone and Dronabinol. Nabilone (Cesamet) is structurally similar to THC and was initially approved by the FDA for the treatment of chemotherapy-induced nausea and vomiting. Still, its role in the treatment of neuropathic pain and noncancer chronic pain has been reviewed in a metaanalysis. According to this study, oral Nabilone can be used as an adjunct in the treatment of different forms of chronic pain, with low abuse potential (Tsang and Giudice, 2016). Dronabinol is a synthetic form of THC available as oral formulations. It is also approved for use in chemotherapy induced nausea and vomiting and has shown promise in patients with noncancer pain. A clinical trial demonstrated that the long-term use of oral Dronabinol was safe for neuropathic pain secondary to multiple sclerosis after 16 weeks of use (Schimrigk et al., 2017). The latest guidelines from the Canadian Pain Society state that cannabinoids can be used as a third line option for the management of chronic neuropathic pain (Moulin et al., 2014).
3. **Oromucosal formulations**: Nabiximols (Sativex) is an oromucosal spray containing THC and CBD in a 1:1 ratio. In this formulation, CBD is intended to attenuate the psychomimetic effects caused by THC. Each dose of a 100 μL spray, containing 2.7 mg of THC and 2.5 mg of CBD, was tested to treat spasticity in patients with multiple sclerosis with minimal side effects (Keating, 2017). In Europe, Nabiximols showed efficacy in chronic neuropathic pain (Überall, 2020). Similar results were seen in a metaanalysis pooling results from 9 randomized controlled trials: Nabiximols spray in patients with chronic noncancer neuropathic pain refractory to usual treatments was associated with a statistically significant benefit compared to placebo (Dykukha et al., 2021).

4. **Intranasal preparation**: No intranasal formulations of cannabinoids are recommended. This route of administration is being researched, given the rapid absorption through the nasal mucosa. Matarazzo et al. developed nanostructured lipid carriers (NLCs) for the intranasal administration of CBD in animal models. These NLCs are colloidal particles made of a biodegradable lipid matrix and can encapsulate lipophilic drugs. The main aim of using these carriers is to increase the bioavailability of drugs. Their initial animal study showed that the intranasal delivery method produced a more significant and lasting antinociceptive effect with neuropathic pain than the oral or nasal administration of a CBD solution. However, human studies are needed to confirm these findings before recommending clinical use (Matarazzo et al., 2021).

Hundreds of animal studies have measured the efficacy of synthetic cannabinoids for the treatment of chronic and neuropathic pain, and positive results from these studies have encouraged human trials and metaanalyses. In this chapter, we review the most significant studies assessing the efficacy and dosage of cannabinoids for the treatment of pain (Table 2.1).

Conclusion

Understanding the basics of the ECS has helped us explore its role in treating different components of neuropathic pain. With increasing cannabis-related products available for consumption in the market and the legalization of cannabis use in 36 states of the United States, it is critical to understand the indications and adverse effects of synthetic cannabinoids. Only oral formulations like Nabilone and Dronabinol are FDA-approved for use in patients with chemotherapy induced nausea and vomiting. There is encouraging data for these compounds in treating neuropathic pain states. Though the FDA disapproves of Nabiximols for clinical use, it is being explored to treat spasticity in patients with Multiple Sclerosis, and studies have found beneficial effects in patients with neuropathic pain. Positive results for the role of synthetic cannabinoids in treating neuropathic pain have encouraged multiple pain societies to include these preparations as adjuncts to treat resistant pain in their guidelines, but further human trials are required to test the safety and adverse effect profile of these drugs in long-term use.

References

Abate, G., Uberti, D., Tambaro, S., 2021. Potential and limits of cannabinoids in Alzheimer's disease therapy. Biology. 2079-7737. 10 (6), 542. doi:10.3390/biology10060542.

Acuna-Goycolea, C., Obrietan, K., Pol, A.N.V.D., 2010. Cannabinoids excite circadian clock neurons. J. Neurosci. 15292401. 30 (30), 10061–10066. Available from: http://www.jneurosci.org/cgi/reprint/30/30/10061. doi:10.1523/JNEUROSCI.5838-09.2010.

Arseneault, L., Cannon, M., Witton, J., Murray, R.M., 2004. Causal association between cannabis and psychosis: examination of the evidence. Br. J. Psychiatry 184, 110–117. doi:10.1192/bjp.184.2.110.

Babson, K.A., Sottile, J., Morabito, D., 2017. Cannabis, cannabinoids, and sleep: a review of the literature. Curr. Psychiatry Rep. 15351645. 19 (4). Available from: http://www.springerlink.com/content/1523-3812/. doi:10.1007/s11920-017-0775-9.

Baratta, F., Pignata, I., Enri, L.R., Brusa, P., 2022. Cannabis for medical use: analysis of recent clinical trials in view of current legislation. Front. Pharmacol. 16639812. 13. Available from: http://www.frontiersin.org/Pharmacology. doi:10.3389/fphar.2022.888903.

Bolla, K.I., Brown, K., Eldreth, D., Tate, K., Cadet, J.L., 2002. Dose-related neurocognitive effects of marijuana use. Neurology. 00283878. 59 (9), 1337–1343. Available from: http://www.neurology.org. doi:10.1212/01.WNL.0000031422.66442.49.

Cohen, S.P., Mao, J., Neuropathic pain: mechanisms and their clinical implications. BMJ. 17561833. 348 (2014), Available from: http://www.bmj.com/highwire/filestream/685378/field_highwire_article_pdf/0/bmj.f7656. doi:10.1136/bmj.f7656United.

Crean, R.D., Crane, N.A., Mason, B.J., 2011. An evidence-based review of acute and long-term effects of cannabis use on executive cognitive functions. J. Addict. Med. 19353227. 5 (1), 1–8. doi:10.1097/ADM.0b013e31820c23fa.

Dykukha, I., Malessa, R., Essner, U., Überall, M.A., 2021. Nabiximols in chronic neuropathic pain: a meta-analysis of randomized placebo-controlled trials. Pain Med. 15264637. 22 (4), 861–874. Available from: http://oxfordjournals.org/en/press/announcements-from-oup/pain-medicine.html. doi:10.1093/pm/pnab050.

Fezza, F., Bari, M., Florio, R., Talamonti, E., Feole, M., Maccarrone, M., Endocannabinoids, 2014. Related compounds and their metabolic routes. Molecules. 1420-3049. 19 (11), 17078–17106. doi:10.3390/molecules191117078.

Haleem, R., Wright, R., 2020. A scoping review on clinical trials of pain reduction with cannabis administration in adults. J. Clin. Med. Res. 1918-3003. 12 (6), 344–351. doi:10.14740/jocmr4210.

Howlett, A.C., Barth, F., Bonner, T.I., Cabral, G., Casellas, P., Devane, W.A., Felder, C.C., Herkenham, M., Mackie, K., Martin, B.R., Mechoulam, R., Pertwee, R.G., 2002. International union of pharmacology. XXVII. Classification of cannabinoid receptors. Pharmacol. Rev. 00316997. 54 (2), 161–202. doi:10.1124/pr.54.2.161.

Ibrahim, M.M., Porreca, F., Lai, J., Albrecht, P.J., Rice, F.L., Khodorova, A., Davar, G., Makriyannis, A., Vanderah, T.W., Mata, H.P., Malan, T.P., 2005. CB2 cannabinoid receptor activation produces antinociception by stimulating peripheral release of endogenous opioids. Proc. Natl. Acad. Sci. USA 102 (8), 3093–3098. doi:10.1073/pnas.0409888102.

Karst, M., Salim, K., Burstein, S., Conrad, I., Hoy, L., Schneider, U., 2003. Analgesic effect of the synthetic cannabinoid CT-3 on chronic neuropathic pain: a randomized controlled trial. JAMA. 00987484. 290 (13), 1757–1762. Available from: http://jama.jamanetwork.com/journal.aspx. doi:10.1001/jama.290.13.1757.

Kathmann, M., Flau, K., Redmer, A., Tränkle, C., Schlicker, E., 2006. Cannabidiol is an allosteric modulator at mu- and delta-opioid receptors. Naunyn Schmiedebergs Arch. Pharmacol. 00281298. 372 (5), 354–361. doi:10.1007/s00210-006-0033-x.

Kaul, M., Zee, P.C., Sahni, A.S., 2021. Effects of cannabinoids on sleep and their therapeutic potential for sleep disorders. Neurotherapeutics. 18787479. 18 (1), 217–227. Available from: http://www.springerlink.com/content/1933-7213/. doi:10.1007/s13311-021-01013-w.

Keating, G.M., 2017. Delta-9-tetrahydrocannabinol/cannabidiol oromucosal spray (Sativex®): a review in multiple sclerosis-related spasticity. New Zealand Drugs. 11791950. 77 (5),

563–574. Available from: http://rd.springer.com/journal/40265. doi:10.1007/s40265-017-0720-6.

Lau, T., Schloss, P., 2008. The cannabinoid CB1 receptor is expressed on serotonergic and dopaminergic neurons. Eur. J. Pharmacol. 00142999. 578 (2-3), 137–141. doi:10.1016/j.ejphar.2007.09.022.

Lawn, W., Freeman, T.P., Pope, R.A., Joye, A., Harvey, L., Hindocha, C., Mokrysz, C., Moss, A., Wall, M.B., Bloomfield, M.A., Das, R.K., Morgan, C.J., Nutt, D.J., Curran, H.V., 2016. Acute and chronic effects of cannabinoids on effort-related decision-making and reward learning: an evaluation of the cannabis 'amotivational' hypotheses. Psychopharmacology. 14322072. 233 (19-20), 3537–3552. Available from: link.springer.de/link/service/journals/00213/index.htm. doi:10.1007/s00213-016-4383-x.

Lawn, W., Mokrysz, C., Lees, R., Trinci, K., Petrilli, K., Skumlien, M., Borissova, A., Ofori, S., Bird, C., Jones, G., Bloomfield, M.A.P., Das, R.K., Wall, M.B., Freeman, T.P., Curran, H.V., 2022. The CannTeen study: cannabis use disorder, depression, anxiety, and psychotic-like symptoms in adolescent and adult cannabis users and age-matched controls. J. Psychopharmacol. 14617285. 36 (12), 1350–1361. Available from: https://journals.sagepub.com/home/JOP. doi:10.1177/02698811221108956.

Lim, G., Sung, B., Ji, R.R., Mao, J., 2003. Upregulation of spinal cannabinoid-1-receptors following nerve injury enhances the effects of Win 55,212-2 on neuropathic pain behaviors in rats. Pain. 03043959. 105 (1-2), 275–283. Available from: www.elsevier.com/locate/painonline. doi:10.1016/S0304-3959(03)00242-2.

Maldonado, R., Baños, J.E., Cabañero, D., 2016. The endocannabinoid system and neuropathic pain. Pain 18726623. 157, S23–S32. www.elsevier.com/locate/painonline. doi:10.1097/j.pain.0000000000000428.

Malfitano, A.M., Basu, S., Maresz, K., Bifulco, M., Dittel, B.N., 2014. What we know and do not know about the cannabinoid receptor 2 (CB2). Semin. Immunol. 10963618. 26 (5), 369–379. Available from: http://www.elsevier.com/inca/publications/store/6/2/2/9/4/5/index.htt. doi:10.1016/j.smim.2014.04.002.

Marsicano, G., Wotjak, C.T., Azad, S.C., Bisogno, T., Rammes, G., Cascioll, M.G., Hermann, H., Tang, J., Hofmann, C., Zieglgänsberger, W., Di Marzo, V., Lutz, B., 2002. The endogenous cannabinoid system controls extinction of aversive memories. Nature. 00280836. 418 (6897), 530–534. doi:10.1038/nature00839.

Matarazzo, A.P., Elisei, L.M.S., Carvalho, F.C., Bonfílio, R., Ruela, A.L.M., Galdino, G., Pereira, G.R., 2021. Mucoadhesive nanostructured lipid carriers as a cannabidiol nasal delivery system for the treatment of neuropathic pain. Eur. J. Pharm. Sci. 18790720. 159. Available from: www.elsevier.com/locate/ejps. doi:10.1016/j.ejps.2020.105698.

Mills, B., Yepes, A., Nugent, K., 2015. Synthetic cannabinoids. Am. J. Med. Sci. 15382990. 350 (1), 59–62. Available from: http://www.journals.elsevier.com/the-american-journal-of-the-medical-sciences/. doi:10.1097/MAJ.0000000000000466.

Moulin, D.E., Boulanger, A., Clark, A.J., Clarke, H., Dao, T., Finley, G.A., Furlan, A., Gilron, I., Gordon, A., Morley-Forster, P.K., Sessle, B.J., Squire, P., Stinson, J., Taenzer, P., Velly, A., Ware, M.A., Weinberg, E.L., Williamson, O.D., 2014. Pharmacological management of chronic neuropathic pain: revised consensus statement from the Canadian pain society. Pain Res. Manag. 1203-6765. 19 (6), 328–335. doi:10.1155/2014/754693.

Nurmikko, T.J., Serpell, M.G., Hoggart, B., Toomey, P.J., Morlion, B.J., Haines, D., 2007. Sativex successfully treats neuropathic pain characterised by allodynia: a randomised, double-blind, placebo-controlled clinical trial. Pain. 03043959. 133 (1-3), 210–220. doi:10.1016/j.pain.2007.08.028.

Pertwee, R.G., 2006. Cannabinoid pharmacology: the first 66 years. Br. J. Pharmacol. 14765381. 147 (1), S163. doi:10.1038/sj.bjp.0706406.

Pertwee, R.G., 2009. Emerging strategies for exploiting cannabinoid receptor agonists as medicines. Br. J. Pharmacol. 14765381. 156 (3), 397–411. doi:10.1111/j.1476-5381.2008.00048.x.

Racz, I., Nadal, X., Alferink, J., Baños, J.E., Rehnelt, J., Martín, M., Pintado, B., Gutierrez-Adan, A., Sanguino, E., Manzanares, J., Zimmer, A., Maldonado, R., 2008. Crucial role of CB2 cannabinoid receptor in the regulation of central immune responses during neuropathic pain. J. Neurosci. 02706474. 28 (46), 12125–12135. Available from: http://www.jneurosci.org/cgi/reprint/28/46/12125. doi:10.1523/JNEUROSCI.3400-08.2008.

Schimrigk, S., Marziniak, M., Neubauer, C., Kugler, E.M., Werner, G., Abramov-Sommariva, D., 2017. Dronabinol is a safe long-term treatment option for neuropathic pain patients. Eur. Neurol. 14219913. 78 (5-6), 320–329. Available from: www.karger.com/journals/ene/ene_jh.htm. doi:10.1159/000481089.

Schouten, M., Dalle, S., Mantini, D., Koppo, K., 2024. Cannabidiol and brain function: current knowledge and future perspectives. Front. Pharmacol. 1663-9812. 14. doi:10.3389/fphar.2023.1328885.

Shah, S., Schwenk, E.S., Sondekoppam, R.V., Clarke, H., Zakowski, M., Rzasa-Lynn, R.S., Yeung, B., Nicholson, K., Schwartz, G., Hooten, W.M., Wallace, M., Viscusi, E.R., Narouze, S., 2023. ASRA Pain Medicine consensus guidelines on the management of the perioperative patient on cannabis and cannabinoids. Reg. Anesth. Pain Med. 15328651. 48 (3), 97–117. Available from: https://rapm.bmj.com/content/by/year/2019. doi:10.1136/rapm-2022-104013.

Sideris, A., Lauzadis, J., Kaczocha, M., 2024. The basic science of cannabinoids. Anesth. Analg. 15267598. 138 (1), 42–53. Available from: http://journals.lww.com/anesthesia-analgesia/toc/publishahead. doi:10.1213/ANE.0000000000006472.

Spindle, T.R., Cone, E.J., Schlienz, N.J., Mitchell, J.M., Bigelow, G.E., Flegel, R., Hayes, E., Vandrey, R., 2018. Acute effects of smoked and vaporized cannabis in healthy adults who infrequently use cannabis: a crossover trial. JAMA Netw. Open. 25743805. 1 (7), e184841. doi:10.1001/jamanetworkopen.2018.4841.

Starowicz, K., Makuch, W., Osikowicz, M., Piscitelli, F., Petrosino, S., Di Marzo, V., Przewlocka, B., 2012. Spinal anandamide produces analgesia in neuropathic rats: possible CB 1- and TRPV1-mediated mechanisms. Neuropharmacology. 18737064. 62 (4), 1746–1755. doi:10.1016/j.neuropharm.2011.11.021.

Surendran, Q., Surendran, L., Heinrich , 2021. Myrcene-what are the potential health benefits of this flavouring and aroma agent? Front. Nutr. 8. PMID: 34350208. doi:10.3389/fnut.2021.699666.

Tsang, C.C., Giudice, M.G., 2016. Nabilone for the management of pain. Pharmacotherapy. 18759114. 36 (3), 273–286. Available from: http://onlinelibrary.wiley.com/journal/10.1002/(ISSN)1875-9114/issues. doi:10.1002/phar.1709.

Turcotte, D., Doupe, M., Torabi, M., Gomori, A., Ethans, K., Esfahani, F., Galloway, K., Namaka, M., 2015. Nabilone as an adjunctive to gabapentin for multiple sclerosis-induced neuropathic pain: a randomized controlled trial. Pain Med. 15264637. 16 (1), 149–159. Available from: http://onlinelibrary.wiley.com/journal/10.1111/(ISSN)1526-4637. doi:10.1111/pme.12569.

Überall, M.A., 2020. A review of scientific evidence for THC:CBD oromucosal spray (Nabiximols) in the management of chronic pain. J. Pain Res. 1178-7090. 13, 399–410. doi:10.2147/jpr.s240011.

Van Hecke, O., Austin, S.K., Khan, R.A., Smith, B.H., Torrance, N., 2014. Neuropathic pain in the general population: a systematic review of epidemiological studies. Pain. 18726623. 155 (4), 654–662. Available from: www.elsevier.com/locate/painonline. doi:10.1016/j.pain.2013.11.013.

Vučković, S., Srebro, D., Savić Vujović, K., Vučetić, Č., Prostran, M., 2018. Cannabinoids and pain: new insights from old molecules. Front. Pharmacol. 1663-9812. 9. doi:10.3389/fphar.2018.01259.

Ware, M.A., Wang, T., Shapiro, S., Robinson, A., Ducruet, T., Huynh, T., Gamsa, A., Bennett, G.J., Collet, J.P., 2010. Smoked cannabis for chronic neuropathic pain: a randomized controlled trial. CMAJ. 14882329. 182 (14), E694. Available from: http://www.cmaj.ca/cgi/reprint/182/14/E694. doi:10.1503/cmaj.091414.

Wei, F., Dubner, R., Zou, S., Ren, K., Bai, G., Wei, D., Guo, W., 2010. Molecular depletion of descending serotonin unmasks its novel facilitatory role in the development of persistent pain. J. Neurosci. 15292401. 30 (25), 8624–8636. Available from: http://www.jneurosci.org/cgi/reprint/30/25/8624. doi:10.1523/JNEUROSCI.5389-09.2010.

Wilsey, B., Marcotte, T., Deutsch, R., Gouaux, B., Sakai, S., Donaghe, H., 2013. Low-dose vaporized cannabis significantly improves neuropathic pain. J. Pain. 15288447. 14 (2), 136–148. doi:10.1016/j.jpain.2012.10.009.

Young-Wolff, K.C., Chi, F.W., Campbell, C.I., Does, M.B., Brown, Q.L., Alexeeff, S.E., Ansley, D., Wang, X., Lapham, G.T., 2024. Association of psychiatric and substance use disorders with cannabis use and cannabis use disorder during early pregnancy in northern California. Addiction. 13600443. 119 (11), 1987–1997. Available from: http://onlinelibrary.wiley.com/journal/10.1111/(ISSN)1360-0443. doi:10.1111/add.16622.

Zhang, J., Hoffert, C., Vu, H.K., Groblewski, T., Ahmad, S., O'Donnell, D., 2003. Induction of CB2 receptor expression in the rat spinal cord of neuropathic but not inflammatory chronic pain models. Eur. J. Neurosci. 17 (12), 2750–2754. doi:10.1046/j.1460-9568.2003.02704.x.

CHAPTER

Cannabinoid receptors 3

Royce Copland, and Matthew Chung

Department of Pain Medicine, Division of Anesthesia, Critical Care and Pain Medicine, The University of Texas, MD Anderson Cancer Center, Houston, TX, United States

Introduction to cannabinoid receptors

The endocannabinoid system (ECS) is a complex neuromodulatory network integral to human physiology, influencing a wide array of physiological processes. The ECS is recognized as a crucial endogenous antinociceptive system, exerting modulatory actions at all stages of pain processing, including peripheral sites of injury, the spinal cord, and supraspinal brain regions (Fine and Rosenfeld, 2013). Its primary role in pain homeostasis is protective, dampening excessive neuronal activity and limiting the transmission of pain signals. This system can downregulate stress-related signals that lead to chronic inflammation and certain types of pain. However, it is also involved in causing inflammation-associated symptoms, depending on the physiological context (Fine and Rosenfeld, 2013).

The ECS comprises four main components: endogenous cannabinoids, receptors, metabolic enzymes, and downstream signaling pathways. Cannabinoid receptors form the cornerstone of this system, mediating the effects of both endogenous and exogenous compounds when bound to a cannabinoid receptor. Within the cannabinoid receptor system, the biological effects of cannabinoids are mediated by two primary cannabinoid receptors: cannabinoid receptor type 1 (CB1) and cannabinoid receptor type 2 (CB2) (Zou and Kumar, 2018). These are cell membrane receptors belonging to the G protein-coupled receptor (GPCR) superfamily, also known as seven-transmembrane receptors (Zou and Kumar, 2018).

These classic cannabinoid receptors, CB1 and CB2, exhibit a distinct anatomical expression pattern throughout the human body. CB1 receptors are predominantly found in the central nervous system, including brain regions such as the cortex, hippocampus, basal ganglia, and cerebellum, as well as in peripheral tissues of presynapses of axonal nerve endings, which mediate the inhibition of neurotransmitter release (Fine and Rosenfeld, 2013; Zou and Kumar, 2018). Their abundance in the nervous system explains the well-known psychoactive, neuromodulatory effects and pain perception related to cannabinoids. In contrast, CB2 receptors are mainly located in the immune system, including immune cells such as lymphocytes, macrophages, and microglia, as well as in peripheral tissues (Zou and Kumar, 2018). The

Cannabinoids for Chronic Pain. https://doi.org/10.1016/B978-0-443-36773-1.00019-2

distribution of CB2 receptors highlights their crucial role in modulating cytokine-mediated neuroinflammation.

CB1 and CB2 are integral components of the ECS that establish and maintain homeostasis in the human body. The ECS acts as a bridge between the body and mind, using the interaction between cannabinoid ligands and receptors to regulate various bodily functions, cognitive processes, and pain processes. These processes include brain plasticity, learning and memory, neuronal development, appetite regulation, lipid metabolism, mood regulation, reward processing, cardiovascular function, and modulating nociception (Matei et al., 2023). Consequently, dysregulation within the ECS can cause an imbalance in receptor expression or function, which has been linked to many human diseases, including neurological disorders (stroke, multiple sclerosis, Parkinson's disease), psychiatric conditions (major depression, generalized anxiety, schizophrenia, post traumatic stress disorder, cardiovascular disease, metabolic syndrome, and chronic painsyndromes (Matei et al., 2023; Gallego-Landin et al., 2021; Kendall and Yudowski, 2017).

Over the past two decades, significant research on endocannabinoids, such as anandamide (AEA) and 2-arachidonoylglycerol (2-AG), has led to the discovery of non-classical cannabinoid receptors. These non-classicial cannabinoid receptors include specific receptors within the following classifications: orphan G-protein-coupled receptors, transient receptor potential (TRP) channels, and PPARs (Zou and Kumar, 2018). While CB1 and CB2 are the classical and most extensively studied members, the landscape of cannabinoid-responsive receptors is expanding, and it has become evident that the effects of cannabinoids are not solely mediated by CB1 and CB2 receptors but also involve a complex interplay with other receptor systems. This broad pathophysiological involvement underscores the critical importance of the ECS for physicians to understand these processes in greater detail. This chapter reviews the fundamentals of cannabinoid receptor types, structure, signaling mechanisms, physiological roles, therapeutic potential, and interplay with chronic pain.

History and discovery of cannabinoid receptors

The historical use of cannabis for medicinal purposes, dating back to 2700 B.C., laid the groundwork for the modern discovery and understanding of cannabinoid receptors (Zuardi, 2006). Ancient civilizations across the globe, including China, India, and the Middle East, recognized the therapeutic potential of cannabis and integrated it into their medical practices. These early applications targeted various conditions, including insomnia, headaches, pain relief, and appetite stimulation (Zuardi, 2006). The empirical knowledge accumulated over centuries through the medicinal use of cannabis hinted at the existence of a complex biological system that mediated its effects. The scientific quest to understand this complex system gained momentum in the 1900s. Cannabidiol (CBD) was first isolated from cannabis in 1940, and its structure was reported in 1963 (Crocq, 2020). In 1964, Raphael Mechoulam and his research group successfully isolated and characterized delta-9-tetrahydrocannabinol

(THC), the principal psychoactive constituent of cannabis (Crocq, 2020). This seminal discovery by Mechoulam served as a catalyst for subsequent investigations into the cannabinoid receptor system, a pivotal element of the ECS.

In 1988, Allyn Howlett and William Devane, working at St. Louis University Medical School, identified cannabinoid receptor type 1 (CB1) (Crocq, 2020). This discovery was made through studies conducted on rats, revealing specialized protein molecules embedded within the cell membrane of neural tissue in the brain that THC activated. In 1993, a second cannabinoid receptor (CB2) was identified. Dr. Sean Munro and his team at the University of Cambridge made this discovery using a PCR-based strategy, enabling them to isolate and identify the CB2 receptor (Crocq, 2020). CB1 and CB2 were initially identified as the primary components of the ECS responsible for mediating the effects of cannabinoids within the body. This understanding prevailed until 1999, when further research and discoveries expanded the scope of the ECS and its interactions with various receptors.

Early 21st-century research revealed that the ECS interacts with a broader range of receptors, including orphan G protein-coupled receptors, TRPy potential channels, and peroxisome proliferator-activated receptors. In 1999, researchers at the University of Toronto identified and cloned orphan G protein-coupled receptor 55 (GPR55), a specific receptor in the superfamily of protein-coupled receptors (Morales and Jagerovic, 2016). This discovery led to the understanding that some endogenous and synthetic cannabinoid ligands can bind to receptors other than CBR1 and CBR2, which are non-cannabinoid receptors. In the same year, a group of European researchers discovered that endogenous cannabinoid anandamide (AEA) can interact with transient receptor potential vanilloid 1 (TRPV1) ion channels (Zygmunt et al., 1999). This finding was part of a larger wave of discoveries at the beginning of the 21st century that revealed how endogenous cannabinoids interact with various receptors via intracellular binding pathways. In 2005, Life Sciences published an article reporting the first instance of cannabinoid compounds binding to peroxisome proliferator-activated receptors situated on the surface of the cell's nucleus (Burstein, 2005). These findings demonstrate that cannabinoid signaling extends beyond the classical CB1 and CB2 pathways and encompasses a broader spectrum of receptor systems.

Cannabinoid receptor 1 and 2

Structure and characteristics

Cannabinoid receptors, specifically CB1 and CB2, are integral components of the endocannabinoid system, playing crucial roles in numerous physiological processes. These receptors are classified as G protein-coupled receptors (GPCRs), a large family of cell surface receptors responding to various extracellular stimuli. In humans, the genetic blueprints for CB1 and CB2 are found on different chromosomes: the *CNR1* gene, encoding the CB1 receptor, is located on chromosome 6, while the *CNR2* gene, responsible for the CB2 receptor, resides on chromosome 1 (Laprairie et al., 2012). The human CB1 receptor is composed of 472 amino acids, making it a larger protein than the CB2 receptor, which consists of 360 amino acids (Laprairie et al., 2012). Despite these differences in size, there is a notable degree of sequence similarity between the two receptors. Their overall amino acid sequence identity is approximately 44%, a similarity that increases to about 68% when considering only the transmembrane regions (Laprairie et al., 2012).

At a protein level, both CB1 and CB2 receptors belong to the rhodopsin-like (Class A) family of GPCRs. This classification implies a typical architecture characterized by seven transmembrane domains (7-TM). These domains are helical stretches of amino acids that span the cell membrane, creating a specific three-dimensional structure crucial for receptor function. These 7-TM domains are interconnected by three extracellular loops, which interact with ligands in the extracellular space, and three intracellular loops, which facilitate interactions with intracellular signaling molecules. IN as Addition, both receptors possess an extracellular N-terminal tail and an intracellular C-terminal tail, which can also contribute to ligand binding and downstream signaling (Shao et al., 2016).

Like other GPCRs, the fundamental function of cannabinoid receptors relies heavily on their 7-TM-spanning domains. These domains are not only critical for the specific binding of ligands but also for mediating the conformational changes necessary for interacting with intracellular signaling molecules, primarily G proteins. This interaction initiates a cascade of intracellular events, leading to a cellular response. A unique aspect of cannabinoid receptors, distinguishing them from many other Class A GPCRs, is the nature of their endogenous ligands. While many Class A GPCRs bind hydrophilic cations, cannabinoid receptors are activated by neutral, highly lipophilic ligands derived from arachidonic acid, a fatty acid (Reggio, 2010).

Furthermore, cannabinoid receptor ligands can be classified into four categories based on their interaction nature: agonists, antagonists, partial agonists, and inverse agonists (Shahbazi et al., 2020). Research has uncovered another type of agonist, a biased agonist. Unlike traditional agonists, which activate all downstream signaling pathways equally, biased agonists preferentially activate specific downstream signaling pathways over others following receptor binding (Khajehali et al., 2015; Ibsen et al., 2017). This biased signaling enables the modulation of specific cellular responses while avoiding undesirable side effects associated with activating other pathways. The ongoing research on biased agonism related to cannabinoid receptors holds immense potential for developing novel pharmacological interventions with enhanced therapeutic efficacy and reduced side effects (Ibsen et al., 2017).

Given these foundational characteristics of cannabinoid receptors and their ligand interactions, a closer examination of the structural differences between CB1 and CB2 receptors reveals how these variations contribute to their distinct pharmacological profiles. The most crucial sequence differences between CB1 and CB2 are in the N-terminal extracellular loop II (ECL2), C-terminus, various isomers, and the number of toggle switches on the receptor (Shahbazi et al., 2020). The second extracellular loop near the first transmembrane domain on CB1 plays a critical regulatory role, functioning as a molecular lid that overlays the main orthosteric binding pocket. The disparity in the ECL2 structure between CB1 and CB2 selectively governs which endocannabinoids or synthetic ligands will preferentially interact with and activate each receptor subtype (Shahbazi et al., 2020). This selectivity in ligand binding dictates the downstream physiological responses and therapeutic potential of targeting either CB1 or CB2, making the ECL2 a key determinant in cannabinoid receptor pharmacology. Variations in the C-terminus regions also distinguish CB1 from CB2. The intracellular C-terminus, particularly the juxtamembrane helix 8 (H8) domain in CB1, is crucial for G-protein interaction and activation (Shahbazi et al., 2020).

Another key feature in cannabinoid receptors is the presence of a toggle switch, whose activation leads to G protein binding: CB1 has two toggle switches (amino acid residues) involving two transmembranes (TM3 and TM6). In contrast, CB2 has a single toggle switch (amino acid residue) on TM6 (Shahbazi et al., 2020; Malfitano et al., 2014). These amino acid residues undergo conformational changes upon agonist binding, pivotal in receptor activation and subsequent G protein engagement (Shahbazi et al., 2020). Beyond the monomeric form in CB2, the CB1 receptor can exist as a homodimer, comprising two identical units, or form heterodimers with other GPCRs (Malfitano et al., 2014; Smith and Milligan, 2010). This ability to form dimers suggests a potential for modulation by other receptor systems. Furthermore, the CB1 receptor uniquely features an allosteric modulatory binding site, distinct from the primary orthosteric cannabinoid binding site (Tham et al., 2019). These structural differences dictate ligand affinity, receptor activation, and signal transduction profiles, underscoring their relevance for drug development and therapeutic targeting.

Signaling transduction pathways and mechanisms of action

Due to the lipophilic nature of cannabinoids, it was initially thought that these compounds exert various biological effects by disrupting the cell membrane nonspecifically. However, it was discovered through the cloning of CB1 and CB2 receptors that these receptors are primarily located within the cell membrane, and their activation occurs when ligands bind to them (Krishna Kumar et al., 2019). The binding of a ligand to the receptor induces conformational changes in the receptor protein, which, in turn, trigger the activation of various intracellular signaling pathways. The primary signaling mechanism of CB1 and CB2 involves the coupling of inhibitory G proteins (Gi) with pertussis toxin-sensitive (PTX) protein, leading to the inhibition of cyclic AMP production (Zou and Kumar, 2018; Krishna Kumar et al., 2019). However, there are multiple other signaling pathways that these classic cannabinoid receptors modulate, including the activation of phospholipase C (PLC), Mitogen-Activated Protein Kinase (MAPK), and G-protein-coupled inwardly rectifying potassium (GIRK) channels, as well as β-arrestin recruitment (Zou and Kumar, 2018; Howlett and Shim, 2013).

The primary signal transduction pathway of CB1 and CB2 involves binding cannabinoid ligands to their respective receptors, triggering a conformational change in the receptor protein. PTX protein couples with inhibitory G proteins (Gi). These Gi proteins inhibit the activity of adenylyl cyclase, a key enzyme in cAMP synthesis.

Consequently, the intracellular concentration of cAMP is decreased, and this reduction in cAMP levels subsequently downregulates the activity of protein kinase A (PKA), affecting downstream phosphorylation events and cellular responses (Howlett et al., 2010). The cannabinoid receptor-mediated inhibition of adenylyl cyclase and the subsequent reduction in PKA activity result in significant alterations in cellular signaling pathways, ultimately influencing a diverse range of cellular responses, including inflammation, immune modulation, and pain sensation.

Beyond its primary role in regulating cAMP, CB1 and CB2 have multiple intracellular signaling cascades to fine-tune cellular activity. The cannabinoid receptors can activate Phospholipase C (PLC) via Gβγ subunits or Gq/11 coupling (Howlett and Shim, 2013). This PLC activation hydrolyzes phosphatidylinositol 4,5-bisphosphate (PIP2) into inositol triphosphate (IP3) and diacylglycerol (DAG) (Ibsen et al., 2017). IP3 binds to its receptors on the endoplasmic reticulum, mobilizing intracellular Ca^{2+} stores, thereby increasing cytosolic calcium concentrations and activating Protein Kinase C (PKC) (Howlett and Shim, 2013). Furthermore, the activation of cannabinoid receptors can directly modulate ion channels. The dissociation of the Gαi/o subunit from the Gβγ dimer upon receptor activation allows the Gβγ subunits to activate G-protein-coupled inwardly rectifying potassium (GIRK) channels, causing membrane hyperpolarization and reduced excitability while simultaneously inhibiting voltage-gated calcium channels to decrease neurotransmitter release at presynaptic terminals (Zou and Kumar, 2018; Howlett and Shim, 2013). One key difference between CB1 and CB2 is that CB2 fails to partake in the modulation of calcium and potassium ion channels, even though inhibition of adenylyl cyclase still occurs (Zou and Kumar, 2018; Reggio, 2010). In addition to these signaling pathways, CB1 activation stimulates the Mitogen-Activated Protein Kinase (MAPK) cascades through complex mechanisms involving Gβγ and receptor tyrosine kinases, influencing long-term processes like gene expression and cell survival (Howlett and Shim, 2013).

Regulation of receptor signaling is achieved through desensitization and internalization of the receptor. These processes ensure cellular responsiveness and prevent overstimulation upon prolonged or excessive agonist binding (Howlett et al., 2010). Upon the binding of agonist molecules to CB1 or CB2 receptors, a cascade of events is initiated that leads to the recruitment of β-arrestin proteins. This recruitment is facilitated by the phosphorylation of the receptors' intracellular loops and C-terminal tail by G protein-coupled receptor kinases (GRKs) (Howlett et al., 2010). The phosphorylation of CB1 and CB2 receptors creates high-affinity binding sites for β-arrestins, specifically β-arrestin 1 and β-arrestin 2 (Howlett and Shim, 2013). The binding of β-arrestins to the phosphorylated receptor triggers two significant consequences: desensitization and internalization. β-arrestins sterically hinder the interaction between the activated receptor and the downstream G proteins, causing G protein uncoupling. This uncoupling effectively terminates G protein-mediated signaling, leading to a diminished cellular response despite the continued presence of the agonist (Howlett et al., 2010). The internalization process involves β-arrestins functioning as adaptor proteins that link the receptors to the endocytic machinery of the cell. The β-arrestin-receptor complex interacts with a protein called clathrin and initiates endocytosis, removing the receptor from the cell membrane to further reduce cellular responses (Howlett et al., 2010).

Retrograde endocannabinoid signaling and ligand formation

Endocannabinoids are not stored in vesicles like classical neurotransmitters but are synthesized "on demand" from membrane phospholipid precursors in response to elevated intracellular calcium or receptor activation (Burston and Woodhams, 2014). Unlike classic neurotransmission, the ECS employs a retrograde signaling mechanism where the neurotransmitter moves from the postsynaptic to the presynaptic neuron. This process can be broken down into four steps: postsynaptic synthesis, diffusion across the synaptic cleft, presynaptic receptor activation, and degradation. The first evidence supporting retrograde endocannabinoid signaling came from the observation of depolarization-induced suppression of inhibition (DSI) and excitation (DSE) (Ohno-Shosaku and Kano, 2014; Diana and Marty, 2004). DSI and DSE are two related forms of short-term plasticity in GABAergic and glutamatergic transmission. Still, it was later discovered that the ECS is involved in long-term plasticity at excitatory and inhibitory synapses (Diana and Marty, 2004; Kano et al., 2009; Heifets and Castillo, 2009). This retrograde process begins with postsynaptic depolarization, causing a calcium influx via G-protein-coupled receptors and the synthesis of two main endocannabinoids: 2-arachidonoylglycerol (2-AG) and anandamide (AEA) (Burston and Woodhams, 2014). Diacylglycerol lipase A mainly produces 2-arachidonoylglycerol (2-AG), and anandamide (AEA) synthesis involves N-acylphosphatidylethanolamine-hydrolyzing phospholipase D. When these lipid-soluble endocannabinoids are synthesized (Di Marzo and De Petrocellis, 2012), they are released post synaptically and diffuse across the synaptic cleft to the presynaptic terminal. At the presynaptic terminal, newly synthesized endocannabinoids bind to CB1, leading to the inhibition of neurotransmitter release by either blocking voltage-gated calcium channels or activating potassium channels. These endocannabinoids modulate the release of neurotransmitters, including acetylcholine, noradrenaline, dopamine, 5-hydroxytryptamine, gamma-aminobutyric acid (GABA), and glutamate (Lu and Mackie, 2016). These receptors play a pivotal role in modulating neurotransmission throughout the nervous system as they inhibit the release of both excitatory and inhibitory neurotransmitters. Scientists have discovered that retrograde signaling by endocannabinoids plays a significant role in suppressing neurotransmitter release, specifically gamma-aminobutyric acid (GABA) and glutamate at excitatory and inhibitory synapses, leading to short and long-term synaptic plasticity. This phenomenon occurs within various brain structures, including the nucleus accumbens, amygdala, hippocampus, visual cortex, somatosensory cortex, and prefrontal cortex, all critical structures involved in pain perception and its downregulation (Lu and Mackie, 2016). This mechanism is fundamental to synaptic plasticity and the homeostatic regulation of neuronal excitability in the brain, which could be a target for centralized pain disorders. Lastly, endocannabinoid action concludes with cellular uptake and enzymatic breakdown. Specifically, monoacylglycerol lipase (MAGL) on the presynapse primarily degrades 2-AG into arachidonic acid and glycerol (Zou and Kumar, 2018). Anandamide, conversely, is hydrolyzed by fatty acid amide hydrolase (FAAH), which is located on the postsynapse, into arachidonic acid and ethanolamine (Katona and Freund, 2008). This on-demand, retrograde signaling mechanism allows the ECS to precisely regulate synaptic activity spatially and temporally, particularly during heightened neuronal activity such as that seen in chronic pain states. Endocannabinoids can effectively act as a "brake," contributing to pain suppression and homeostatic balance by inhibiting the release of pain-transmitting neurotransmitters from presynaptic terminals in pain pathways.

Ligand affinity

The endocannabinoids 2-arachidonoylglycerol (2-AG) and anandamide (AEA) are key endogenous ligands of the cannabinoid receptors, CB1 and CB2, each exhibiting distinct binding affinities and activation profiles that contribute to diverse physiological effects. 2-AG acts as a full agonist with a strong affinity for CB1 and CB2 receptors, producing a maximal biological response upon binding. Research indicates that 2-AG is present in considerably higher amounts in the brain and is essential for multiple types of synaptic plasticity, including depolarization-induced suppression of inhibition (DSI) and depolarization-induced suppression of excitation (DSE), which are vital in pain modulation throughout the neuraxis (Heifets and Castillo, 2009). 2-AG can also function as a positive allosteric modulator of GABA-A receptors, further impacting neuronal inhibition (Tham et al., 2019; Bakas et al., 2017).

AEA binds strongly to CB1 receptors, acting as a partial agonist, but exhibits little to no activity at CB2 receptors (Di Marzo and De Petrocellis, 2012). This implies that even when AEA occupies the CB1 receptor, the resulting biological response is submaximal, falling short of the full effect elicited by a full agonist, such as 2-AG. While AEA can also bind to CB2 receptors, it is often considered CB1-selective due to its considerably higher affinity for CB1 compared to CB2 and its more prominent involvement in the modulation of nerve cell activity mediated by CB1 receptors (Lu and Mackie, 2016; Clapper et al., 2010). It is considered a key endogenous ligand for CB1 receptors in the CNS (Katona and Freund, 2008). Levels of AEA are dynamically regulated in pain states, with research showing that reduced AEA levels in the cerebrospinal fluid have been associated with chronic migraine, suggesting a potential deficiency contributing to pain (Greco et al., 2018). Conversely, peripheral AEA levels can increase in response to inflammation or nerve injury, possibly as an endogenous response to analgesia (Clapper et al., 2010). AEA plays a significant role in peripheral pain gating through the activation of CB1 receptors on sensory nerve terminals, controlling the access of nociceptive inputs to the CNS.

Anatomical distribution

CB1 was first discovered in the brain, but it was through immunohistochemistry that CB1 receptors were located throughout various tissue structures in the human body. The CB1 receptor exhibits a widespread distribution throughout the central nervous system (CNS) and peripheral nervous system (PNS), with varying levels of expression across different anatomical regions of the body. CB1 receptors are primarily localized to the presynaptic terminals and axons of central and peripheral neurons, mediating neurotransmitter release inhibition (Howlett and Abood, 2017). High levels of CB1 expression are observed in the olfactory bulb, neocortex, pyriform cortex, hippocampus, basal ganglia, brainstem (nucleus of the solitary tract and area postrema), and cerebellum (Zou and Kumar, 2018; Kendall and Yudowski, 2017; Mackie, 2005; Howlett and Abood, 2017). Other regions with moderate to high CB1 presence include the periaqueductal gray, hypothalamus, and amygdala (Mackie, 2005). Within the spinal cord, CB1 receptors are notably present in the dorsal primary afferent regions. Areas of low CB1 expression in the central nervous system include the thalamus, ventral horn, and medullary respiratory center (Mackie, 2005). The distribution of CB1 within the central nervous system correlates with its role in controlling motor function, cognition, memory, and analgesia. CB1 is also expressed in the peripheral nervous system, as many studies suggest a high concentration of expression on presynaptic terminals, where it mediates the retrograde signaling of endocannabinoids (Ohno-Shosaku and Kano, 2014; Kano et al., 2009; Lu and Mackie, 2016). In addition, there is a high level of CB1 expression in the trigeminal ganglion, dorsal root ganglion, and dermal nerve endings of primary sensory neurons, where it regulates nociception from afferent nerve fibers (Zou and Kumar, 2018). CB1 is expressed in low amounts in astrocytes, oligodendrocytes, and microglial cells (Castillo et al., 2012). Beyond the CNS and PNS, CB1 receptors are also present in various tissues throughout the body. The expression of CB1 has been reported in adipose tissue, skeletal tissue, bone, eye, gastrointestinal tract, hepatic cells, vascular and cardiac tissue, as well as reproductive tissues (testes, ovaries, uterus) (Maccarrone et al., 2015).

Cannabinoid receptor 2 (CB2) is predominantly found outside the central nervous system, with its highest expression in cells and tissues of the immune system, such as T cells, B cells, macrophages, monocytes, and mast cells, as well as in hematopoietic cells and immune organs like the spleen, tonsils, and thymus (Zou and Kumar, 2018; Kendall and Yudowski, 2017; Howlett and Abood, 2017). Beyond the immune system, CB2 receptors are also located on peripheral nerve terminals, in keratinocytes of the skin, bone, liver, and the gastrointestinal tract (Zou and Kumar, 2018; Howlett and Abood, 2017). While considered primarily peripheral, CB2 receptors are also present in the central nervous system, though at lower levels and mainly in glial cells (microglia and astrocytes) (Howlett and Abood, 2017). CB1 functions by suppressing the release of cytokines and chemokines, as well as hindering the migration of neutrophils and macrophages. This contributes to the reduction of chronic inflammation and the regulation of chronic pain. Growing evidence suggests that

CB2 receptors may be upregulated in microglia after peripheral nerve damage to modulate neuroimmune interaction with inflammation (Vučković et al., 2018).

Physiological pain: Role of cannabinoid receptors 1 and 2 activation

Activation of cannabinoid receptors CB1 and CB2 modulates a wide range of physiological processes, particularly pain, inflammation, and neurotransmission. While they share some structural similarities, their distinct anatomical locations and signaling pathways lead to different primary functions. The physiological functions of CB1 receptor activation are predominantly linked to its high expression throughout the central nervous system (CNS). The primary function of CB1 is its role as a presynaptic inhibitor via modulating synaptic transmission through retrograde signaling. CB1 receptors are densely located on presynaptic nerve terminals in key pain-processing areas like the spinal cord, periaqueductal gray (PAG), rostral ventral medulla (RVM), and thalamus. When activated by cannabinoids, these presynaptic receptors inhibit the release of neurotransmitters, including acetylcholine, noradrenaline, dopamine, 5-hydroxytryptamine, D-aspartate, and cholecystokinin (Mackie, 2005; Johnson et al., 2025). Through the regulation of neurotransmitter release, CB1 receptors have a significant role in influencing neuronal excitability, leading to analgesia.

CB1 receptors regulate pain perception by modulating descending pain pathways and cognitive aspects of pain. CB1 receptors are also highly concentrated in brainstem regions, such as the periaqueductal gray (PAG) and rostral ventromedial medulla (RVM), which serve as control centers for the descending inhibitory pain system (Zhang et al., 2024). Activation of CB1 receptors in these areas enhances the activity of this descending pathway, which then travels down the spinal cord to suppress pain signals at the dorsal horn further. Beyond simply blocking the sensory component of pain, CB1 receptors in higher brain regions, such as the amygdala, prefrontal cortex, and limbic system, modulate the emotional and cognitive aspects of the pain experience, including pain-related anxiety and fear (Mackie, 2005; Zhang et al., 2024; Wang et al., 2023).

The physiological role of the CB2 receptor in pain regulation is fundamentally different from that of the CB1 receptor. While CB1 receptors primarily exert direct control over neuronal excitability, CB2 receptors modulate pain largely through indirect mechanisms, primarily by regulating inflammation and the activity of immune cells. CB2R is a crucial "brake" on the immune system, preventing excessive inflammation and maintaining immune homeostasis. Activation suppresses proinflammatory cytokine release (TNF-alpha, IL-1β, IL-6) and inhibits immune cell activation, proliferation, and migration (Vučković et al., 2018). A key feature of the CB2 receptor is its dynamic expression and capacity to shift microglial phenotypes. Under normal conditions, CB2s are present at very low levels in the CNS. However, in response to pathological states such as inflammation or peripheral nerve injury, their expression is significantly upregulated in microglia and astrocytes (Vučković et al., 2018; Kangtai et al., 2023). Activation of the upregulated CB2 on microglia encourages them to shift from a pro-inflammatory (M1) state to an anti-inflammatory, neuroprotective (M2) state (Xu et al., 2023).

Beyond the CNS, CB2 on peripheral immune cells like macrophages and mast cells is also crucial for pain regulation. Their activation can inhibit the migration of these cells to sites of injury and reduce the release of inflammatory substances, thereby contributing to an overall anti-inflammatory and analgesic effect (Vučković et al., 2018; Johnson et al., 2025). CB2 receptors are also found on peripheral nerve terminals and keratinocytes. Activation of CB2 receptors on keratinocytes can stimulate the release of the endogenous opioid β-endorphin, which then acts on nearby sensory neurons to inhibit nociception, thereby providing an additional layer of peripheral pain control (Manzanares et al., 2006).

Atypical cannabinoid receptors

The classical view of cannabinoid action being mediated solely through CB1 and CB2 receptors has been challenged by observations that specific cannabinoid ligands, such as cannabidiol, can elicit physiological effects without activating either CB1 or CB2 (Crocq, 2020; Morales and Jagerovic, 2016). This has led to advanced research into identifying additional cannabinoid-responsive receptors. The orphan G protein-coupled receptor family, meaning a GPCR receptor that lacks a specific ligand, has been identified as binding to both cannabinoid and non-cannabinoid ligands. GPR55 is the most well-studied orphan GPCR related to the ECS, and due to its sequence homology with classical CB1 and CB2, it has been proposed that GPR55 should be classified as a third cannabinoid receptor. GPR55 exhibits a low sequence amino acid identity with CB1 receptors (13%) and CB2 receptors (14%) and is widely expressed in the brain (Davis, 2022). GPR55 can form complexes with CB1 and CB2 receptors, known as heterodimers, which influence signaling pathways. Lysophosphatidylinositol (LPI) is considered a key endogenous ligand activating GPR55 but has yet to be identified as the primary ligand associated with GPR55 (Davis, 2022). THC, AEA, and 2-AG have a high affinity and act as agonists, whereas cannabidiol has a high affinity for GPR55 and acts as an antagonist (Morales and Jagerovic, 2016). GPR55 is found throughout the body but is particularly concentrated in the brain (caudate nucleus, putamen, hippocampus), peripheral tissue (dorsal root ganglion), gastrointestinal tract, and bone (Morales and Jagerovic, 2016).

Transient Receptor Potential (TRP) is a group of membrane-associated protein channels involved in transducing chemical and physical stimuli (Davis, 2022). Transient Receptor Potential Vanilloid 1 (TRPV1), known as the capsaicin receptor, is one of the six non-selective cation channels. While structurally distinct from GPCRs, TRPV1 is directly modulated by several endocannabinoids and phytocannabinoids, making some consider it an "ionotropic cannabinoid receptor" or a key component of the expanded ECS (Zygmunt et al., 1999). Anandamide is a full agonist at TRPV1 channels (Clapper et al., 2010). Phytocannabinoids such as CBD and THC also interact with and modulate TRPV1 activity (Clapper et al., 2010). Activation of TRPV1 leads to the influx of cations, particularly calcium, and results in neuronal depolarization. Prolonged activation can lead to receptor desensitization, causing the receptor to remain closed and inactive for an extended period. This desensitization process, resulting from the protracted activation of TRPV1, is believed to be the primary mechanism behind the analgesic and anti-inflammatory effects of TRPV1 agonists (Vučković et al., 2018; Johnson et al., 2025; Davis, 2022).

Peroxisome Proliferator-Activated Receptors (PPARs) are nuclear receptor proteins that function as ligand-activated transcription factors (Burstein, 2005). Distinct from cell surface receptors like CB1 and CB2, PPARs reside within the nucleus of the cell. Upon ligand activation, they operate as transcription factors, binding to DNA and directly controlling the expression of targeted genes. Ligands that activate PPARs include the endocannabinoid anandamide (AEA) and phytocannabinoids like CBD and CBG (Burstein, 2005). The activation of PPARs is primarily linked to potent anti-inflammatory effects. By altering gene expression, PPAR activation can suppress the synthesis of pro-inflammatory cytokines (like TNF-α and IL-6) in immune cells such as monocytes and macrophages (Vučković et al., 2018).

Identifying GPR55, TRPV1, and PPARs as cannabinoid-responsive receptors has significantly expanded the scope of the endocannabinoid system. These findings help explain why certain cannabinoids—especially CBD—exhibit pharmacologic effects despite low affinity for CB1 and CB2. They also highlight the complexity and potential of cannabinoid-based therapies. Recognizing the roles of these atypical receptors is essential for understanding drug efficacy, predicting side effects, and designing next-generation cannabinoid therapeutics that target pain through novel mechanisms beyond classical CB1/CB2 pathways.

References

Bakas, T., van Nieuwenhuijzen, P.S., Devenish, S.O., McGregor, I.S., Arnold, J.C., Chebib, M., 2017. The direct actions of cannabidiol and 2-arachidonoyl glycerol at GABAA receptors. Pharmacol. Res. 10961186. 119, 358–370. Available from: http://www.elsevier.com/inca/publications/store/6/2/2/9/3/1/index.htt. doi:10.1016/j.phrs.2017.02.022.

Burstein, S., 2005. PPAR-γ: a nuclear receptor with affinity for cannabinoids. Life Sci. 00243205. 77 (14), 1674–1684. doi:10.1016/j.lfs.2005.05.039.

Burston, J.J., Woodhams, S.G., 2014. Endocannabinoid system and pain: an introduction. Proc. Nutr. Soc. 0029-6651. 73 (1), 106–117. doi:10.1017/s0029665113003650.

Castillo, P.E., Younts, T.J., Chávez, A.E., Hashimotodani, Y., 2012. Endocannabinoid signaling and synaptic function. Neuron. 08966273. 76 (1), 70–81. doi:10.1016/j.neuron.2012.09.020.

Clapper, J.R., Moreno-Sanz, G., Russo, R., Guijarro, A., Vacondio, F., Duranti, A., Tontini, A., Sanchini, S., Sciolino, N.R., Spradley, J.M., Hohmann, A.G., Calignano, A., Mor, M., Tarzia, G., Piomelli, D., 2010. Anandamide suppresses pain initiation through a peripheral endocannabinoid mechanism. Nat. Neurosci. 1097-6256. 13 (10), 1265–1270. doi:10.1038/nn.2632.

Crocq, M.-A., 2020. History of cannabis and the endocannabinoid system. Dialogues Clin. Neurosci. 1958-5969. 22 (3), 223–228. doi:10.31887/dcns.2020.22.3/mcrocq.

Davis, M.P., 2022. Overview of the Endocannabinoid System and Endocannabinoidome. Springer Science and Business Media LLC, pp. 1–40. doi:10.1007/978-3-030-89918-9_1.

Di Marzo, V., De Petrocellis, L., 2012. Why do cannabinoid receptors have more than one endogenous ligand? Philos. Trans. R. Soc. Lond. B Biol. Sci. 0962-8436. 367 (1607), 3216–3228. doi:10.1098/rstb.2011.0382.

Diana, M.A., Marty, A., 2004. Endocannabinoid-mediated short-term synaptic plasticity: depolarization-induced suppression of inhibition (DSI) and depolarization-induced suppression of excitation (DSE). Br. J. Pharmacol. 0007-1188. 142 (1), 9–19. doi:10.1038/sj.bjp.0705726.

Fine, P.G., Rosenfeld, M.J., 2013. The endocannabinoid system, cannabinoids, and pain. Rambam Maimonides Med. J. 20769172. 4 (4). doi:10.5041/RMMJ.10129.

Gallego-Landin, I., García-Baos, A., Castro-Zavala, A., Valverde, O., 2021. Reviewing the role of the endocannabinoid system in the pathophysiology of depression. Front. Pharmacol. 1663-9812. 12. doi:10.3389/fphar.2021.762738.

Greco, R., Demartini, C., Zanaboni, A.M., Piomelli, D., Tassorelli, C., 2018. Endocannabinoid system and migraine pain: an update. Front. Neurosci. 1662-453X. 12. doi:10.3389/fnins.2018.00172.

Heifets, B.D., Castillo, P.E., 2009. Endocannabinoid signaling and long-term synaptic plasticity. Annu. Rev. Physiol. 0066-4278. 71 (1), 283–306. doi:10.1146/annurev.physiol.010908.163149.

Howlett, A., Blume, L., Dalton, G., 2010. CB1 Cannabinoid receptors and their associated proteins. Curr. Med. Chem. 09298673. 17 (14), 1382–1393. doi:10.2174/092986710790980023.

Howlett, A.C., Abood, M.E., 2017. CB1 and CB2 receptor pharmacology. Adv. Pharmacol. 80, 169–206. Available from: http://www.elsevier.com/wps/find/bookdescription.cws_home/700817/description#description. doi:10.1016/bs.apha.2017.03.007.

Howlett, A.C., Shim, J.-Y., Cannabinoid receptors and signal transduction. Madame Curie Bioscience Database. (2013), Available from: https://www.ncbi.nlm.nih.gov/books/NBK6154/.

Ibsen, M.S., Connor, M., Glass, M., 2017. Cannabinoid CB 1 and CB 2 Receptor Signaling and Bias. Cannabis Cannabinoid Res. 2378-8763. 2 (1), 48–60. doi:10.1089/can.2016.0037.

Johnson, B.W., Strand, N.H., Raynak, J.C., Jara, C., Habtegiorgis, K., Hand, B.A., Hong, S., Maloney, J.A., 2025. Cannabinoids in chronic pain management: a review of the history, efficacy, applications, and risks. Biomedicines. 2227-9059. 13 (3), 530. doi:10.3390/biomedicines13030530.

Kano, M., Ohno-Shosaku, T., Hashimotodani, Y., Uchigashima, M., Watanabe, M., 2009. Endocannabinoid-mediated control of synaptic transmission. Physiol. Rev. 0031-9333. 89 (1), 309–380. doi:10.1152/physrev.00019.2008.

Katona, I., Freund, T.F., 2008. Endocannabinoid signaling as a synaptic circuit breaker in neurological disease. Nat. Med. 1078-8956. 14 (9), 923–930. doi:10.1038/nm.f.1869.

Kendall, D.A., Yudowski, G.A., 2017. Cannabinoid receptors in the central nervous system: their signaling and roles in disease. Front. Cell. Neurosci. 1662-5102. 10. doi:10.3389/fncel.2016.00294.

Khajehali, E., Malone, D.T., Glass, M., Sexton, P.M., Christopoulos, A., Leach, K., 2015. Biased agonism and biased allosteric modulation at the CB1 cannabinoid receptor. Mol. Pharmacol. 0026895X. 88 (2), 368–379. doi:10.1124/mol.115.099192.

Krishna Kumar, K., Shalev-Benami, M., Robertson, M.J., Hu, H., Banister, S.D., Hollingsworth, S.A., Latorraca, N.R., Kato, H.E., Hilger, D., Maeda, S., Weis, W.I., Farrens, D.L., Dror, R.O., Malhotra, S.V., Kobilka, B.K., Skiniotis, G., 2019. Structure of a signaling cannabinoid receptor 1-G protein complex. Cell. 10974172. 176 (3), 448. Available from: https://www.sciencedirect.com/journal/cell. doi:10.1016/j.cell.2018.11.040.

Laprairie, R.B., Kelly, M.E.M., Denovan-Wright, E.M., 2012. The dynamic nature of type 1 cannabinoid receptor CB1 transcription. Br. J. Pharmacol. 14765381. 167 (8), 1583–1595. doi:10.1111/j.1476-5381.2012.02175.x.

Lu, H.-C., Mackie, K., 2016. An introduction to the endogenous cannabinoid system. Biol. Psychiatry. 00063223. 79 (7), 516–525. doi:10.1016/j.biopsych.2015.07.028.

Maccarrone, M., Bab, I., Bíró, T., Cabral, G.A., Dey, S.K., Di Marzo, V., Konje, J.C., Kunos, G., Mechoulam, R., Pacher, P., Sharkey, K.A., Zimmer, A., 2015. Endocannabinoid signaling at the periphery: 50 years after THC. Trends Pharmacol. Sci. 01656147. 36 (5), 277–296. doi:10.1016/j.tips.2015.02.008.

Mackie, K., 2005. Distribution of cannabinoid receptors in the central and peripheral nervous system. Handb. Exp. Pharmacol. 18650325. 168, 299–325. Available from: http://www.springer.com/series/164. doi:10.1007/3-540-26573-2_10.

Malfitano, A.M., Basu, S., Maresz, K., Bifulco, M., Dittel, B.N., 2014. What we know and do not know about the cannabinoid receptor 2 (CB2). Semin. Immunol. 10445323. 26 (5), 369–379. doi:10.1016/j.smim.2014.04.002.

Manzanares, J., Julian, M., Carrascosa, A., 2006. Role of the cannabinoid system in pain control and therapeutic implications for the management of acute and chronic pain episodes. Curr. Neuropharmacol. 1570159X. 4 (3), 239–257. doi:10.2174/157015906778019527.

Matei, D., Trofin, D., Iordan, D.A., Onu, I., Condurache, I., Ionite, C., Buculei, I., 2023. The endocannabinoid system and physical exercise. Int. J. Mol. Sci. 1422-0067. 24 (3), 1989. doi:10.3390/ijms24031989.

Morales, P., Jagerovic, N., 2016. Advances towards the discovery of GPR55 ligands. Curr. Med. Chem. 09298673. 23 (20), 2087–2100. doi:10.2174/0929867323666160425113836.

Ohno-Shosaku, T., Kano, M., 2014. Endocannabinoid-mediated retrograde modulation of synaptic transmission. Current Opinion in Neurobiology. 09594388 29, 1–8. doi:10.1016/j.conb.2014.03.017.

Reggio, P., 2010. Endocannabinoid binding to the cannabinoid receptors: what is known and what remains unknown. Curr. Med. Chem. 09298673. 17 (14), 1468–1486. doi:10.2174/092986710790980005.

Shahbazi, F., Grandi, V., Banerjee, A., Trant, J.F., 2020. Cannabinoids and cannabinoid receptors: the story so far. iScience. 25890042. 23 (7), 101301. doi:10.1016/j.isci.2020.101301.

Shao, Z., Yin, J., Chapman, K., Grzemska, M., Clark, L., Wang, J., Rosenbaum, D.M., 2016. High-resolution crystal structure of the human CB1 cannabinoid receptor. Nature. 0028-0836. 540 (7634), 602–606. doi:10.1038/nature20613.

Smith, N.J., Milligan, G., 2010. Allostery at G protein-coupled receptor homo- and heteromers: uncharted pharmacological landscapes. Pharmacol. Rev. 00316997. 62 (4), 701–725. doi:10.1124/pr.110.002667.

Tham, M., Yilmaz, O., Alaverdashvili, M., Kelly, M.E.M., Denovan-Wright, E.M., Laprairie, R.B., 2019. Allosteric and orthosteric pharmacology of cannabidiol and cannabidiol-dimethylheptyl at the type 1 and type 2 cannabinoid receptors. Br. J. Pharmacol. 0007-1188. 176 (10), 1455–1469. doi:10.1111/bph.14440.

Vučković, S., Srebro, D., Savić Vujović, K., Vučetić, Č., Prostran, M., 2018. Cannabinoids and pain: new insights from old molecules. Front. Pharmacol. 1663-9812. 9. doi:10.3389/fphar.2018.01259.

Wang, Y., Wang, Q., Tang, L., Zhang, X., 2023. Cannabinoid modulations of pain- and stress-related circuits. Ageing Neurodegener. Dis. 27695301. 3 (3), 16. doi:10.20517/and.2023.19.

Xu, K., Yifei, W., Tian, Z., Yuanfan, X., Chaoran, W., Wang, Z., 2023. Microglial cannabinoid CB2 receptors in pain modulation. Int. J. Mol. Sci. 1422-0067. 24 (3), 2348. doi:10.3390/ijms24032348.

Zhang, M., Wang, T., Meng, F., Jiang, M., Shengxi, W., Hui, X., 2024. The endocannabinoid system in the brain undergoes long-lasting changes following neuropathic pain. iScience. 25890042. 27 (12), 111409. doi:10.1016/j.isci.2024.111409.

Zou, S., Kumar, U., 2018. Cannabinoid receptors and the endocannabinoid system: signaling and function in the central nervous system. Int. J. Mol. Sci. 1422-0067. 19 (3), 833. doi:10.3390/ijms19030833.

Zuardi, A.W., 2006. History of cannabis as a medicine: a review. Revista Brasileira de Psiquiatria. 1516-4446. 28 (2), 153–157. doi:10.1590/s1516-44462006000200015.

Zygmunt, P.M., Petersson, J., Andersson, D.A., Chuang, H.H., Sørgård, M., Di Marzo, V., Julius, D., Högestätt, E.D., 1999. Vanilloid receptors on sensory nerves mediate the vasodilator action of anandamide. Nature. 00280836. 400 (6743), 452–457. doi:10.1038/22761.

CHAPTER

Phytocannabinoids: Cannabidiol (CBD)

4

Orlando Florete Jr.
Orles Pain Management and Regenerative Medicine Group, Jacksonville, FL, United States

Introduction

Cannabidiol (CBD) is a nonpsychoactive cannabinoid derived from *Cannabis sativa* that has gained widespread medical interest due to its diverse pharmacological properties. It is the second most common phytocannabinoid and is extracted from both the hemp and marijuana plants. The most common phytocannabinoid, delta-9-tetrahydrocannabinol (THC), is derived from the female cannabis plant, while CBD is extracted from both the male cannabis plant (hemp) and the female plant. Unlike delta-9-THC, CBD does not produce intoxication, making it attractive for therapeutic use. CBD exhibits anti-inflammatory, anxiolytic, anticonvulsant, antipsychotic, and neuroprotective effects, among others (Mechoulam and Shvo, 1963). These characteristics have spurred research into its application in conditions ranging from epilepsy and anxiety to chronic pain and neurodegenerative disorders.

CBD gained significant popularity in the wellness circle after consumers recognized its potential. It became mainstream in the United States following the passage of the 2018 Farm Bill, which legalized hemp-derived products containing less than 0.3% THC (U.S. Congress, n.d.). This chapter explores the history and evolution of CBD, its science-backed benefits, its potential and future challenges, and the drug's clinical, legal, and ethical aspects.

Historical use of CBD

The medicinal use of cannabis dates back thousands of years, with early documentation in ancient China, India, and the Middle East. In these cultures, cannabis was employed to treat a variety of ailments, including pain, seizures, and inflammatory conditions. Although ancient healers did not isolate individual cannabinoids, the therapeutic effects attributed to cannabis likely resulted, in part, from CBD, which is nonintoxicating and possesses calming and anticonvulsant properties now supported by modern pharmacology (Mechoulam and Shvo, 1963).

Cannabis entered Western medicine more formally in the 19th century through the work of Irish physician William B. O'Shaughnessy, who observed its use in India

Cannabinoids for Chronic Pain. https://doi.org/10.1016/B978-0-443-36773-1.00005-2

and introduced it to Europe for treating spasms and seizures. Cannabis extracts became common in pharmacies as over-the-counter remedies for neuralgia, insomnia, and epilepsy. These preparations were chemically crude, containing varying ratios of THC, CBD, and other phytochemicals. Nonetheless, their popularity hinted at the plant's broad therapeutic utility (VanDolah et al., 2019).

CBD was first isolated in 1940 by chemist Roger Adams, but its molecular structure was not determined until 1963 by Raphael Mechoulam and Yechiel Gaoni. This discovery was pivotal as it allowed scientists to begin studying CBD in isolation from THC (Mechoulam and Shvo, 1963). A year later, Mechoulam's team also identified the structure of THC, making it possible to distinguish between the psychoactive and non-psychoactive effects of cannabis. Despite these breakthroughs, CBD remained overlooked for decades.

Interest in CBD reemerged in the early 2000s, with growing public and scientific attention to its therapeutic potential, particularly for refractory epilepsy. Case reports, including high-profile examples such as Charlotte Figi—who experienced dramatic reductions in seizures using a CBD-rich extract—sparked media attention and policy changes. Several US states enacted CBD-specific legislation, enabling limited access for medical use even where cannabis remained otherwise restricted (Devinsky et al., 2017).

The turning point in CBD's modern medical history came in 2018, when the US FDA approved Epidiolex, a purified CBD formulation, for the treatment of Dravet syndrome and Lennox–Gastaut syndrome (Devinsky et al., 2017). This regulatory approval affirmed the therapeutic validity of CBD and catalyzed global research interest. From its ancient medicinal use to modern pharmaceutical development, CBD's trajectory highlights the enduring relevance of natural compounds and the importance of rigorously evaluating traditional therapies.

Legal landscape of CBD

The legal classification of CBD has evolved dramatically over the past two decades, reflecting changing societal attitudes, emerging medical evidence, and legislative reform. Historically, CBD was subject to the same restrictions as other cannabis derivatives due to its association with the *Cannabis sativa* plant, despite lacking psychoactive effects. Under the Controlled Substances Act of 1970 in the United States, all cannabinoids, including CBD, were designated as Schedule I substances, indicating high abuse potential and no accepted medical use.

This rigid classification began to be challenged in the 2010s as scientific studies and anecdotal evidence increasingly demonstrated CBD's potential therapeutic value, particularly in pediatric epilepsy. In response to public demand, numerous US states passed laws permitting the use of low-THC, high-CBD products for medical purposes. These "CBD-only" laws created a legal paradox in which CBD was simultaneously allowed at the state level and prohibited federally, complicating research, manufacturing, and clinical application.

A significant shift occurred with the passage of the Agricultural Improvement Act of 2018, commonly referred to as the 2018 Farm Bill (U.S. Congress, n.d.).

This federal legislation redefined "hemp" as *Cannabis sativa* containing no more than 0.3% THC by dry weight and removed hemp and its derivatives—including CBD—from the list of controlled substances. While this move federally legalized hemp-derived CBD, it did not regulate it as a drug or dietary supplement, leaving oversight in a regulatory gray area managed by the US FDA.

The FDA's position on CBD remains cautious. While it approved Epidiolex as a prescription medication, it has consistently warned manufacturers against marketing unapproved CBD products with disease treatment claims. Over-the-counter CBD products are considered unlawful when advertised for medical use, and the FDA has issued multiple warning letters to companies for mislabeling, unsafe manufacturing practices, and unsubstantiated health claims. As a result, consumers and clinicians face a market filled with unregulated, often inaccurately labeled products.

Internationally, CBD's legal status varies widely. Countries like Canada and the United Kingdom allow CBD products with limited restrictions, while others enforce stricter controls or outright bans. In 2020, the United Nations Commission on Narcotic Drugs voted to reclassify cannabis and remove it from the most restrictive international drug control schedule, acknowledging its medicinal value. However, the global regulatory landscape for CBD remains fragmented, with considerable inconsistency in standards, permissible concentrations, and clinical guidance.

Mechanism of action

CBD exerts its effects through a complex and diverse range of molecular targets, many of which are distinct from the canonical cannabinoid receptors. Unlike THC, which acts primarily as a partial agonist at the CB1 and CB2 receptors of the endocannabinoid system (ECS), CBD has a low affinity for these receptors. Instead, CBD functions predominantly as an indirect modulator of the ECS. One of its key actions involves inhibiting fatty acid amide hydrolase (FAAH), the enzyme that breaks down anandamide, an endogenous cannabinoid. By reducing anandamide degradation, CBD increases its availability in the synaptic cleft, indirectly enhancing CB1 and CB2 receptor activity and contributing to mood stabilization, neuroprotection, and analgesia.

Beyond its effects on the ECS, CBD interacts with various non-cannabinoid receptors, further enhancing its therapeutic versatility. One prominent target is the transient receptor potential vanilloid type 1 (TRPV1) channel, which regulates pain perception, inflammation, and body temperature. CBD acts as an agonist at TRPV1, modulating nociceptive signaling and contributing to its analgesic and anti-inflammatory effects. In addition, CBD activates TRPA1 and desensitizes TRPM8 channels, which play roles in sensory nerve signaling and cold perception. These actions help explain CBD's role in managing neuropathic pain and inflammatory conditions.

Another vital receptor influenced by CBD is the 5-hydroxytryptamine 1A (5-HT1A) serotonin receptor, which is implicated in anxiety, depression, and stress responses. CBD acts as a partial agonist at this receptor, enhancing serotonergic transmission. This mechanism is believed to underlie many of CBD's anxiolytic and antidepressant-like effects observed in preclinical and some clinical studies. CBD's interaction with the 5-HT1A receptor also influences cardiovascular function and neuroendocrine responses to stress, suggesting a broader role in homeostatic regulation beyond mood disorders.

CBD also engages nuclear receptors such as the PPAR-γ, which modulates gene expression involved in inflammation, lipid metabolism, and insulin sensitivity. Activation of PPAR-γ by CBD may contribute to neuroprotective effects in neurodegenerative diseases and provide metabolic benefits in conditions such as type 2 diabetes. Furthermore, CBD has been shown to inhibit adenosine reuptake, thereby enhancing adenosine receptor signaling, which may further contribute to its anti-inflammatory and cardioprotective actions. CBD's multifaceted pharmacodynamics reflect its potential as a therapeutic agent across a broad spectrum of medical conditions, though many of these mechanisms require further validation in human studies.

Pharmacokinetics

The pharmacokinetics of CBD are influenced by its physicochemical properties, route of administration, and metabolic pathways. CBD is a highly lipophilic compound with low water solubility, which limits its oral bioavailability. CBD undergoes extensive first-pass hepatic metabolism when administered orally, resulting in an estimated bioavailability of 6% to 19%. Absorption may be delayed and variable, particularly when taken on an empty stomach. However, co-administration with a high-fat meal significantly enhances absorption by up to four- or five-fold, highlighting the importance of administration conditions in therapeutic use.

Following absorption, CBD distributes widely throughout the body due to its lipophilicity, with preferential accumulation in adipose tissue and in highly perfused organs such as the brain, liver, and lungs. Plasma protein binding is high (approximately 94%), primarily to albumin. The volume of distribution (Vd) is large, and steady-state concentrations can take several days to achieve with chronic dosing. Peak plasma concentrations (Tmax) are typically observed 1–4 hours after oral administration, though this varies by formulation. Inhaled CBD exhibits a more rapid onset of action (within minutes) and higher bioavailability (up to 31%), making it suitable for conditions requiring fast symptom relief.

CBD is metabolized extensively in the liver, predominantly via the cytochrome P450 enzyme system. The major isoenzymes involved are CYP3A4 and CYP2C19, with minor contributions from CYP1A1, CYP2D6, and others. Metabolism yields a range of hydroxylated and carboxylated metabolites, including 7-hydroxy-CBD (7-OH-CBD), which retains some biological activity, and 7-carboxy-CBD (7-COOH-CBD), the major inactive metabolite. Interindividual variability in CYP enzyme activity—whether due to genetic polymorphisms, liver disease, or drug-drug interactions—can significantly influence CBD plasma levels and therapeutic response.

Excretion of CBD and its metabolites occurs primarily via the fecal route, with minimal renal clearance. The elimination half-life of CBD varies based on the route and duration of administration. After a single dose, the terminal half-life is approximately

18–32 hours in humans; however, with repeated dosing, accumulation occurs, and the half-life can extend to several days. These pharmacokinetic characteristics support the rationale for both acute and chronic therapeutic regimens. Notably, CBD's metabolism via CYP450 enzymes poses a risk of pharmacokinetic drug interactions, particularly with medications with narrow therapeutic indices or those metabolized by the same pathways.

Therapeutic applications

Therapeutic use of CBD in epilepsy

The strongest clinical evidence supporting the use of CBD lies in its application for treatment-resistant epilepsy, particularly in pediatric populations. Randomized, placebo-controlled trials have demonstrated that CBD significantly reduces seizure frequency in individuals with Lennox-Gastaut syndrome, Dravet syndrome, and tuberous sclerosis complex—three forms of epilepsy often refractory to conventional antiepileptic drugs (AEDs). These results led to the 2018 FDA approval of Epidiolex, a purified, plant-derived oral CBD solution, marking the first cannabis-derived medication approved for prescription use in the United States (Devinsky et al., 2017). In pivotal trials, CBD at doses of 10–20 mg/kg/day, when used as an adjunct to existing AEDs, resulted in a median seizure reduction of up to 44% compared to placebo (Devinsky et al., 2017). Notably, patients with a high baseline seizure burden often experienced substantial functional improvements, including increased alertness and quality of life.

Mechanistically, CBD's anticonvulsant effects appear to be multifactorial. Unlike traditional AEDs, CBD does not act solely via voltage-gated sodium channels or GABAergic pathways. Instead, it modulates several neurochemical targets, including the TRPV1 channel, GPR55, and the 5-HT1A serotonin receptor, which influence neuronal excitability and synaptic transmission (Ibeas Bih et al., 2015). CBD also increases levels of the endogenous cannabinoid anandamide by inhibiting fatty acid amide hydrolase (FAAH), thereby enhancing the tone of the endocannabinoid system. These interactions may work synergistically with other AEDs, which partly explains CBD's success as an adjunctive therapy in drug-resistant epilepsy.

While generally well tolerated, the use of CBD in epilepsy is not without clinical considerations. Adverse effects observed in trials include somnolence, decreased appetite, diarrhea, and elevated hepatic transaminases, mainly when used with valproic acid (Devinsky et al., 2017; Chesney et al., 2020). Interactions with clobazam—another frequently prescribed AED—are also notable; CBD increases the levels of clobazam's active metabolite (N-desmethylclobazam), necessitating potential dose adjustments (Zendulka et al., 2016). Despite these challenges, the risk-benefit profile for patients with refractory seizures is favorable. Importantly, studies continue to evaluate CBD monotherapy and its utility in other forms of epilepsy, including generalized and focal epilepsies, where data remain less robust. Future research is expected to refine dosing strategies, explore age- and subtype-specific responses, and assess long-term neurodevelopmental outcomes in pediatric users.

Therapeutic use of CBD in chronic pain and inflammation

CBD has been widely explored for its potential role in the treatment of chronic pain conditions, particularly those involving neuropathic and inflammatory mechanisms. Preclinical studies in rodent models have demonstrated that CBD reduces hyperalgesia and mechanical allodynia through actions on multiple pain-related targets, including vanilloid receptors and the inhibition of inflammatory cytokine release (Ibeas Bih et al., 2015; Häuser et al., 2018). These mechanisms suggest that CBD can influence central and peripheral sensitization processes in chronic pain. In humans, observational and patient-reported outcomes suggest benefits in conditions such as fibromyalgia, diabetic neuropathy, and osteoarthritis. However, clinical trials evaluating CBD monotherapy remain limited, and many available studies have focused on formulations combining CBD with THC, complicating the interpretation of CBD-specific effects.

Mechanistically, CBD's analgesic and antiinflammatory properties are believed to stem from its ability to modulate several signaling pathways. It acts on TRPV1 channels to alter nociceptive signaling and may indirectly influence endocannabinoid tone by inhibiting the breakdown of anandamide. CBD has also been shown to downregulate pro-inflammatory mediators, including tumor necrosis factor-alpha (TNF-α), interleukin-6 (IL-6), and cyclooxygenase-2 (COX-2) (Ibeas Bih et al., 2015). These effects may be especially beneficial in inflammatory conditions such as rheumatoid arthritis and inflammatory bowel disease. CBD's interactions with PPARs and adenosine signaling further support its potential as a modulator of immune responses and oxidative stress, both of which are critical contributors to chronic inflammatory pain syndromes.

Despite promising mechanistic and early clinical data, the evidence base supporting the routine use of CBD in chronic pain remains incomplete. A 2018 overview of systematic reviews noted modest benefits for chronic pain, particularly when CBD is used as part of a combined cannabinoid therapy (Häuser et al., 2018). Challenges include heterogeneity in study designs, variability in dosing and formulations, and a lack of standardized endpoints. Nonetheless, CBD's favorable safety profile—especially its lack of intoxicating or addictive properties—makes it an attractive option for patients seeking alternatives to opioids or nonsteroidal antiinflammatory drugs (NSAIDs). Further large-scale, randomized, placebo-controlled trials are needed to clarify the role of CBD monotherapy in chronic pain and identify optimal patient populations and dosing strategies.

Therapeutic use of CBD in neurodegenerative disorders

CBD has shown promise as a potential neuroprotective agent in several neurodegenerative diseases, including Alzheimer's disease, Parkinson's disease, Huntington's disease, and amyotrophic lateral sclerosis (ALS). Preclinical studies have demonstrated that CBD reduces oxidative stress, modulates inflammatory signaling, and attenuates excitotoxicity—three central pathophysiological processes involved in neurodegeneration (Ibeas Bih et al., 2015; Cassano et al., 2020). In animal models of Alzheimer's disease, CBD has been observed to reduce neuroinflammation, preserve synaptic integrity, and attenuate memory impairment. Similarly, in Parkinsonian models, CBD has been shown to protect dopaminergic neurons and improve motor function, possibly through anti-inflammatory and antioxidant effects mediated by PPAR-γ activation and inhibition of reactive oxygen species.

The potential benefits of CBD in Parkinson's disease (PD) are further supported by small-scale human studies. In one randomized pilot study, CBD improved sleep and quality of life in PD patients without worsening motor symptoms. Its anxiolytic and antipsychotic properties may also be valuable in addressing nonmotor symptoms such as REM sleep behavior disorder, psychosis, and anxiety, which are common complications in advanced PD. Although larger trials are lacking, these early results suggest that CBD could be used as an adjunct to traditional dopaminergic therapies to address the multifaceted symptom burden in PD patients (Cassano et al., 2020). In Huntington's disease, CBD's anti-inflammatory and mitochondrial-stabilizing effects have shown promise in animal models, though human trials have yielded mixed outcomes.

Alzheimer's disease (AD) remains an area of active investigation. In transgenic mouse models, CBD administration reduced microglial activation and amyloid-beta-induced neurotoxicity, potentially slowing disease progression. These effects appear to involve both CB2 receptor-dependent and independent mechanisms, such as modulation of toll-like receptor signaling and reduction of oxidative stress (Ibeas Bih et al., 2015). Although no human clinical trials have definitively established CBD as a treatment for Alzheimer's or other neurodegenerative diseases, its low toxicity and multi-targeted actions justify continued exploration. Future studies must clarify appropriate dosing regimens, long-term safety, and the specific disease stage at which CBD may confer the greatest benefit.

Therapeutic use of CBD in anxiety disorders and depression

CBD has shown increasing promise in the treatment of anxiety-related disorders, including generalized anxiety disorder (GAD), social anxiety disorder (SAD), panic disorder, and post-traumatic stress disorder (PTSD). Preclinical studies in rodent models have consistently demonstrated anxiolytic effects, with reductions in autonomic arousal and avoidance behaviors. Human trials have yielded encouraging results, though limited in number and scale. In a randomized, double-blind study involving healthy participants subjected to a simulated public speaking test, oral CBD at a dose of 300 mg significantly reduced self-reported anxiety compared to placebo, with effects comparable to conventional anxiolytics (Bergamaschi et al., 2011). Neuroimaging data support these findings, showing that CBD modulates activity in brain regions implicated in anxiety, including the amygdala, hippocampus, and prefrontal cortex.

The proposed mechanisms underlying CBD's anxiolytic properties are complex and not fully elucidated, but are thought to involve multiple receptor systems. CBD acts as a partial agonist at the 5-HT1A receptor, a serotonin subtype associated with mood and anxiety regulation. This receptor interaction is thought to be central to its antianxiety and potential antidepressant effects, as evidenced by both preclinical and human data (Ibeas Bih et al., 2015). In addition, CBD modulates endocannabinoid tone by increasing anandamide levels, an endogenous cannabinoid that influences emotion and stress responses. Other targets include GABAergic transmission, TRPV1 receptors, and PPARs, contributing to neuroplasticity and inflammatory balance. This multimodal action profile may offer advantages over selective serotonin reuptake inhibitors, particularly in treatment-resistant cases.

The evidence supporting CBD's role in major depressive disorder (MDD) is less robust than for anxiety, but preliminary data are promising. Animal models of depression have shown that CBD produces rapid-acting antidepressant-like effects, similar to ketamine, possibly through the enhancement of synaptogenesis and neurotrophic factors such as brain-derived neurotrophic factor (BDNF) (Cassano et al., 2020). Small observational studies and open-label trials have reported improvements in mood, sleep, and emotional regulation among users with depressive symptoms, though placebo-controlled trials remain limited. Notably, CBD appears to lack the sedative and dependency risks associated with benzodiazepines and may be better tolerated than some antidepressants. Nonetheless, clinical guidance currently recommends CBD as an adjunctive, rather than first-line, treatment in mood and anxiety disorders until further high-quality research becomes available.

Therapeutic use of CBD in substance use disorders

CBD has gained attention as a potential adjunctive treatment for substance use disorders, including opioid, tobacco, alcohol, and cannabis use disorders. Unlike other cannabinoids, such as THC, CBD is nonreinforcing and nonintoxicating, which positions it uniquely as a therapeutic agent that does not carry significant abuse liability. CBD's ability to modulate craving, anxiety, impulsivity, and memory reconsolidation processes—key factors in addiction cycles—makes it a promising candidate for relapse prevention. Preliminary studies have suggested that CBD may interfere with the reward circuitry involved in drug-seeking behaviors by acting on both the serotonergic and endocannabinoid systems (Ibeas Bih et al., 2015; Hurd et al., 2019).

One of the most studied areas is the use of CBD in opioid use disorder (OUD). In a double-blind, randomized, placebo-controlled trial, acute administration of CBD (400–800 mg/day) significantly reduced cue-induced craving and anxiety in individuals abstinent from heroin. Notably, these effects persisted up to one week after the final dose, suggesting durable neurobehavioral modulation. The proposed mechanism includes attenuation of amygdala activation and normalization of salience attribution during cue exposure, mediated by 5-HT1A receptor activity and indirect endocannabinoid signaling. CBD's nonsedating, nonintoxicating nature enhances its suitability as an adjunct to medications for opioid use disorder (MOUD), such as buprenorphine and methadone.

In the context of tobacco and nicotine dependence, CBD has demonstrated potential to reduce cigarette consumption and cravings. One pilot study found that smokers who used a CBD inhaler decreased the number of cigarettes smoked by approximately 40% over one week compared to a placebo. Although the mechanisms are not yet fully understood, CBD's anxiolytic and anti-compulsive properties are believed to disrupt habitual use patterns and mitigate withdrawal-related symptoms. Similar interest has emerged regarding alcohol use disorder, where CBD has been shown in animal studies to reduce ethanol intake, prevent relapse-like behavior, and protect against alcohol-induced neurodegeneration. However, robust human trials in these areas remain scarce.

Beyond its primary effects on drug craving and consumption, CBD may support recovery by addressing comorbid psychiatric symptoms such as anxiety, insomnia, and depression—conditions that commonly co-occur with addiction and increase relapse risk. This multifaceted pharmacological profile may allow CBD to stabilize the emotional and cognitive dysregulation that underlies many substance use patterns. Despite these encouraging findings, the current evidence base remains preliminary. Larger, long-term randomized trials are needed to determine optimal dosing, duration, and formulation. Until then, CBD may be cautiously considered as a complementary approach within comprehensive treatment plans for substance use disorders, particularly where conventional pharmacotherapy is ineffective or contraindicated.

Esthetic and dermatological applications of CBD

The use of CBD in esthetics has grown rapidly, driven by its purported anti-inflammatory, antioxidant, sebostatic, and anti-aging effects. Topically applied CBD is incorporated into various skincare products, including creams, serums, and facial oils. Mechanistically, CBD interacts with cutaneous endocannabinoid receptors (primarily CB2 and transient receptor potential channels) and exerts local effects without systemic absorption. Studies have demonstrated that CBD can reduce the production of pro-inflammatory cytokines in keratinocytes and sebocytes, potentially benefiting patients with inflammatory skin disorders such as acne vulgaris and rosacea. In particular, CBD's ability to inhibit lipid synthesis in sebaceous glands via TRPV4 channel activation suggests a promising role in reducing sebum production and skin oiliness.

In addition to acne treatment, CBD's antioxidant properties may contribute to its utility in anti-aging skincare. Oxidative stress and free radical formation play central roles in the degradation of collagen and elastin, leading to skin laxity, fine lines, and pigmentation changes. CBD scavenges reactive oxygen species and may reduce oxidative damage to skin cells, particularly when combined with other established antioxidants such as vitamins C and E. Furthermore, anecdotal and preliminary clinical reports suggest CBD may improve skin hydration and barrier function by supporting keratinocyte homeostasis and modulating local inflammation. While consumer interest is high and initial findings are encouraging, most esthetic applications of CBD remain under-investigated in rigorous clinical trials, and dermatologic use should be guided by evidence-based practice as research evolves.

Regulatory considerations for topical CBD in esthetic medicine

The regulation of topical CBD products used in esthetic medicine remains fragmented and inconsistent across jurisdictions, posing challenges for both clinicians and consumers. In the United States, the 2018 Farm Bill legalized hemp-derived CBD at the federal level, provided it contains less than 0.3% THC by dry weight; however, this law did not extend specific regulatory oversight to cosmetic products containing CBD (U.S. Congress, n.d.). The FDA maintains that while CBD may be legally present in cosmetics, it must not be adulterated or misbranded, and no therapeutic or disease-treatment claims can be made without formal drug approval. Product quality, purity, and labeling accuracy can vary widely without FDA premarket approval or standardization. Independent studies have found that many over-the-counter CBD skincare products are mislabeled concerning CBD concentration and may contain contaminants such as heavy metals or residual solvents. As a result, esthetic practitioners must exercise caution when recommending or applying CBD-containing formulations, ideally favoring products with third-party testing, certificates of analysis, and good manufacturing practice certification. Internationally, regulatory frameworks differ significantly; while countries like Canada and the European Union allow topical CBD in cosmetics with certain restrictions, others, such as parts of Asia and the Middle East, maintain strict prohibitions. Given the growing popularity of CBD in esthetics, harmonized guidelines and clinical-grade product standards are urgently needed to ensure safety, efficacy, and professional accountability in dermatologic practice.

General tolerability of CBD

CBD is generally well-tolerated across a range of doses and formulations, making it a favorable candidate for therapeutic use in both pediatric and adult populations. Clinical trials, particularly those involving Epidiolex—a pharmaceutical-grade, purified CBD oral solution—have demonstrated a relatively mild adverse event profile even at high doses (up to 20 mg/kg/day). Commonly reported side effects include somnolence, fatigue, decreased appetite, and gastrointestinal symptoms such as diarrhea and nausea. Most of these effects are dose-dependent and tend to resolve with dose adjustment or gradual titration. Importantly, CBD does not produce the euphoric or psychoactive effects associated with THC, and it has not demonstrated abuse liability in human or animal studies.

Hepatotoxicity, while rare, is among the more serious potential adverse effects of CBD. In some studies—especially those involving higher doses or concomitant use with hepatically metabolized drugs—transient elevations in liver transaminases (ALT and AST) have been observed. This is particularly relevant in patients taking valproic acid, with which CBD appears to have a synergistic risk for hepatic enzyme elevation. As a result, periodic monitoring of liver function tests is recommended during long-term or high-dose CBD therapy, particularly in populations with preexisting liver disease or polypharmacy.

CBD may also interact with other medications by influencing cytochrome P450 enzymes, especially CYP3A4 and CYP2C19. These interactions can lead to altered plasma concentrations of drugs such as clobazam, warfarin, and certain antidepressants or antiepileptics. Clinically significant drug-drug interactions, while not frequent, underscore the importance of evaluating a patient's medication profile before initiating CBD therapy. In most cases, interactions can be managed by dose adjustments and close monitoring, but caution is warranted, particularly in older adults and individuals on complex pharmacologic regimens.

Tolerability may also vary by formulation and route of administration. Oral preparations tend to have more systemic effects and a delayed onset, while topical formulations are typically well tolerated due to limited systemic absorption. Inhaled CBD products may cause mild respiratory irritation but are not commonly associated with serious adverse events. Overall, the tolerability profile of CBD supports its use in diverse clinical contexts, though patient-specific factors such as age, liver function, and concurrent medications must be considered. As CBD use continues to expand, long-term safety data and standardized reporting of adverse events will be essential for optimizing clinical guidance.

Drug interactions with CBD

CBD has the potential to interact with a variety of prescription medications, primarily due to its influence on hepatic drug-metabolizing enzymes. CBD is a known inhibitor of several cytochrome P450 (CYP) isoenzymes, most notably CYP3A4 and CYP2C19, which metabolize approximately 60% of pharmaceutical agents. Inhibiting these enzymes can result in elevated plasma concentrations of co-administered drugs, potentially enhancing therapeutic effects or increasing the risk of toxicity. This interaction profile necessitates careful evaluation of concurrent medications when initiating CBD therapy, particularly in patients on polypharmacy regimens.

One of the most well-documented drug interactions involves CBD and clobazam, an antiepileptic drug metabolized by CYP2C19. Studies have shown that co-administration of CBD can significantly increase the serum levels of clobazam's active metabolite, N-desmethylclobazam, leading to increased sedation and other central nervous system effects. In clinical trials of Epidiolex, this interaction required dose adjustments and close monitoring. Similarly, CBD may enhance the effects of other central nervous system depressants, such as benzodiazepines and barbiturates, through additive pharmacodynamic interactions, even in the absence of direct metabolic interference.

CBD has also been shown to interact with drugs that are CYP3A4 substrates, including certain statins (e.g., atorvastatin), calcium channel blockers, immunosuppressants (e.g., tacrolimus, cyclosporine), and anticoagulants such as warfarin. In the case of warfarin, co-administration with CBD has resulted in elevated international normalized ratio (INR) values, necessitating dose adjustments to avoid bleeding complications. In patients on narrow therapeutic index medications, such as antiepileptics, chemotherapeutic agents, and immunosuppressants, clinicians should monitor for potential changes in efficacy or toxicity when CBD is introduced or discontinued.

Beyond hepatic enzyme inhibition, CBD may influence drug transport proteins, such as P-glycoprotein (P-gp), which can affect the absorption and distribution of various medications. While the clinical significance of these transporter-level interactions remains under investigation, this further underscores the complexity of CBD's pharmacokinetic profile. Given this broad interaction potential, clinicians should obtain a thorough medication history, initiate CBD at conservative doses, and perform therapeutic drug monitoring when appropriate. As the availability and use of CBD expand, comprehensive guidelines and cannabinoid-specific drug interaction databases will be essential to ensuring safe clinical practice.

Use of CBD in special populations

Pediatric population

CBD has been most extensively studied in pediatric patients with drug-resistant epilepsy, such as those with Lennox-Gastaut syndrome, Dravet syndrome, and tuberous sclerosis complex. The FDA-approved CBD formulation, Epidiolex, has demonstrated significant reductions in seizure frequency in children aged 2 years and older. Dosing in pediatric populations is typically weight-based (e.g., 10–20 mg/kg/day) and titrated based on therapeutic response and tolerability. While CBD is generally well tolerated, common adverse effects include somnolence, decreased appetite, diarrhea, and elevated liver enzymes, especially when used concurrently with valproic acid (Devinsky et al., 2017; Chesney et al., 2020). Long-term safety data in children remain limited, and its use outside of epilepsy treatment should be approached cautiously and under medical supervision.

Geriatric population

In older adults, CBD presents unique pharmacologic considerations due to age-related changes in drug metabolism, an increased likelihood of comorbid conditions, and higher rates of polypharmacy. Hepatic and renal function decline with age, potentially altering CBD metabolism and excretion. Moreover, the inhibitory effect of CBD on cytochrome P450 enzymes increases the risk of drug interactions, especially with commonly prescribed medications such as anticoagulants, antihypertensives, and sedatives (Zendulka et al., 2016; VanDolah et al., 2019). Although anecdotal reports suggest benefits for chronic pain, sleep disturbances, and anxiety in geriatric patients, high-quality, age-specific clinical trials are lacking. Lower starting doses, slow titration, and close monitoring for adverse effects and interactions are recommended in this population.

Pregnancy and lactation

Data on the safety of CBD during pregnancy and lactation are sparse and mainly limited to animal models. Preclinical studies have raised concerns about potential teratogenicity and impaired neurodevelopment in offspring exposed to cannabinoids in utero. While CBD is nonintoxicating, it crosses the placental barrier and has been detected in breast milk. Due to the lack of controlled human studies and potential risks to fetal and neonatal development, the use of CBD is not currently recommended during pregnancy or breastfeeding (VanDolah et al., 2019).

Patients with hepatic or renal impairment

Patients with impaired liver function are at greater risk of altered CBD pharmacokinetics and increased systemic exposure due to decreased metabolic clearance. Clinical studies of Epidiolex have shown that hepatic impairment, particularly moderate to severe, necessitates dose adjustment and frequent liver function monitoring (Devinsky et al., 2017; Chesney et al., 2020). Although renal clearance plays a minor role in CBD elimination, patients with concurrent renal and hepatic disease may experience compounded pharmacologic variability. As such, clinicians should initiate treatment at reduced doses and adjust the dose based on tolerability and laboratory findings. In all special populations, individualized assessment and careful risk-benefit analysis are critical to the safe and effective use of CBD.

Product formulations

CBD is available in a wide range of formulations that differ in route of administration, pharmacokinetics, and clinical utility. Oral formulations, such as oils, capsules, soft gels, and edibles, are among the most commonly used. These products are convenient and allow for precise dosing, but they are subject to significant first-pass metabolism, resulting in delayed onset and variable bioavailability (typically 6%–19%). The presence of food, especially high-fat meals, can significantly enhance absorption.

Pharmaceutical-grade oral solutions, like Epidiolex, offer standardized dosing, making them preferable in clinical settings, particularly for conditions like epilepsy.

Sublingual preparations, including tinctures and sprays, offer an alternative to oral ingestion by enabling partial absorption through the oral mucosa. This route bypasses some hepatic metabolism and allows for a faster onset of action—often within 15–45 minutes—while preserving ease of administration. Sublingual delivery is commonly used for symptoms requiring more immediate relief, such as anxiety or breakthrough pain. However, absorption rates can vary depending on formulation composition and mucosal contact time, and standardized dosing may be challenging in nonpharmaceutical products.

Topical formulations, including creams, balms, and transdermal patches, are increasingly used in dermatologic and musculoskeletal conditions. These products are designed for localized delivery and typically result in minimal systemic absorption, which reduces the risk of systemic side effects and drug interactions. Topical CBD may be useful in managing inflammatory skin disorders, localized pain, and muscle soreness. Transdermal systems, which deliver CBD through the skin into the bloodstream, are under investigation and may offer controlled, sustained delivery for systemic effects. However, more research is needed to validate efficacy, permeability, and dosage accuracy in clinical use.

Inhaled formulations, such as vaporized CBD or smoked hemp flower, offer the most rapid onset of action (within minutes) and higher systemic bioavailability (up to 31%) compared to oral products. These are sometimes preferred for acute symptom relief, including anxiety or breakthrough pain. However, concerns about pulmonary safety—especially with unregulated vaping products—limit the recommendation of this route for long-term or routine use. In addition, variability in delivered dose and lack of product standardization can complicate clinical application. Product selection should be based on the patient's therapeutic goals, comorbidities, lifestyle factors, and need for precise dosing.

Research and future directions

Research into CBD has expanded significantly in the past decade, driven by its therapeutic potential across neurological, psychiatric, inflammatory, and metabolic conditions. While epilepsy remains the most well-studied and clinically validated indication, ongoing investigations are exploring CBD's efficacy in anxiety disorders, post-traumatic stress disorder (PTSD), schizophrenia, autism spectrum disorder (ASD), and substance use disorders. Preliminary findings suggest CBD's effects on serotonin receptors, the endocannabinoid system, and neuroinflammation may offer meaningful symptom relief in these conditions. Large-scale, placebo-controlled trials are underway to establish dosing protocols, safety profiles, and long-term outcomes (Bergamaschi et al., 2011; Hurd et al., 2019; VanDolah et al., 2019).

In addition to neuropsychiatric applications, research is examining the role of CBD in pain management, particularly as an adjunct or alternative to opioids. The opioid crisis has spurred interest in nonaddictive analgesics, and CBD's interaction with pain pathways and inflammatory mediators positions it as a potential candidate (Häuser et al., 2018). Studies are also underway evaluating CBD in inflammatory and autoimmune diseases such as rheumatoid arthritis, multiple sclerosis, and inflammatory bowel disease (Ibeas Bih et al., 2015; Cassano et al., 2020). These efforts aim to clarify symptom control and disease-modifying potential, given CBD's immunomodulatory properties.

Pharmaceutical and formulation research is focused on enhancing CBD's bioavailability and therapeutic consistency. Novel delivery systems, such as nanotechnology-based emulsions, transdermal patches, and inhalable aerosols, are being developed to overcome limitations in absorption and metabolism (VanDolah et al., 2019). Synthetic analogs and combination therapies with other cannabinoids or conventional drugs are also being explored to optimize efficacy and reduce adverse effects. Biomarker-driven studies may soon enable personalized cannabinoid therapy, guiding clinicians toward specific formulations and doses based on individual genetic and metabolic profiles.

Despite increasing evidence and growing interest, challenges remain in integrating CBD into mainstream medical practice. Regulatory inconsistencies, lack of standardized dosing guidelines, and variable product quality limit its widespread adoption. Future research must emphasize rigorous methodology, long-term safety data, and population-specific outcomes to ensure clinical validity. In addition, harmonizing international legal frameworks and establishing comprehensive educational resources for healthcare professionals will be essential to realizing CBD's full therapeutic potential.

Clinical and ethical considerations

The clinical use of CBD presents a unique set of considerations that extend beyond pharmacology and efficacy. One of the foremost challenges is the lack of standardized dosing guidelines across indications, age groups, and formulations. Because CBD products vary in concentration, bioavailability, and purity, clinicians often rely on empirical titration rather than evidence-based protocols. This variability complicates efforts to predict therapeutic outcomes and manage adverse effects. Moreover, many patients self-initiate CBD use without medical supervision, leading to unrecognized interactions, subtherapeutic dosing, or the use of unverified products.

Another key issue is product quality and labeling accuracy. While pharmaceutical-grade CBD, such as Epidiolex, undergoes rigorous testing and regulatory oversight, most over-the-counter (OTC) CBD products do not. Independent analyses have shown significant discrepancies between labeled and actual CBD content, as well as contamination with THC, heavy metals, pesticides, and residual solvents. These quality control failures have both clinical and legal implications, particularly for patients subject to drug testing or with conditions sensitive to THC exposure. Clinicians must remain vigilant and encourage the use of products that are third-party lab-verified and certified under good manufacturing practices.

Ethical considerations also arise in the context of informed consent and vulnerable populations. Many patients turn to CBD after exhausting conventional therapies, sometimes driven by anecdotal reports or marketing claims that lack scientific backing. In such cases, physicians must ensure that patients understand the current evidence base, limitations, and uncertainties associated with CBD use. Special attention is warranted when recommending CBD for pediatric patients, pregnant individuals, or those with cognitive impairments. Clinicians must balance hope and innovation with critical appraisal and evidence-based care.

Finally, access and affordability present important socioeconomic considerations. Pharmaceutical CBD can be prohibitively expensive and is not consistently covered by insurance, while OTC products vary widely in price and quality. In addition, patients living in jurisdictions with restrictive cannabis laws may face legal barriers or stigma when pursuing CBD therapy. These disparities raise questions of equity and justice in access to potentially beneficial treatments. Moving forward, equitable regulation, insurance reimbursement models, and education of providers and patients will be essential to the responsible integration of CBD into clinical practice.

Conclusion

CBD has emerged as a versatile and promising compound within the therapeutic landscape, offering potential benefits across a broad spectrum of neurological, psychiatric, dermatologic, and inflammatory conditions. Its nonintoxicating profile, multifaceted mechanisms of action, and favorable tolerability make it an attractive candidate for primary and adjunctive therapy. While the strongest clinical evidence currently supports its use in pediatric epilepsy, expanding research continues to explore CBD's utility in anxiety, pain, neurodegeneration, and dermatologic health.

Despite its potential, the integration of CBD into evidence-based medical practice is still evolving. Regulatory inconsistencies, lack of standardized dosing, product quality concerns, and limited long-term safety data pose significant barriers to routine clinical adoption. Clinicians must remain cautious and informed, balancing patient interest with the current evidence base and guiding therapy while emphasizing safety, transparency, and shared decision-making.

Special populations—including pediatric, geriatric, pregnant, and medically complex patients—require individualized assessment and heightened vigilance, particularly in the context of polypharmacy and organ dysfunction. Ethical considerations regarding accessibility, product labeling, and patient autonomy are also central to the responsible use of CBD. As research deepens and clinical experience grows, these challenges can be addressed through robust clinical trials, regulatory oversight, and practitioner education.

In summary, cannabidiol represents both an opportunity and a responsibility for modern medicine. Continued interdisciplinary collaboration among clinicians, researchers, regulators, and patients will be essential in defining CBD's precise role in clinical care. As our understanding advances, CBD may evolve from a complementary alternative to a cornerstone of personalized, integrative therapy.

References

Bergamaschi, M.M., Queiroz, R.H.C., Chagas, M.H.N., De Oliveira, D.C.G., De Martinis, B.S., Kapczinski, F., Quevedo, J., Roesler, R., Schröder, N., Nardi, A.E., Martín-Santos, R., Hallak, J.E.C., Zuardi, A.W., Crippa, J.A.S., 2011. Cannabidiol reduces the anxiety induced by simulated public speaking in treatment-nave social phobia patients. Neuropsychopharmacology. 0893133X. 36 (6), 1219–1226. doi:10.1038/npp.2011.6.

Cassano, T., Villani, R., Pace, L., Carbone, A., Bukke, V.N., Orkisz, S., Avolio, C., Serviddio, G., 2020. From cannabis sativa to cannabidiol: promising therapeutic candidate for the treatment of neurodegenerative diseases. Front. Pharmacol. 16639812. 11. Available from http://www.frontiersin.org/Pharmacology. doi:10.3389/fphar.2020.00124.

Chesney, E., Oliver, D., Green, A., Sovi, S., Wilson, J., Englund, A., Freeman, T.P., McGuire, P., 2020. Adverse effects of cannabidiol: a systematic review and meta-analysis of randomized clinical trials. Neuropsychopharmacology. 0893-133X. 45 (11), 1799–1806. doi:10.1038/s41386-020-0667-2.

Devinsky, O., Cross, J.H., Laux, L., Marsh, E., Miller, I., Nabbout, R., Scheffer, I.E., Thiele, E.A., Wright, S., 2017. Trial of cannabidiol for drug-resistant seizures in the dravet syndrome. N. Engl. J. Med. 15334406. 376 (21), 2011–2020. Available from: http://www.nejm.org/doi/pdf/10.1056/NEJMoa1611618. doi:10.1056/NEJMoa1611618.

Häuser, W., Petzke, F., Fitzcharles, M.A., 2018. Efficacy, tolerability and safety of cannabis-based medicines for chronic pain management—an overview of systematic reviews. Eur. J. Pain. 1090-3801. 22 (3), 455–470. doi:10.1002/ejp.1118.

Hurd, Y.L., Spriggs, S., Alishayev, J., Winkel, G., Gurgov, K., Kudrich, C., Oprescu, A.M., Salsitz, E., 2019. Cannabidiol for the reduction of cue-induced craving and anxiety in drug-abstinent individuals with heroin use disorder: a double-blind randomized placebo-controlled trial. Am. J. Psychiatry. 0002-953X. 176 (11), 911–922. doi:10.1176/appi.ajp.2019.18101191.

Ibeas Bih, C., Chen, T., Nunn, A.V.W., Bazelot, M., Dallas, M., Whalley, B.J., 2015. Molecular targets of cannabidiol in neurological disorders. Neurotherapeutics. 18787479. 12 (4), 699–730. Available from: http://www.springerlink.com/content/1933-7213/. doi:10.1007/s13311-015-0377-3.

Mechoulam, R., Shvo, Y., 1963. Hashish—I. Tetrahedron. 00404020. 19 (12), 2073–2078. doi:10.1016/0040-4020(63)85022-x.

U.S. Congress, n.d. Agriculture Improvement Act of 2018. 2.

VanDolah, H.J., Bauer, B.A., Mauck, K.F., 2019. Clinicians' guide to cannabidiol and hemp oils. Mayo Clin. Proc. 19425546. 94 (9), 1840–1851. Available from http://www.journals.elsevier.com/mayo-clinic-proceedings. doi:10.1016/j.mayocp.2019.01.003.

Zendulka, O., Dovrtělová, G., Nosková, K., Turjap, M., Šulcová, A., Hanuš, L., Juřica, J., 2016. Cannabinoids and cytochrome P450 interactions. Curr. Drug Metab. 13892002. 17 (3), 206–226. doi:10.2174/1389200217666151210142051.

CHAPTER

Phytocannabinoids: Tetrahydrocannabinol (THC)

5

Roshni Patel

RNP Regenerative Medicine, Center of Excellence in Pain and Regenerative Medicine, Quantum Brain Health, Farmington, CT, United States

Introduction

The primary psychoactive phytocannabinoid found in *Cannabis sativa* is Δ9-tetrahydrocannabinol (THC), first discovered and structurally characterized by Gaoni and Mechoulam in 1964 at the Hebrew University of Jerusalem (Gaoni and Mechoulam, 1964).

This discovery was the catalyst for cannabinoid research and helped in understanding the therapeutic and psychoactive effects of THC (Mechoulam et al., 2014).

With over 100 identified cannabinoids in the cannabis plant, THC is the most researched due to its clinical implications and unique pharmacological profile (Appendino, 2020).

THC's therapeutic effects mainly arise from its interactions with the endocannabinoid system (ECS), primarily the cannabinoid receptors CB1 and CB2, naturally occurring cannabinoids in the body, and the enzymes involved in their metabolism (Zou and Kumar, 2018).

CB1 receptors are mainly found in the central nervous system, especially in areas of the brain linked to cognition, memory, movement coordination, and pain processing, while CB2 receptors are primarily located in immune and peripheral tissues (Howlett and Abood, 2017).

THC is a partial agonist at both receptor subtypes, but with very different affinities and functional outcomes, which account for its diverse pharmacological effects, including analgesia, appetite stimulation, and psychoactive responses (Pertwee, 2008).

Systematic reviews and metaanalyses have shown that THC effectively reduces nausea and vomiting related to chemotherapy. The FDA-approved medications like dronabinol and nabilone, both THC-based formulations, have been found to work better at preventing nausea compared to standard treatments (Sallan et al., 1975; Tramèr et al., 2001).

In chronic pain management, cannabinoid medications containing THC have been shown to provide modest yet meaningful pain relief in chronic neuropathic

Cannabinoids for Chronic Pain. https://doi.org/10.1016/B978-0-443-36773-1.00002-7

pain. Several randomized controlled trials have found improvements in pain levels and overall quality of life (Whiting et al., 2015; McDonagh et al., 2022).

Also, in HIV-related wasting or cancer cachexia, THC has been effective in stimulating appetite (Badowski and Yanful, 2018).

In addition, THC has been beneficial in reducing muscle spasticity symptoms in individuals with multiple sclerosis (Zajicek et al., 2003).

Chemical and pharmacokinetic profile

Δ^9-tetrahydrocannabinol (THC) is characterized by a 21-carbon terpenophenolic skeleton composed of polyketide and terpenoid biosynthetic precursors derived from acetyl-CoA and isoprenyl pyrophosphate pathways, respectively (Grotenhermen, 2003).

The molecular formula $C_{21}H_{30}O_2$ has multiple stereoisomeric configurations. The therapeutic and psychoactive effects are mediated mainly by the naturally occurring (−)-trans-Δ^9-tetrahydrocannabinol isomer, which is stereoselective in receptor binding and essential for cannabinoid receptor activation (Lucas et al., 2018).

The tricyclic dibenzopyran structure has three key ligand-receptor interactions that elicit a specific biological response: the phenolic hydroxyl-bearing aromatic A-ring, the oxygen-containing pyran B-ring, and the cyclohexenyl C-ring with the characteristic pentyl side chain at the C3 position (Hansen et al., 2024).

The pharmacokinetic interactions of THC are due to its lipophilic interactions across biological membranes, contributing to rapid tissue penetration and presenting formulation challenges for pharmaceutical development (Sharma et al., 2012). The hydrophobic nature requires specialized delivery systems to achieve sufficient aqueous solubility for systemic absorption, especially for oral administration routes where solubility represents a significant pharmacokinetic limitation (Lucas et al., 2018).

The molecular geometry and conformational flexibility of the dibenzopyran structure allow THC to function as a partial agonist at both CB1 and CB2 cannabinoid receptors, with lower intrinsic activity compared to full agonists while maintaining enough receptor affinity to elicit a therapeutic response (Grotenhermen, 2003).

Absorption kinetics and route-dependent bioavailability

THC absorption depends on the administration route due to the complex interactions between anatomical, physical, chemical, and metabolic properties (Hansen et al., 2024).

The most rapid systemic absorption is via the pulmonary route: inhalation. THC crosses the extensive alveolar-capillary system to achieve peak plasma concentrations within 6–10 min postinhalation, corresponding to the cardiovascular circulation time from lungs to brain (Lucas et al., 2018). With inhalation, the bioavailability ranges from 10% to 35%, mainly due to variability between individuals, primarily due to differences in smoking behaviors, such as how deeply one inhales, holding one's breath, puff frequency, as well as variations in lung function and alveolar surface area (Grotenhermen, 2003).

With oral administration, THC absorption is very different due to undergoing extensive first-pass hepatic metabolism mediated by the cytochrome P450 enzyme systems, substantially reducing systemic bioavailability to 4%–12% and delaying peak plasma concentrations, which occur 1.5–3 h postingestion (Hansen et al., 2024).

The gastrointestinal absorption of THC depends on specific food interactions, with lipid-containing meals enhancing solubility through bile salt micelle formation and facilitating lymphatic transport, potentially increasing bioavailability by 2-3 fold compared to fasting conditions (Lucas et al., 2018).

The significant interindividual oral bioavailability is due to the differential expression of intestinal and hepatic cytochrome P450 enzymes, gastric emptying rates, and gastrointestinal transit times, requiring individualized clinical application dosing strategies (Wright and Katamesh, 2024).

Sublingual and oral mucosal delivery systems bypass the first-pass metabolism through direct absorption via the oral mucosa, which more closely follows the pharmacokinetic profiles of oral administration (Grotenhermen, 2003).

These delivery routes have bioavailability rates of 15%–25%, with onset times of 15–45 min; however, absorption may be affected by saliva production and swallowing reflexes that redirect the drug to gastrointestinal absorption pathways, subject to hepatic metabolism (Lucas et al., 2018).

Distribution patterns

After systemic absorption, THC has a biphasic distribution: first, there is an initial rapid equilibration with highly perfused organs during the distribution phase (α-phase), followed by slower redistribution into lipophilic tissue compartments during the elimination phase (β-phase) (Hansen et al., 2024).

THC's lipophilicity results in extensive binding within body tissues, with a distribution volume ranging from about 3.4 to 9.6 L/kg, reflecting its wide distribution beyond just the bloodstream and significantly surpassing the total amount of water in the body (Lucas et al., 2018).

THC enters the brain quickly after inhalation or intravenous administration, reaching similar concentrations in the brain and blood plasma within a few minutes. This rapid entry into the brain occurs because THC easily crosses the blood-brain barrier by passively diffusing through lipid-rich membranes (Grotenhermen, 2003).

THC's extensive adipose tissue distribution represents an important pharmacokinetic consideration, especially in chronic users, where prolonged accumulation in lipophilic compartments creates deep tissue reservoirs that undergo slow-release kinetics (Sharma et al., 2012). That is why the extended elimination half-life is observed in frequent users, where elimination half-lives may extend to 5–13 days compared to 1–3 days in occasional users, reflecting the gradual redistribution from deep tissue compartments back into systemic circulation (Hansen et al., 2024).

Protein binding of THC to albumin and lipoproteins in plasma is approximately 95%–97%, leaving the unbound fraction available for pharmacological activity and tissue distribution (Lucas et al., 2018).

Also, the placental transfer of THC represents a significant clinical concern during pregnancy, with its lipophilic characteristics allowing for transplacental passage and potential accumulation in fetal tissues (Lucas et al., 2018).

Similarly, THC demonstrates substantial excretion into breast milk, with milk-to-plasma ratios attributed to its lipophilic properties, raising concerns about infant exposure during breastfeeding (Sharma et al., 2012).

Hepatic metabolism and cytochrome P450 enzymatic pathways

The hepatic cytochrome P450 and CYP2C9 are the primary metabolic pathways responsible for the initial hydroxylation of THC, forming the pharmacologically active metabolite 11-hydroxy-THC (11-OH-THC) (Lucas et al., 2018).

The predominant metabolite, 11-OH-THC, has significant psychoactive potency, with equivalent or potentially increased CB1 receptor binding affinity compared to the parent compound, alongside superior blood-brain barrier penetration characteristics that potentially contribute to the pharmacological profile observed after oral administration (Grotenhermen, 2003).

Subsequent oxidative metabolism converts 11-OH-THC to the pharmacologically inactive carboxylated metabolite 11-nor-9-carboxy-THC (THC-COOH), which undergoes phase II glucuronidation reactions catalyzed by UDP-glucuronosyltransferase enzymes before biliary and renal elimination (Lucas et al., 2018).

The metabolism demonstrates capacity-limited kinetics at higher doses, with evidence of saturable metabolism contributing to nonlinear pharmacokinetics and disproportionate increases in systemic exposure with dose escalation (Hansen et al., 2024).

Genetic polymorphisms in CYP2C9 significantly impact THC clearance, with variant alleles *2 and *3 encoding enzymes with reduced catalytic activity compared to the wild-type *1 allele (Wright, 2024). Individuals carrying the CYP2C93/*3 genotype (poor metabolizers, approximately 1%–3% of Caucasian populations) demonstrate threefold higher systemic THC exposure and 70% lower THC-COOH formation compared to extensive metabolizers, necessitating dose adjustments to prevent excessive pharmacological effects (Sachse-Seeboth et al., 2009).

Intermediate metabolizers (heterozygous *1/*3 or *1/*2 genotypes) exhibit intermediate metabolic capacity with corresponding intermediate systemic exposure profiles (Wright, 2024).

Extrahepatic metabolism contributes substantially to THC clearance, particularly in tissues relevant to specific administration routes (Lucas et al., 2018).

Intestinal CYP2C9 expression mediates significant first-pass metabolism following oral administration, while pulmonary CYP450 activity may contribute to local metabolism following inhalation exposure (Hansen et al., 2024).

Brain tissue demonstrates limited but measurable CYP450 expression, potentially contributing to local THC metabolism and metabolite formation within the central nervous system (Grotenhermen, 2003).

Elimination kinetics

The elimination kinetics of THC are complicated, with an initial rapid decline due to tissue distribution, followed by slower terminal elimination due to redistribution from deep tissue compartments (Hansen et al., 2024).

The initial distribution half-life is 6 min, representing the rapid equilibration with highly perfused tissues, while the terminal elimination half-life exhibits pronounced interindividual variability, ranging from 20 to 30 h in occasional users to several days in chronic consumers (Lucas et al., 2018).

This extended elimination phase reflects the slow release of THC from adipose tissue reservoirs, where accumulation following chronic exposure creates a persistent source for systemic re-exposure long after cessation of use (Grotenhermen, 2003).

Elimination occurs primarily through biliary excretion (65%–80%), with secondary renal elimination (20%–35%) of glucuronidated metabolites (Sharma et al., 2012).

The predominant urinary metabolite, THC-COOH-glucuronide, serves as the primary biomarker for cannabis exposure in forensic and clinical toxicology applications, with detection windows extending weeks to months in chronic users due to the extensive tissue reservoir and slow elimination kinetics (Lucas et al., 2018).

Fecal elimination primarily involves unchanged THC and hydroxylated metabolites, reflecting biliary excretion pathways and limited enterohepatic recirculation (Hansen et al., 2024).

The prolonged elimination kinetics contribute to significant accumulation potential with repeated dosing, particularly in chronic users, where steady-state concentrations may require weeks to achieve due to the long terminal half-life (Grotenhermen, 2003).

This has important implications for therapeutic dosing regimens, workplace drug testing protocols, and forensic interpretation of analytical results, where the detection of THC metabolites may not correlate with use or acute impairment (Wright, 2024).

Mechanisms of action: Tetrahydrocannabinol

ECS architecture and receptor-mediated signal transduction

Δ^9-tetrahydrocannabinol produces its wide-ranging pharmacological and psychotropic effects through interactions with the ECS. This neuromodulatory network regulates many physiological processes, including nociceptive processing, metabolic regulation, emotional modulation, motor coordination, and immunological surveillance (Pertwee, 2008).

The ECS is made up of two primary G-protein-coupled receptor subtypes—cannabinoid receptor type 1 (CB1) and cannabinoid receptor type 2 (CB2)—which have a 68% sequence homology within their seven-transmembrane domains but distinct anatomical distribution patterns and functional profiles (Munro et al., 1993; Brust et al., 2023).

CB1 receptors are the most abundantly expressed G-protein-coupled receptors within the central nervous system, with the highest density concentrations in regions critical for cognitive processing, emotional regulation, and motor control. CB2 receptors have the highest concentration in peripheral immune tissues and undergo significant upregulation under pathophysiological inflammatory conditions (Howlett and Abood, 2017).

The partial agonist characteristics of THC result in maximal receptor activation levels substantially below those achieved by full agonists, creating a distinct signature characterized by ceiling effects for specific biological responses while maintaining the capacity for functional antagonism in high receptor density environments (Di Marzo, 2009). This pharmacological property contributes significantly to THC's therapeutic window and may explain the development of tolerance phenomena that paradoxically enhance therapeutic utility by mitigating adverse psychoactive effects while preserving clinical benefits.

CB1 receptor

CB1 receptors are found at presynaptic terminals throughout the central and peripheral nervous systems, functioning as critical modulators of neurotransmitter release through retrograde signaling mechanisms (Pertwee, 2008).

Once THC binds to the CB1 receptor, it couples with the Gi/o proteins, initiating a cascade of intracellular events, including the inhibition of adenylyl cyclase activity, which results in decreased cyclic adenosine 3',5'-monophosphate (cAMP) concentrations, modulation of voltage-gated calcium channel conductance, and activation of potassium channels (Howlett and Abood, 2017). This leads to presynaptic inhibition of neurotransmitter release, affecting glutamatergic and GABAergic transmission, thereby altering neuronal excitability and synaptic plasticity mechanisms.

The anatomical distribution of CB1 receptors correlates with regions highly susceptible to THC-induced alterations in cognitive function, emotional processing, and motor coordination (Pertwee, 2008).

Within the hippocampal formation, CB1 receptor activation by THC disrupts long-term potentiation mechanisms important for memory consolidation, providing the molecular basis for the characteristic short-term memory impairment associated with acute THC intoxication. In the amygdala, CB1 receptor activation can lead to anxiety and paranoid ideation, which are frequently observed in predisposed individuals. Basal ganglia CB1 receptor activation affects the motor coordination deficits and altered temporal perception characteristic of THC intoxication (Pertwee, 2008).

CB2 receptor

CB2 receptors are expressed within the immune cells: monocytes, macrophages, T and B lymphocytes, dendritic cells, and microglial cells. They are essential modulators of inflammatory responses and immune system homeostasis (Li et al., 2023).

THC activation of CB2 receptors starts an inflammatory signaling cascade characterized by the inhibition of proinflammatory cytokine production, including tumor necrosis factor-α, interleukin-1β, and interleukin-6. It also simultaneously promotes the production of antiinflammatory mediators such as interleukin-10 (Cabral and Staab, 2005). These immunomodulatory effects contribute significantly to THC's therapeutic efficacy in inflammatory conditions, including multiple sclerosis, inflammatory bowel disease, and rheumatoid arthritis.

CB2 receptor activation by THC modulates immune cell migration patterns within the central nervous system and peripheral tissues, influencing inflammatory cell populations' recruitment and activation states (Walter and Stella, 2004). This mechanism is essential in neuroinflammatory conditions, where microglial CB2 receptor activation by THC may contribute to neuroprotective effects by suppressing inflammatory cytokine cascades and promoting tissue repair mechanisms.

Noncannabinoid receptor interactions and polypharmacological mechanisms

The complicated therapeutic and side effect profiles are due to interactions beyond those of CB1 and CB2 receptors, at multiple noncannabinoid receptor systems (Di Marzo, 2009). THC modulates dopaminergic neurotransmission through indirect mechanisms involving CB1 receptor-mediated disinhibition of dopaminergic neurons in the mesolimbic reward pathway, resulting in increased dopamine concentrations within the nucleus accumbens and contributing to both the reinforcing properties and therapeutic mood-elevating effects of THC (Spiller et al., 2019). This dopaminergic modulation occurs through indirect network effects rather than direct dopamine receptor interaction, involving CB1 receptor activation on GABAergic interneurons that typically provide tonic inhibition of dopaminergic cell firing.

THC interacts with the endogenous opioidergic system by involving μ-opioid receptors and shared downstream signaling pathways, including mitogen-activated protein kinase cascades and β-arrestin recruitment mechanisms (Pertwee, 2008). This pharmacological synergy enables the potentiation of endogenous and exogenous opioid analgesic effects, with possible roles in opioid dose reduction strategies in pain management protocols, while minimizing opioid-related adverse effects and dependency risks.

Another noncannabinoid target is the transient receptor potential vanilloid 1 (TRPV1) channels, which mediate nociceptive, thermoregulatory, and inflammatory responses, and peroxisome proliferator-activated receptor γ (PPARγ), which influences gene transcription programs related to metabolic homeostasis and immune function regulation (Di Marzo et al., 2009). THC's interaction with TRPV1 channels contributes to its analgesic properties and neuropathic pain states.

THC also has significant modulatory effects on arachidonic acid metabolism pathways through the suppression of cyclooxygenase-2 (COX-2) enzymatic activity, reducing proinflammatory prostaglandin E_2 synthesis (Pertwee, 2008). This interaction contributes substantially to THC's antiinflammatory and analgesic properties in models of neuropathic and inflammatory pain, where prostaglandin-mediated sensitization represents a crucial pathophysiological component.

THC-mediated suppression of COX-2 activity occurs through direct enzymatic inhibition and transcriptional downregulation of COX-2 expression, resulting in the suppression of inflammatory prostaglandin cascades (Di Marzo et al., 2009). This antiinflammatory effect is essential in chronic inflammatory conditions, where sustained prostaglandin production contributes to tissue damage and pain sensitization. The COX-2 suppression by THC is dose-dependent, with prolonged antiinflammatory effects extending beyond the duration of acute psychoactive responses.

Pharmacological effects of tetrahydrocannabinol

Psychoactive effects

The psychoactive properties of Δ^9-tetrahydrocannabinol are the most recognized manifestation of its pharmacological activity, mediated primarily through CB1 receptor activation within neuroanatomical regions important for cognitive processing, emotional regulation, and sensorimotor integration (Grotenhermen, 2003).

The "high" associated with THC comprises multiple subjective experiences, including euphoria, altered temporal perception, enhanced sensory acuity, cognitive disruption, and mood alterations that directly correlate with CB1 receptor density and distribution patterns across functionally specialized brain regions (Ashton, 2001). These psychotropic effects emerge through THC-induced disruption of normal endocannabinoid signaling within key neuronal circuits, resulting in alterations in neurotransmitter release and synaptic plasticity mechanisms that control higher-order cognitive functions.

In the hippocampal formation, CB1 receptor activation by THC disrupts long-term potentiation and synaptic plasticity processes essential for memory consolidation and retrieval processes, which is the neurobiological basis for the characteristic short-term memory impairment observed during acute cannabis intoxication (Grotenhermen, 2003).

Electrophysiological studies demonstrate that THC suppresses both GABAergic and glutamatergic neurotransmission within hippocampal CA1 and CA3 regions, disrupting the excitatory-inhibitory balance required for optimal information processing and storage for memory processing (Ashton, 2001).

Amygdala CB1 receptor activation contributes to the emotional disturbances of THC intoxication, including both anxiolytic and anxiogenic responses, which are dose and individual-dependent (Weizman et al., 2018).

Functional neuroimaging studies reveal that acute THC administration increases amygdala reactivity to emotional stimuli while simultaneously disrupting prefrontal cortical control mechanisms, creating a neurobiological environment for the increased emotional lability and anxiety susceptibility observed in specific individuals (Weizman et al., 2018).

Activation of prefrontal cortical CB1 receptors by THC produces disruptions in executive function, working memory, and attention, which manifest as cognitive clouding, impaired decision-making, and altered risk assessment capabilities (Grotenhermen, 2003). These cognitive impairments result from THC-induced suppression of glutamatergic neurotransmission within prefrontal-subcortical circuits critical for cognitive control and behavioral regulation. The dose-dependent nature of these executive function deficits has a threshold effect, with low-dose exposures producing subtle attentional changes, while higher doses result in pronounced cognitive dysfunction that significantly impairs complex task performance.

Analgesic mechanisms

Δ^9-tetrahydrocannabinol has significant analgesic properties mediated through central nervous system CB1 receptor activation and peripheral CB2 receptor-mediated antiinflammatory response, effective against neuropathic and inflammatory pain conditions refractory to conventional analgesics (Russo, 2008).

The ECS has tonic activity in pain modulation, as evidenced by the hyperalgesic responses produced by CB1 receptor antagonist administration, indicating continuous endogenous pain suppression that can be pharmacologically enhanced through exogenous THC administration (Russo, 2008).

Central analgesic mechanisms involve THC activation of CB1 receptors located throughout the neuraxis, including critical pain-processing regions such as the periaqueductal gray, rostral ventromedial medulla, and spinal dorsal horn, where cannabinoid receptor activation produces antinociceptive effects up to ten-fold more potent than morphine in electrophysiological assays (Russo, 2008).

In the spinal cord dorsal horn regions, CB1 receptor activation by THC produces presynaptic inhibition of nociceptive neurotransmitter release, particularly substance P and glutamate, while also modulating GABAergic and glycinergic inhibitory mechanisms that gate nociceptive transmission to supraspinal processing centers (Ware et al., 2008).

A neuroimaging study documented that THC analgesia involves a functional disconnection between the anterior cingulate cortex and the dorsolateral prefrontal cortex, two critical regions involved in cognitive-emotional pain regulation, suggesting that cannabinoid analgesia operates through a separate mechanism from opioid pathways (Weizman et al., 2018). This functional decoupling correlates with subjective pain relief and possibly represents a novel neurobiological target for analgesic intervention that circumvents the tolerance and dependence associated with opioid receptor-mediated analgesia.

Peripheral analgesia involves CB2 receptor activation on immune cells within injured tissues, producing localized antiinflammatory effects through the suppression of proinflammatory cytokine cascades, which include tumor necrosis factor-α, interleukin-1β, and interleukin-6, while also promoting the release of antiinflammatory mediators such as interleukin-10 (Ware et al., 2008). This peripheral antiinflammatory function is essential in neuropathic pain conditions, where immune system activation contributes to persistent pain sensitization and tissue damage.

Antiemetic effects

The antiemetic properties of THC are among the earliest recognized and clinically validated therapeutic applications, with synthetic THC formulations, such as dronabinol and nabilone, receiving FDA approval for chemotherapy-induced nausea and vomiting (CINV). Clinical studies have shown superiority over the antiemetic agents metoclopramide and prochlorperazine (5-HT_3 receptor antagonists) in CINV (Ng and Keshock, 2023). THC's antiemetic activity involves CB1 receptor activation within the dorsal vagal complex of the brainstem, targeting the area postrema and nucleus tractus solitarius, the essential components of the central emetic reflex circuitry responsible for nausea and vomiting responses (Grotenhermen, 2003).

THC also affects serotonergic transmission by inhibiting 5-HT_3 receptor-mediated signaling pathways, the primary triggers for chemotherapy-induced emetic response (Grotenhermen, 2003), which is essential for breakthrough nausea when combined with other antiemetics.

THC's antiemetic activity has a steep dose-response curve with a narrow therapeutic window, which requires careful titration to achieve optimal antiemetic benefit while minimizing psychoactive adverse effects that may affect patient compliance (Ng and Keshock, 2023).

Appetite stimulation

THC has potent orexigenic effects mediated through complex hypothalamic mechanisms involving CB1 receptor activation within the arcuate nucleus and related feeding control centers. Cannabinoid signaling affects critical neuropeptide systems: neuropeptide Y, agouti-related peptide, and melanocortin pathways, which influence appetite regulation and energy homeostasis (Grotenhermen, 2003). The appetite-stimulating effects of THC not only cause hunger but also involve enhanced food palatability, increased meal frequency, and altered macronutrient preferences that contribute to significant weight gain in appropriate clinical contexts.

The neurobiological mechanisms underlying THC-induced appetite stimulation involve activating the mesolimbic reward pathways by enhancing dopaminergic signaling within the nucleus accumbens, increasing the value of food consumption, and promoting food-seeking behaviors (Wenzel and Cheer, 2017). This reward system activation synergizes with hypothalamic feeding circuits, resulting in increased motivation to consume food and enhanced satisfaction from eating experiences. This contributes to its effectiveness in clinical wasting syndromes.

THC also has peripheral orexigenic effects, seen on gastrointestinal hormones via enhancement of ghrelin signaling and suppression of satiety signals of leptin and cholecystokinin, creating a coordinated central and peripheral drive for increased food intake (Grotenhermen, 2003). These hormonal effects occur through direct CB1 receptor activation in gastrointestinal tissues and indirect impact through vagal and sympathetic nervous system pathways that coordinate feeding behavior with metabolic status.

Clinical applications of THC appetite stimulation have significant therapeutic value in HIV-associated wasting syndrome, cancer cachexia, and anorexia nervosa, especially when traditional appetite stimulants demonstrate limited efficacy (Ng and Keshock, 2023).

THC appetite stimulation has a rapid onset with sustained duration, usually producing increased food intake within 1–2 h of administration, maintaining enhanced appetite for 4–6 h depending on dosing parameters and individual pharmacokinetic factors (Grotenhermen, 2003). This pharmacokinetic effect benefits meal-timing strategies that optimize nutritional intake while minimizing interference with regular daily activities.

Muscle relaxation

THC has significant muscle relaxant and antispasmodic effects, mediated through CB1 receptor activation in the spinal cord motor circuits and supraspinal motor control regions, reducing muscle tone, spasm frequency, and movement-related pain in neurological conditions with pathological muscle hyperactivity (Lynch and Campbell, 2011).

In multiple sclerosis, spinal cord injury, and other neurological conditions where traditional antispasmodic medications demonstrate limited efficacy or produce intolerable adverse effects, THC serves as an option.

THC produces muscle relaxation through CB1 receptor-mediated motor neuron excitability within spinal cord ventral horn regions: cannabinoid signaling inhibits excessive cholinergic transmission at neuromuscular junctions and modulates spinal reflex pathways that contribute to pathological muscle hyperactivity (Lynch and Campbell, 2011).

THC's antispasmodic effects also involve modulation of motor cortex and basal ganglia circuits that control voluntary movement and muscle tone regulation (Grotenhermen, 2003). CB1 receptor activation within these regions affects motor control, resulting in reduced involuntary movements, decreased muscle rigidity, and improved coordination in appropriate clinical populations.

The dose-response characteristics of THC muscle relaxation have a gradual onset with sustained duration, usually requiring several days to weeks of consistent dosing to achieve maximal therapeutic benefits (Lynch and Campbell, 2011). This points to the complexity of neuroplasticity in therapeutic muscle relaxation and separates THC's antispasmodic effects from the immediate but short-lived effects of traditional muscle relaxants.

Medical indications: FDA-approved THC-based pharmaceutical formulations

Dronabinol

Dronabinol, marketed as Marinol (capsule formulation) and Syndros (oral solution), represents the first successful pharmaceutical development of synthetic Δ^9-tetrahydrocannabinol, receiving initial FDA approval in 1985 for CINV and HIV/AIDS-related anorexia, following comprehensive clinical investigation demonstrating superior efficacy compared to conventional therapeutic alternatives (O'Donnell et al., 2023).

Clinical efficacy data supporting dronabinol approval demonstrate significant antiemetic benefits in patients experiencing breakthrough nausea and vomiting despite optimized conventional antiemetic regimens, with particular effectiveness observed in delayed-phase CINV that proves refractory to 5-HT_3 receptor antagonists and neurokinin-1 receptor antagonists (Badowski, 2017). Randomized controlled trials comparing dronabinol with prochlorperazine and metoclopramide revealed superior patient-reported outcomes and reduced rescue medication requirements (O'Donnell et al., 2023).

The appetite stimulation indication for dronabinol emerged from controlled clinical investigations in HIV/AIDS patients experiencing progressive weight loss refractory to conventional nutritional interventions, with trials demonstrating statistically significant increases in body weight, appetite scores, and nutritional markers compared to placebo controls (Badowski, 2017). Average weight gain was 2–4 kg over 6–12 week treatment periods, with clinically meaningful improvements in patient functional status and quality of life measures (Badowski, 2017).

Nabilone

Nabilone (Cesamet) is a synthetic cannabinoid analog of THC, created to address the pharmacokinetic limitations of dronabinol. It has increased bioavailability (approximately 60% compared to 10%–20% for dronabinol) and improved dose-linear pharmacokinetics, which allow for a more predictable clinical response (Haney et al., 2013). Nabilone has an extended half-life and reduced first-pass hepatic metabolism, which provides therapeutic advantages for patients requiring consistent cannabinoid exposure (McGilveray, 2005).

FDA approval for nabilone in CINV management, in controlled clinical trials, demonstrated equivalent or superior antiemetic efficacy compared to dronabinol and traditional antiemetics, with advantages in patients requiring flexible dosing regimens or experiencing tolerance to alternative interventions (Badowski, 2017). The increased bioavailability of nabilone allows effective antiemetic activity at lower doses when compared to dronabinol, reducing dose-limiting psychoactive adverse effects while maintaining therapeutic benefits (McGilveray, 2005).

Nabiximols (Sativex)

Nabiximols, Sativex, approved for multiple sclerosis-related spasticity, is the first standardized botanical drug derived from cannabis extracts to receive regulatory approval. It combines equal ratios of THC and cannabidiol (CBD) in an oromucosal spray formulation that bypasses first-pass hepatic metabolism, providing precise dosing control (Barnes, 2006). Each spray delivers 2.7 mg of THC and 2.5 mg of CBD in a standardized botanical matrix of minor cannabinoids and terpenes.

Clinical effectiveness data from a Belgian study with 276 patients showed sustained therapeutic benefits, with mean spasticity numerical rating scale scores improving from 8.1 at baseline to 5.2 at week 4, 4.6 at week 8, and 4.1 at week 12, representing clinically significant improvements exceeding the minimal clinically important difference threshold of a 20% reduction (D'hooghe et al., 2021).

Contraindications

Psychiatric disorders

Δ^9-tetrahydrocannabinol is an absolute contraindication in individuals with established psychotic disorders, prodromal psychotic symptoms, or an elevated genetic predisposition for schizophrenia spectrum conditions due to epidemiological evidence demonstrating dose-dependent relationships between cannabis exposure and psychotic disorder development (Chesney et al., 2024). There is a significant association between cannabis use disorder and the development of schizophrenia, accounting for 10% to 50% of new schizophrenia cases (Tandon et al., 2024).

A meta-analysis of 83 longitudinal studies revealed that cannabis use precipitates psychotic symptom onset 2.7 years earlier than observed in nonusers. There is also a significant dose-response relationship, where heavy cannabis users had a 3.9-fold higher odds of developing psychotic symptoms compared to nonusers (Ortiz-Medina et al., 2018).

Neuroimaging has shown that, with acute THC exposure, there is an increased activation of the amygdala and anterior cingulate cortical regions, which are areas associated with fear processing and paranoid ideation. Chronic exposure during adolescence is associated with decreased cortical thickness in the prefrontal regions, particularly in males with high genetic risk scores for schizophrenia. These alterations provide the neurobiological evidence supporting clinical observations of THC-induced psychotic exacerbations and long-term vulnerability to psychotic disorder development.

Populations with elevated risks have up to 50% of individuals experiencing cannabis-induced psychosis subsequently developing either schizophrenia or bipolar disorder, pointing to the prognostic importance of identifying cannabis-related psychotic episodes (Starzer et al., 2018). Individuals with prodromal symptoms or ultra-high-risk states for psychosis have a 2.09-fold increased odds of lifetime cannabis use and a 5.49-fold increased prevalence of cannabis use disorder compared to healthy controls (Carney et al., 2017).

Cardiovascular disease

THC has a significant contraindication in patients with cardiovascular disease. The acute hemodynamic effects of dose-dependent increases in heart rate, blood pressure elevation, and orthostatic hypotension, which substantially increase myocardial oxygen demand and arrhythmogenic potential, are well documented (Subramaniam et al., 2019).

There is a 4.8-fold increased risk of myocardial infarction within 1 year of cannabis use, with the elevated risk persisting for several hours postexposure, and an annual myocardial infarction risk of 1.5%–3% among daily cannabis users (Subramaniam et al., 2019).

There are case reports of severe cardiovascular complications, including paroxysmal sustained ventricular tachycardia, cardiac arrest, and acute coronary syndrome, in young adults without cardiovascular risk factors after cannabis exposure, particularly with high-potency preparations (Yahud et al., 2020).

THC activates the CB1 receptors found throughout cardiovascular tissues, stimulating the sympathetic nervous system with parasympathetic system inhibition, resulting in tachycardia, increased contractility, and altered coronary perfusion dynamics (Singh, 2017).

Chronic THC exposure is associated with endothelial dysfunction, oxidative stress promotion, and atherosclerotic progression. In contrast, acute exposure may precipitate coronary artery vasospasm, platelet activation, and prothrombotic states, which are particularly dangerous in individuals with underlying coronary artery disease (Subramaniam et al., 2019).

Specific contraindications due to THC-induced hemodynamic stress, causing life-threatening cardiovascular events, include unstable angina pectoris, myocardial infarction or acute coronary syndrome, significant cardiac arrhythmias, including atrial fibrillation and ventricular tachycardia, uncontrolled hypertension, and advanced heart failure with reduced ejection fraction (Singh, 2017).

Pregnancy and lactation

Cannabis use during pregnancy and lactation represents an absolute contraindication. There is extensive evidence of placental transfer of THC and its metabolites, with accumulation in fetal neural tissues during critical neurodevelopmental windows (Mulligan and Hamre, 2023). THC has high lipophilicity, allowing for easy placental passage and preferential distribution to developing fetal brain regions, where ECS signaling modulates developmental pathways of neuronal migration, synaptogenesis, axonal pathfinding, and glial cell specification (Mulligan and Hamre, 2023).

There are significant associations between prenatal cannabis exposure and adverse neurodevelopmental outcomes, including deficits in executive function, attention regulation, working memory, and behavioral control, that persist throughout childhood and adolescence (Baranger et al., 2024).

Structural brain alterations found in children with documented prenatal cannabis exposure include reduced cortical thickness, compromised white matter integrity, and altered functional connectivity patterns in networks affecting cognitive and emotional regulation (Baranger et al., 2024).

Prenatal THC exposure disrupts genetic expressions essential for brain development, affecting neuronal proliferation, differentiation, and synaptic development by interfering with endogenous cannabinoid signaling cascades, which maintain developmental homeostasis (Mulligan and Hamre, 2023). CB1 receptor expression reaches adult levels in the human fetal brain by 24 weeks of gestation, with increased vulnerability during the second and third trimesters, when cortical development and synaptic maturation occur most rapidly.

THC is excreted into breast milk, and detection persists up to one-week post-consumption due to extensive lipophilic distribution and prolonged elimination kinetics. This creates a potential for protracted neonatal exposure during critical postnatal neurodevelopmental periods (Baranger et al., 2024).

Guidelines from the American College of Obstetricians and Gynecologists, the American Academy of Pediatrics, and the Society of Obstetricians and Gynecologists of Canada uniformly recommend complete cannabis abstinence during pregnancy and breastfeeding based on established evidence of developmental risks (Mulligan and Hamre, 2023).

Substance use disorder history

Individuals with present or a history of substance use disorders represent an essential contraindication for THC consumption due to shared neurobiological pathways mediating addiction vulnerability and demonstrated risks of cross-sensitization, relapse precipitation, and polysubstance abuse development (Ortiz-Medina et al., 2018).

THC activates mesolimbic dopaminergic reward circuitry through CB1 receptor-mediated increased dopamine release in the nucleus accumbens, with identical neurobiological substrates involved in alcohol, opioid, and stimulant addiction processes.

There are strong associations between cannabis use and relapse to other substances of abuse, especially in individuals in early recovery phases, when neurobiological systems involved in impulse control and decision-making remain compromised (Cheng et al., 2023).

Individuals with opioid use disorder have an increased risk of continued illicit opioid use and treatment dropout when cannabis use occurs concurrently, suggesting interference with recovery-oriented neuroplasticity and behavioral modification processes.

Cannabis use disorder itself has significant heritability and shares genetic risk loci with other psychiatric conditions, including schizophrenia, bipolar disorder, and other substance use disorders, indicating common underlying vulnerability factors that predispose individuals to addiction development across multiple substance classes (Cheng et al., 2023).

Central nervous system depressant coadministration

Concurrent administration of THC with CNS depressant medications, including benzodiazepines, opioid analgesics, barbiturates, sedating antihistamines, and alcohol, is a significant clinical contraindication due to synergistic CNS depression, with potentially fatal respiratory suppression and cardiovascular collapse (Subramaniam et al., 2019).

THC has intrinsic sedative properties that become markedly enhanced through pharmacodynamic interactions with other CNS depressants, creating unpredictable degrees of cognitive impairment, psychomotor dysfunction, and physiological depression.

Severe adverse events, including respiratory arrest, profound sedation requiring mechanical ventilation, and death among individuals combining cannabis with opioid medications, particularly in populations with compromised respiratory function or underlying medical comorbidities, have been reported (Singh, 2017).

The combination demonstrates particular dangers in elderly populations, where baseline cognitive function, drug metabolism, and physiological reserve may be compromised, creating enhanced susceptibility to adverse interactions.

The hepatic cytochrome P450 enzyme systems metabolize THC, the same systems responsible for metabolizing many CNS depressant medications, potentially altering drug clearance and creating unpredictable changes in therapeutic drug concentrations (Subramaniam et al., 2019).

Side effects

Cognitive side effects

The cognitive impairments of THC are dose-dependent, presenting as deficits in working memory, attention, executive function, and psychomotor coordination. There is significant variability between individuals due to age, frequency of use, genetic polymorphisms, and baseline cognitive capacity (Chesney et al., 2024).

Acute cognitive effects usually manifest within minutes to hours of administration and may present with impaired short-term memory consolidation, reduced attention span, altered temporal perception, and decreased psychomotor speed. There is increased difficulty in performing complex tasks requiring sustained attention and cognitive flexibility.

Using standardized cognitive assessment batteries, there are significant impairments in working memory span, verbal learning, and executive function tasks following acute THC exposure, with increased effects correlating directly with plasma THC concentrations and demonstrating partial reversibility upon drug elimination (Singh et al., 2024). Individuals who are chronic users develop a partial tolerance to acute cognitive effects but still exhibit persistent deficits in complex attention, processing speed, and executive function, which may persist even during extended abstinence periods.

Acute THC exposure alters functional connectivity within default mode networks and cognitive control circuits, disrupting neural efficiency and information processing, which correlates with decreased behavioral cognitive performance (Chesney et al., 2024).

Psychiatric adverse effects

THC has a complicated, dose-dependent, variable effect on mood and emotional regulation, with a therapeutic window for anxiolytic effects at low doses potentially transitioning to anxiety, panic, and paranoid ideation at higher dosages or in susceptible individuals (Ortiz-Medina et al., 2018).

Acute psychiatric adverse effects include heightened anxiety, panic attacks, paranoid thoughts, dysphoria, and depersonalization/derealization symptoms that typically occur within 1–3 h of administration and diminish as plasma concentrations decline (Ortiz-Medina et al., 2018). Chronic cannabis use is associated with an increased risk of developing mood and anxiety disorders (Cheng et al., 2023).

Predisposing factors for increased risk of psychiatric adverse effects include a personal or family history of anxiety disorders, previous adverse cannabis experiences, concurrent stressful life circumstances, and genetic variations in ECS function (Singh et al., 2024).

Cardiovascular side effects

Acute THC exposure has dose-dependent cardiovascular side effects, including increases in heart rate (20–50 beats per minute elevation), modest increases in systolic blood pressure, and orthostatic hypotension that may predispose individuals to falls and injuries—an essential consideration in the elderly population (Subramaniam et al., 2019). The changes in hemodynamics peak within 30–60 min of exposure and slowly improve over 2–4 h as plasma concentrations drop.

Development of tolerance to cardiovascular effects occurs relatively rapidly with regular use. However, acute hemodynamic responses may persist even in chronic users and can be particularly pronounced with high-potency preparations or alternative delivery methods (Singh, 2017).

Tolerance, dependence, and withdrawal

Regular use of THC leads to the development of tolerance, which presents as decreasing pharmacological responses, requiring increasingly higher doses to achieve equivalent therapeutic or psychoactive effects. This is due to the downregulation of CB1 receptors and altered ECS function (Ortiz-Medina et al., 2018). Tolerance rapidly develops to psychoactive and cognitive effects, while therapeutic benefits for conditions such as chronic pain and spasticity may be maintained at stable doses over extended periods.

Cannabis use disorder is characterized by compulsive use despite negative consequences, unsuccessful cessation attempts, social/occupational impairment, risky use behaviors, tolerance development, and withdrawal symptoms upon discontinuation. Lifetime users have a prevalence rate of cannabis use disorder of 9%, and up to 17% among adolescent users (Cheng et al., 2023).

Cannabis withdrawal syndrome can present with irritability, anxiety, sleep disturbances, decreased appetite, physical discomfort, and mood alterations. Within 24–72 hs of cessation, these symptoms may persist for days to weeks, depending on the intensity and duration of prior use (Singh et al., 2024).

Other systemic side effects

Cannabinoid hyperemesis syndrome (CHS) is a condition that occurs after long-term, heavy cannabis use. Individuals experiencing CHS have repeated episodes of intense nausea, vomiting, and stomach pain. Many find temporary relief by taking frequent hot showers or baths. The condition emerges after years of regular, potent cannabis use and completely resolves once cannabis use is stopped. In severe cases, symptoms can lead to dangerous outcomes due to dehydration and electrolyte imbalances, requiring immediate medical intervention (Singh et al., 2024).

Gastrointestinal effects include dry mouth due to CB1 receptor-mediated inhibition of salivary gland function, altered gastrointestinal motility that may result in either delayed gastric emptying or accelerated transit, and appetite stimulation, which may lead to significant weight gain in some individuals (Ortiz-Medina et al., 2018).

Respiratory effects affecting individuals utilizing inhalational delivery methods include acute bronchodilation, chronic bronchial irritation, increased respiratory symptoms, including cough and sputum production, and a potential increased risk of respiratory infection (Subramaniam et al., 2019).

Drug-drug interactions with tetrahydrocannabinol: Pharmacokinetic and pharmacodynamic mechanisms

Due to the increasing use and availability of Δ^9-tetrahydrocannabinol (THC), significant drug-drug interactions are occurring as a result of both pharmacokinetic and pharmacodynamic mechanisms (Ho et al., 2024).

Cytochrome p450-mediated interactions

THC undergoes hepatic metabolism via CYP2C9 and CYP3A4, producing the active metabolite 11-hydroxy-THC (Bansal et al., 2023). THC is a high-affinity CYP2C9 substrate, which competitively inhibits the activity of coadministered substrates (Doohan et al., 2021). Clinically significant enzyme inhibition occurs when combined with THC: CYP2C19 activity decreases by 207%, CYP2C9 by 77%, and CYP3A4 by 56% (Bansal et al., 2023).

Critical drug class interactions

Anticoagulants

Cannabidiol potently inhibits CYP2C9, substantially reducing warfarin clearance and causing 1.5–4-fold INR elevations (Grayson et al., 2018). Cases require warfarin dose reductions exceeding 30% to maintain therapeutic anticoagulation (Damkier et al., 2019). Peak drug-drug interaction occurs within 4–7 days of THC use and takes 2–3 weeks to resolve once discontinued (Cortopassi and Brooks, 2020).

Antiepileptic drugs

Cannabidiol significantly alters clobazam pharmacokinetics, increasing parent drug concentrations by 60% and the active metabolite N-desmethylclobazam by 500% (Geffrey et al., 2015). This interaction produces superior seizure control (70% responder rate) but increases sedation incidence to 77% compared to 30%–40% with monotherapy (Anderson et al., 2019). Preemptive clobazam dose reductions of 25%–50% are recommended when initiating cannabidiol (Chang, 2018).

CNs depressants

THC produces synergistic CNS depression with benzodiazepines and opioids through convergent GABAergic modulation, creating unpredictable dose-response relationships and an enhanced respiratory depression risk (Qian et al., 2019). Alcohol-cannabis combinations demonstrate supra-additive cognitive and psychomotor impairment, exceeding individual substance effects (Ho et al., 2024).

Immunosuppressants

When THC is combined with tacrolimus, concentrations increase up to a 2-3 fold due to CYP3A4 inhibition, requiring close therapeutic drug monitoring every 2–3 days during initiation to prevent nephrotoxicity (Ho et al., 2024). In addition, concurrent use with cyclosporine requires close monitoring of the therapeutic window due to shared metabolic pathways (Qian et al., 2019).

FDA-approved THC medications: Clinical evidence and dosing

Dronabinol (Marinol, Syndros) contains synthetic THC and is FDA-approved for CINV after conventional antiemetic failure and AIDS-related anorexia with weight loss (U.S. Food and Drug Administration, 2024).

Nabilone (Cesamet), a synthetic THC analog, is approved solely for refractory CINV (U.S. Food and Drug Administration, 2024).

Chemotherapy-induced nausea and vomiting

A systematic review of 30 randomized trials ($n = 1{,}366$) demonstrated cannabinoids' superior efficacy over traditional antiemetics, including prochlorperazine and metoclopramide (Tramèr et al., 2001). Complete nausea control occurred more frequently with cannabinoids, as did complete vomiting control (Tramèr et al., 2001). Patients significantly preferred cannabinoids over traditional agents (Tramèr et al., 2001; Rocha et al., 2008).

AIDS-related anorexia

A randomized, placebo-controlled trial of 139 AIDS patients with ≥ 2.3 kg weight loss showed that dronabinol 2.5 mg twice daily produced 38% appetite improvement versus 8% with placebo (Beal et al., 1995). Dronabinol patients maintained stable weight, while placebo patients lost 0.4 kg, with 22% of dronabinol patients gaining ≥ 2 kg versus 10.5% with placebo (Beal et al., 1995). Secondary benefits included mood improvement and decreased nausea (Beal et al., 1995).

Clinical dosing protocols

Dronabinol CINV dosing: 5 mg/m^2 initiated 1–3 h pre-chemotherapy, repeated every 2–4 h as needed, with 2.5 mg/m^2 incremental titration to a maximum of 15 mg/m^2 per dose (Tramèr et al., 2001).

Nabilone CINV dosing: 1–2 mg twice daily starting 1–3 h pre-chemotherapy, continued up to 48 h posttreatment, with dose escalation as tolerated (Tramèr et al., 2001).

Dronabinol appetite stimulation: 2.5 mg twice daily for AIDS-related anorexia (Beal et al., 1995).

The FDA governs the regulatory status for these therapeutic indications (U.S. Food and Drug Administration, 2024).

Evidence quality is highest for CINV management, supported by multiple systematic reviews and metaanalyses, while AIDS-related anorexia has moderate evidence from well-designed randomized trials (Tramèr et al., 2001; Rocha et al., 2008; Beal et al., 1995).

References

Anderson, L.L., Absalom, N.L., Abelev, S.V., Low, I.K., Doohan, P.T., Martin, L.J., Chebib, M., McGregor, I.S., Arnold, J.C., 2019. Coadministered cannabidiol and clobazam: preclinical evidence for both pharmacodynamic and pharmacokinetic interactions. Epilepsia 15281167. 60 (11), 2224–2234. Available from: http://onlinelibrary.wiley.com/journal/10.1111/(ISSN)1528-1167. doi:10.1111/epi.16355.

Appendino, G., 2020. The early history of cannabinoid research. Rend. Lincei Sci. Fis. Nat. 2037-4631. 31 (4), 919–929. doi:10.1007/s12210-020-00956-0.

Ashton, C.H., 2001. Pharmacology and effects of cannabis: a brief review. Br. J. Psychiatry 178, 101–106. doi:10.1192/bjp.178.2.101.

Badowski, M.E., 2017. A review of oral cannabinoids and medical marijuana for the treatment of chemotherapy-induced nausea and vomiting: a focus on pharmacokinetic variability and pharmacodynamics. Cancer Chemother Pharmacol. 80 (3), 441–449. PMID: 28780725.

Badowski, M.E., Yanful, P.K., 2018. Dronabinol oral solution in the management of anorexia and weight loss in AIDS and cancer. Ther. Clin. Risk Manag. 1178203X. 14, 643–651. Available from: https://www.dovepress.com/getfile.php?fileID=41371. doi:10.2147/TCRM.S126849.

Bansal, S., Zamarripa, C.A., Spindle, T.R., Weerts, E.M., Thummel, K.E., Vandrey, R., Paine, M.F., Unadkat, J.D., 2023. Evaluation of cytochrome P450-mediated cannabinoid-drug interactions in healthy adult participants. Clin. Pharmacol. Ther. 15326535. 114 (3), 693–703. Available from: http://onlinelibrary.wiley.com/journal/10.1002/(ISSN)1532-6535. doi:10.1002/cpt.2973.

Baranger, D.A., Miller, A.P., Gorelik, A.J., Paul, S.E., Hatoum, A.S., Johnson, E.C., et al., 2024. Prenatal cannabis exposure, the brain, and psychopathology during early adolescence. Nat. Ment. Health 2 (8), 975–986. PMID: 40836962. DOI: 10.1038/s44220-024-00281-7.

Barnes, M.P., 2006. Sativex®: clinical efficacy and tolerability in the treatment of symptoms of multiple sclerosis and neuropathic pain. Expert Opin. Pharmacother. 14656566. 7 (5), 607–615. Available from: http://www.tandfonline.com/loi/ieop20. doi:10.1517/14656566.7.5.607.

Beal, J.E., Olson, R., Laubenstein, L., et al., 1995. Dronabinol as a treatment for anorexia associated with weight loss in patients with AIDS. J. Pain Symptom Manage. 10 (2), 89–97.

Brust, C.A., Swanson, M.A., Bohn, L.M., 2023. Structural and functional insights into the G protein-coupled receptors: CB1 and CB2. Biochem. Soc. Trans. 51 (4), 1533–1543. PMID: 37646476.

Cabral, G.A., Staab, A., 2005. Effects on the immune system. Handb. Exp. Pharmacol. 168, 385–423.

Carney, R., Cotter, J., Firth, J., Bradshaw, T., Yung, A.R., 2017. Cannabis use and symptom severity in individuals at ultra high risk for psychosis: a meta-analysis. Acta Psychiatr. Scand. 136 (1), 5–15. PMID: 28168698.

Chang, B.S., 2018. Cannabidiol and serum antiepileptic drug levels: the ABCs of CBD with AEDs. Epilepsy Curr. 15357511. 18 (1), 33–34. Available from: http://www.epilepsycurrents.org/doi/pdf/10.5698/1535-7597.18.1.33. doi:10.5698/1535-7597.18.1.33.

Cheng, W., Parker, N., Karadag, N., et al., 2023. The relationship between cannabis use, schizophrenia, and bipolar disorder: a genetically informed study. Lancet Psychiatry 10 (6), 441–451. PMID: 37208114. DOI: 10.1016/S2215-0366(23)00143-8.

Chesney, E., Lawn, W., McGuire, P., 2024. Assessing cannabis use in people with psychosis. Cannabis Cannabinoid Res. 2578-5125. 9 (1), 49–58. doi:10.1089/can.2023.0032.

Cortopassi, W.A., Brooks, K.M., 2020. Drug interaction between cannabidiol and tacrolimus: a case report. Am. J. Health Syst. Pharm. 77 (22), 1846–1851. PMID: 33016308.

D'hooghe, M., Willekens, B., Delvaux, V., et al., 2021. Sativex® (nabiximols) cannabinoid oromucosal spray in patients with resistant multiple sclerosis spasticity: the Belgian experience. BMC Neurol. 21 (1), 227. PMID: 34157999. DOI: 10.1186/s12883-021-02246-0.

Damkier, P., Lassen, D., Christensen, M.M.H., Grønkjær Madsen, K., Hellfritzsch, M., Pottegård, A., 2019. Interaction between warfarin and cannabis. Basic Clin. Pharmacol. Toxicol. 1742-7835. 124 (1), 28–31. doi:10.1111/bcpt.13152.

Di Marzo, V., 2009. The endocannabinoid system: its general strategy of action, tools for its pharmacological manipulation and potential therapeutic exploitation. Best Pract. Res. Clin. Endocrinol. Metab. 23 (1), 1–15.

Doohan, P.T., Oldfield, L.D., Arnold, J.C., Anderson, L.L., 2021. Cannabinoid interactions with cytochrome P450 drug metabolism: a full-spectrum characterization. AAPS J. 23 (4), 91. DOI: 10.1208/s12248-021-00616-7.

Gaoni, Y., Mechoulam, R., 1964. Isolation, structure, and partial synthesis of an active constituent of hashish. J. Am. Chem. Soc. 0002-7863. 86 (8), 1646–1647. doi:10.1021/ja01062a046.

Geffrey, A.L., Pollack, S.F., Bruno, P.L., Thiele, E.A., 2015. Drug–drug interaction between clobazam and cannabidiol in children with refractory epilepsy. Epilepsia 56 (8), 1246–1251.

Grayson, L., Vines, B., Nichol, K., Szaflarski J.P., 2018. An interaction between warfarin and cannabidiol, a case report. Epilepsy Behav. Case Rep. 9, 10–11.

Grotenhermen, F., 2003. Pharmacokinetics and pharmacodynamics of cannabinoids. Clin. Pharmacokinet. 0312-5963. 42 (4), 327–360. doi:10.2165/00003088-200342040-00003.

Haney, M., Ramesh, D., Glass, A., Pavlicova, M., Bedi, G., Cooper, Z.D., 2013. Naltrexone maintenance decreases cannabis self-administration and subjective effects in daily cannabis smokers. Neuropsychopharmacology 38 (8), 1557–1565.

Hansen, J.S., Jongen, S., Brouwers, J.R.B.J., et al., 2024. Drug interactions with cannabidiol: a systematic review. Clin. Transl. Sci. 17 (1), e13685.

Ho, J.J.Y., Sheridan, J., Rosengren, R.J., Kennedy, M.A., 2024. Pharmacogenomics, CYP2C9 variation, and cannabinoid pharmacokinetics. Clin. Transl. Sci. 17 (5), e13812. PMID: 38720531.

Howlett, A.C., Abood, M.E., 2017. CB1 and CB2 Receptor Pharmacology. Adv. Pharmacol. 80, 169–206. Available from: http://www.elsevier.com/wps/find/bookdescription.cws_home/700817/description#description. doi:10.1016/bs.apha.2017.03.007.

Lucas, C.J., Galettis, P., Schneider, J., 2018. The pharmacokinetics and the pharmacodynamics of cannabinoids. Br. J. Clin. Pharmacol. 0306-5251. 84 (11), 2477–2482. doi:10.1111/bcp.13710.

Lynch, M.E., Campbell, F., 2011. Cannabinoids for treatment of chronic non-cancer pain; a systematic review of randomized trials. Br. J. Clin. Pharmacol. 13652125. 72 (5), 735–744. doi:10.1111/j.1365-2125.2011.03970.x.

McDonagh, M.S., Morasco, B.J., Wagner, J., Ahmed, A.Y., Fu, R., Kansagara, D., Chou, R., 2022. Cannabis-based products for chronic pain: a systematic review. Ann. Intern. Med. 15393704. 175 (8), 1143–1153. Available from: https://www.acpjournals.org/doi/10.7326/M21-4520. doi:10.7326/M21-4520.

McGilveray, I.J., 2005. Pharmacokinetics of cannabinoids. Pain Res. Manag. 10 (Suppl A), 15A–22A. PMID: 16237477.

Mechoulam, R., Hanuš, L.O., Pertwee, R., Howlett, A.C., 2014. Early phytocannabinoid chemistry to endocannabinoids and beyond. Nat. Rev. Neurosci. 14710048. 15 (11), 757–764. Available from: http://www.nature.com/nrn/. doi:10.1038/nrn3811.

Mulligan, M.K., Hamre, K.M., 2023. Neurobiological and genetic factors in alcohol and cannabis use and co-use. Adv. Drug Alcohol Res. 3, 10981.

Munro, S., Thomas, K.L., Abu-Shaar, M., 1993. Molecular characterization of a peripheral receptor for cannabinoids. Nature 365, 61–65. PMID: 7689702.

Ng, T., Keshock, M.C., 2023. Tetrahydrocannabinol (THC). In: StatPearls [Internet]. StatPearls Publishing, Treasure Island, FL, NBK563174.

O'Donnell, B., Melon, L.C., Bhullar, A., Millington, W., 2023. Dronabinol. In: StatPearls [Internet]. StatPearls Publishing, Treasure Island, FL, NBK557531.

Ortiz-Medina, M.B., Perea, M., Torales, J., Ventriglio, A., Vitrani, G., Aguilar, L., Roncero, C., 2018. Cannabis consumption and psychosis or schizophrenia development. Int. J. Soc. Psychiatry 17412854. 64 (7), 690–704. Available from: http://isp.sagepub.com/. doi:10.1177/0020764018801690.

Pertwee, R.G., 2008. The diverse CB 1 and CB 2 receptor pharmacology of three plant cannabinoids: Δ 9-tetrahydrocannabinol, cannabidiol and Δ 9-tetrahydrocannabivarin. Br. J. Pharmacol. 14765381. 153 (2), 199–215. doi:10.1038/sj.bjp.0707442.

Qian, Y., Gurley, B.J., Markowitz, J.S., 2019. The potential for pharmacokinetic interactions between cannabis products and conventional medications. J. Clin. Psychopharmacol. 39 (5), 462–471.

Rocha, F.C.M., dos Santos Júnior, J.G., Stefano, S.C., da Silveira, D.X., 2008. Systematic review of the literature on clinical and experimental trials on the antitumor effects of cannabinoids in gliomas. Eur. J. Cancer Care 17 (5), 431–443.

Russo, E.B., 2008. Cannabinoids in the management of difficult to treat pain. Ther. Clin. Risk Manag. 11766336. 4 (1), 245–259. Available from: http://www.dovepress.com/therapeutics-and-clinical-risk-management-journal. doi:10.2147/tcrm.s1928.

Sachse-Seeboth, C., Pfeil, J., Seeboth, A., et al. 2009. Interindividual variation in the pharmacokinetics of Δ9-tetrahydrocannabinol as related to genetic polymorphisms in CYP2C9. Clin. Pharmacol. Ther. 85 (3), 273–276.

Sallan, S.E., Zinberg, N.E., Frei, E., 1975. Antiemetic effect of delta-9-tetrahydrocannabinol in patients receiving cancer chemotherapy. N. Engl. J. Med. 0028-4793. 293 (16), 795–797. doi:10.1056/nejm197510162931603.

Sharma, P., Murthy, P., Bharath, M.M.S., 2012. Chemistry, metabolism, and toxicology of cannabis: clinical implications. Iran J. Psychiatry 7 (4), 149-156.

Singh, A., 2017. Role of cannabis in cardiovascular disorders. J. Thorac. Dis. 20721439. 9 (7), 2079–2092. doi:10.21037/jtd.2017.06.104.

Spiller, K.J., Bi, G.-H., He, Y., et al., 2019. Cannabinoid CB1 and CB2 receptor mechanisms underlie cannabis reward and aversion in rats. Br. J. Pharmacol. 176 (9), 1268–1281.

Starzer, M.S.K., Nordentoft, M., Hjorthøj, C., 2018. Rates and predictors of conversion to schizophrenia or bipolar disorder following substance-induced psychosis. Am. J. Psychiatry 175 (4), 343–350. PMID: 29179576.

Subramaniam, V.N., Menezes, A.R., DeSchutter, A., Lavie, C.J., 2019. The cardiovascular effects of marijuana: are the potential adverse effects worth the high? Mo. Med. 00266620. 116 (2), 146–153.

Tandon, R., Nasrallah, H., Akbarian, S., et al., 2024. The schizophrenia syndrome, circa 2024: what we know and how that informs its nature. Schizophr. Res. 264, 1–28.

Tramèr, M.R., Carroll, D., Campbell, F.A., Reynolds, D.J.M., Moore, R.A., McQuay, H.J., 2001. Cannabinoids for control of chemotherapy induced nausea and vomiting: quantitative systematic review. Br. Med. J. 09598146. 323 (7303), 16–21.

U.S. Food and Drug Administration, 2024. FDA and Cannabis: Research and Drug Approval Process. U.S. FDA.

Walter, L., Stella, N., 2004. Cannabinoids and neuroinflammation. Br. J. Pharmacol. 141 (5), 775–785.

Ware, M.A., Daeninck, P., Maida, V., 2008. A review of nabilone in the treatment of chemotherapy-induced nausea and vomiting. Ther. Clin. Risk Manag. 11766336. 4 (1), 99–107.

Weizman, L., Dayan, L., Brill, S., et al., 2018. Cannabis analgesia in chronic neuropathic pain is associated with altered brain connectivity. Neurology 91 (14), e1285–e1294.

Wenzel, J.M., Cheer, J.F., 2017. Endocannabinoid regulation of reward and reinforcement through interaction with dopamine and endogenous opioid signaling. Neuropsychopharmacology 43 (1), 103–115. PMID: 28653666.

Whiting, P.F., Wolff, R.F., Deshpande, S., Di Nisio, M., Duffy, S., Hernandez, A.V., Keurentjes, J.C., Lang, S., Misso, K., Ryder, S., Schmidlkofer, S., Westwood, M., Kleijnen, J., 2015. Cannabinoids for medical use: a systematic review and meta-analysis. American Medical Association, United Kingdom JAMA. JAMA 15383598. 313 (24), 2456–2473. Available from: http://jama.jamanetwork.com/article.aspx?articleid=2338251. doi:10.1001/jama.2015.6358.

Wright, J.A., Katamesh, B.E., et al., 2024. Hypothesized pharmacogenomic and medication influences on tetrahydrocannabinol and cannabidiol metabolism in a cohort of unselected oral cannabis users. J. Cannabis Res. 6, 256. doi:10.1186/s42238-024-00256-6.

Yahud, E., Paul, G., Rahkovich, M., et al., 2020. Cannabis induced cardiac arrhythmias: a case series. Eur. Heart J. Case Rep. 4 (6), 1–9. PMID: 33442601. DOI: 10.1093/ehjcr/ytaa376.

Zajicek, J., Fox, P., Sanders, H., Wright, D., Vickery, J., Nunn, A., Thompson, A., 2003. Cannabinoids for treatment of spasticity and other symptoms related to multiple sclerosis (CAMS study): multicentre randomised placebo-controlled trial. Lancet. 01406736. 362 (9395), 1517–1526. doi:10.1016/S0140-6736(03)14738-1.

Zou, S., Kumar, U., 2018. Cannabinoid receptors and the endocannabinoid system: signaling and function in the central nervous system. Int. J. Mol. Sci. 1422-0067. 19 (3), 833. doi:10.3390/ijms19030833.

CHAPTER

Other phytocannabinoids

6

Roshni Patel

RNP Regenerative Medicine, Center of Excellence in Pain and Regenerative Medicine, Quantum Brain Health, Farmington, CT, United States

Introduction

Beyond the widely recognized psychoactivity of Δ^9-tetrahydrocannabinol (THC), the plant, *Cannabis sativa*, produces a diverse chemical range of over 100 other phytocannabinoids. These other phytocannabinoids offer many therapeutic promises without intoxicating effects (Zagzoog et al., 2020). Cannabidiol (CBD), a nonpsychoactive chemical, has been a focus of scientific inquiry due to its unique pharmacological profile (Russo and Guy, 2006). The early therapeutic rationale for cannabinoid co-administration emerged from findings that CBD could minimize THC's psychoactive side effects, particularly anxiety and intoxication, hence highlighting cannabinoid synergy (Russo and Guy, 2006).

The various compounds found in cannabis, including cannabinoids (like THC and CBD) and terpenes, work together and may offer enhanced therapeutic benefits while minimizing side effects. This is referred to as the "entourage effect." This dual action enhances the therapeutic benefits and overall effects, making the plant's overall impact greater than the sum of its components (Ferber et al., 2020). Specific terpenes, like linalool and limonene, demonstrate anxiolytic properties and synergize with cannabinoids. Beta-caryophyllene functions as a selective CB2 receptor agonist, contributing antiinflammatory effects.

Such interactions may widen the therapeutic window or enhance clinical efficacy, an observation of particular relevance for complex conditions like mood and anxiety disorders (Ferber et al., 2020). This has assisted in broadening the scope of research to probe the distinct properties of minor cannabinoids like cannabigerol (CBG), cannabichromene (CBC), and tetrahydrocannabivarin (THCV). CBG demonstrates potent α2-adrenoceptor agonism and neuroprotective properties. CBC exhibits selective CB2 receptor agonism with antiinflammatory and analgesic effects. THCV presents a unique pharmacological profile as a CB1 antagonist at low doses and a partial agonist at higher concentrations. Each exhibits distinct mechanisms and promising therapeutic utility (Zagzoog et al., 2020).

Cannabinoids for Chronic Pain. https://doi.org/10.1016/B978-0-443-36773-1.00017-9

Mechanisms of action

Unlike THC, which exerts most of its psychoactive effects by activating CB1 and CB2 receptors, other phytocannabinoids have a much more complex story. They operate through polypharmacology—the principle that a single molecule can engage multiple molecular targets to produce therapeutic effects far beyond the traditional endocannabinoid system (Russo and Guy, 2006). This multitarget action is how they achieve these results, often without THC's intoxication (Zagzoog et al., 2020).

The mechanisms of CBD are notably diverse. A key feature is its extremely low affinity for the primary cannabinoid receptors (K_i values > 10,000 nM for both CB1 and CB2), which explains its lack of psychoactivity (Russo and Guy, 2006). Instead of direct activation, CBD acts as a negative allosteric modulator at the CB1 receptor; it changes the receptor's shape to reduce its signaling efficiency, which can blunt THC's psychoactive impact (Laprairie et al., 2015). Beyond this, CBD's therapeutic versatility comes from other systems. It is a potent agonist at serotonin 5-HT_{1a} receptors (EC_{50} = 16 μM), an interaction strong enough at clinical doses to be a likely source of its anxiolytic (antianxiety) and antidepressant effects (Russo and Guy, 2006; Ferber et al., 2020). Furthermore, it targets pain pathways by activating the TRPV1 "capsaicin" receptor, contributing to its analgesic (pain-relieving) properties (Russo and Guy, 2006). CBD can also weakly inhibit the FAAH enzyme, which may boost the brain's anandamide levels and enhance natural endocannabinoid signaling (Russo and Guy, 2006).

Similarly complex, CBG also displays a multireceptor profile. While it is a partial agonist at both CB1 (K_i = 440 nM) and CB2 (K_i = 337 nM) receptors (Zagzoog et al., 2020), its most powerful and clinically relevant interactions lie elsewhere. CBG is a powerful agonist at α_2-adrenoceptors (K_i = 0.2–72.8 nM) (Navarro et al., 2018). This potent adrenergic activity is so significant that it requires caution and close monitoring in patients already receiving antihypertensive therapy. Like CBD, CBG also acts on 5-HT_{1a} receptors, giving it potential for treating mood and anxiety disorders (Zagzoog et al., 2020).

CBC achieves its effects through a more selective pathway, defined by its negligible interaction with CB1 receptors (Zagzoog et al., 2020). This selectivity allows for antiinflammatory effects without any psychoactive risk. Its primary mechanisms are potent and selective agonism at the CB2 receptor, where it shows over 1000-to-1 selectivity compared to CB1 and demonstrates higher efficacy than even THC (Udoh et al., 2019; Zagzoog et al., 2020). CBC is also a potent activator of TRPA1 and TRPV1 channels, which are heavily involved in modulating pain and inflammation signals (Zagzoog et al., 2020).

Perhaps the most unusual profile belongs to THCV, which exhibits a pharmacological "split personality" entirely dependent on its concentration. At low doses (below 1 μM), THCV functions as a neutral CB1 antagonist (K_i = 75.4 nM), meaning it can block the effects of THC or endogenous cannabinoids (Zagzoog et al., 2020). However, at higher concentrations, it flips to become a partial agonist at the CB1 receptor. This dose-dependent switch is a critical consideration for therapy, as low and high doses can produce opposite outcomes. At all concentrations, it remains a partial agonist at CB2 receptors (Zagzoog et al., 2020).

Finally, understanding these individual compounds is incomplete without considering the entourage effect. This is the principle that the full therapeutic impact of cannabis comes from a chemical synergy between its many components, not just from isolated molecules (Ferber et al., 2020). For example, terpenes like linalool and limonene have anxiolytic properties that appear to augment the effects of cannabinoids in treating mood disorders (Ferber et al., 2020). A fascinating case is the common terpene β-caryophyllene, which is itself a selective CB2 receptor agonist (K_i = 155 nM) and provides its antiinflammatory effects, perfectly complementing the other phytocannabinoids in the plant (Russo, 2011).

Indications

The therapeutic applications of non-THC phytocannabinoids are expanding rapidly, but CBD remains the clear frontrunner, with a depth of clinical evidence that has led to major regulatory approval. The journey of its purified oral form, Epidiolex, from research to an FDA-approved medicine sets the stage for the entire field. Landmark clinical trials—specifically for Dravet syndrome, Lennox–Gastaut syndrome, and tuberous sclerosis complex (TSC)—have cemented its role in modern medicine. Table 6.1 provides a snapshot of this evolving therapeutic landscape.

The results from the CBD trials were striking. In patients with Dravet syndrome, CBD delivered a median 39% reduction in convulsive seizures—nearly triple the effect of the placebo (Devinsky et al., 2017). For Lennox–Gastaut syndrome, the outcomes were just as compelling, with CBD significantly cutting the frequency of drop seizures, which are often the most debilitating aspect of the condition (Devinsky et al., 2018). The impact in TSC was equally impressive, producing a significant reduction in seizures that led to clinically meaningful improvements in quality of life for many patients (Thiele et al., 2021).

Beyond CBD, emerging evidence points to exciting applications for other minor cannabinoids. In its first human clinical trial, CBG demonstrated apparent anxiolytic effects. A single 20 mg dose produced a 26.5% reduction in anxiety ratings without causing any intoxication (Cuttler et al., 2024). In addition to its impact on anxiety, CBG exhibits noteworthy preclinical potential by providing neuroprotective qualities in Huntington's disease mouse models (Valdeolivas et al., 2015). Cannabigerolic acid (CBGA) exhibits complex effects in epilepsy models, showing both anticonvulsant and concerning proconvulsant properties (Anderson et al., 2021).

Meanwhile, other compounds are showing promise in early research. The potential of Δ9- THCV in metabolic regulation has drawn attention, as it has been shown to improve insulin sensitivity in animal models (Wargent et al., 2013). CBC shows a different potential, demonstrating analgesic and antiinflammatory properties through its action on CB2 receptors (Sepulveda et al., 2024). However, the path forward is not always straightforward. The CBGA case is a perfect example; in epilepsy models, it

Table 6.1 Therapeutic indications of select non-THC phytocannabinoids.

Cannabinoid	Indication(s)	Level of evidence	Key clinical data	Key references
CBD	Seizures (Dravet, Lennox–Gastaut, TSC)	FDA-approved (Level I)	38.9–48.6% seizure reduction	(Devinsky et al., 2017, 2018; Thiele et al., 2021)
CBG	Anxiety	Level II (Human RCT)	26.5% anxiety reduction at 20 mg.	(Cuttler et al., 2024)
	Neuroprotection (Huntington's disease)	Preclinical	Preserved striatal neurons	(Valdeolivas et al., 2015)
THCV	Metabolic regulation (obesity, diabetes)	Preclinical	Improved insulin sensitivity	(Wargent et al., 2013; Abioye et al., 2020)
	Antipsychotic effects	Preclinical	5-HT_1A receptor activation	(Cascio et al., 2015)
CBC	Antinociceptive, antibacterial, antiseizure	Preclinical	CB2 agonism > THC	(Udoh et al., 2019; Sepulveda et al., 2024)
CBGA	Epilepsy	Preclinical	Complex: anti- and proconvulsant	(Anderson et al., 2021)

exhibits a complex profile with both anticonvulsant and concerning proconvulsant effects (Anderson et al., 2021). This finding is a crucial reminder that, while the field is full of promise, these compounds require rigorous clinical validation before their therapeutic roles can be truly understood.

Contraindications

Unlike Δ9-THC, nonpsychoactive phytocannabinoids such as CBD, CBG, and CBC do not exacerbate psychosis; nevertheless, limited long-term safety data and distinct adverse-effect profiles demand cautious, patient-specific use. Severe hepatic impairment is an absolute contraindication for high-dose oral CBD because systemic exposure increases 2.5–5-fold, and hepatocellular injury risk rises accordingly (Madeo et al., 2023). Pregnancy and lactation are also contraindicated due to insufficient human safety data.

Relative contraindications include moderate hepatic dysfunction, co-administration with valproate or clobazam, and use of narrow-therapeutic-index drugs metabolized by CYP2C19, CYP2C9, or CYP3A4 (Bansal et al., 2023; Lucas et al., 2018). THCV's glucose-lowering effects warrant caution in insulin-treated diabetes, while

Table 6.2 Contraindications for non-THC phytocannabinoids.

Specific contraindication	Affected cannabinoid(s)	Evidence/rationale	Suggested action
Absolute			
Severe hepatic impairment (Child–Pugh C)	CBD (and likely other orally metabolized cannabinoids)	2.5–5.2-fold ↑ CBD exposure; higher ALT elevations (Madeo et al., 2023; Chesney et al., 2020)	Avoid or use only under specialist supervision, with intensive LFT monitoring
Pregnancy/lactation	All	Developmental toxicity in animals; no controlled human data (Huestis et al., 2019)	Avoid unless the benefit outweighs the risk
Hypersensitivity	All	Case reports of cannabinoid allergy and anaphylaxis	Contraindicated
Active seizure disorder (for CBGA)	CBGA	Proconvulsant activity in some models (Anderson et al., 2021)	Avoid until human safety is clarified
Moderate hepatic impairment (Child–Pugh B)	CBD	2-fold ↑ exposure; dosing guidelines recommend a 50% reduction (Madeo et al., 2023; Chesney et al., 2020)	Halve starting dose; monitor LFTs
Relative			
Concomitant valproate	CBD ≥ 20 mg/kg/day	21% ALT > 3 × ULN vs. 0% without valproate (Chesney et al., 2020)	Baseline and q2-4 week LFTs; consider lower CBD dose
Concomitant clobazam or other CYP2C19 substrates	CBD	CBD ↑ N-desmethylclobazam up to 8-fold (Lucas et al., 2018)	Reduce clobazam by 25%–50%; monitor for sedation
Narrow-therapeutic-index CYP2C9/3A4 drugs (e.g., warfarin, tacrolimus)	CBD, THC, possibly others	CBD/THC inhibits CYP2C9/3A4 (Bansal et al., 2023)	Check INR and drug levels; adjust the dose
Diabetes/hypoglycemia risk	THCV	THCV lowers fasting glucose in animals; theoretical additive hypoglycemia with insulin or sulfonylureas (Wargent et al., 2013)	Monitor glucose; consider dose reduction of antidiabetic medications

Continued

Table 6.2 Contraindications for non-THC phytocannabinoids—cont'd

Specific contraindication	Affected cannabinoid(s)	Evidence/rationale	Suggested action
Severe hypotension/ autonomic dysfunction	CBG, CBD	Vasodilatory and blood-pressure-lowering effects were reported in preclinical studies (Nachnani et al., 2021)	Use caution; monitor blood pressure (BP)
Concurrent CNS depressants (benzodiazepines, opioids, alcohol)	CBD, CBG, CBC	Additive sedation possible (Iffland and Grotenhermen, 2017)	Start at a lower dose; counsel on sedation
Poor CYP2C19 metabolizers (genetic)	CBD	Higher plasma CBD and clobazam metabolite levels (Lucas et al., 2018)	Consider pharmacogenetic testing or a lower starting dose

the vasodilatory properties of CBG and CBD require blood-pressure monitoring in hypotension-prone patients (Table 6.2).

Given CBD's dose-dependent risk of hepatocellular injury, baseline and periodic transaminase monitoring are essential—particularly in patients with preexisting hepatic impairment or those receiving concomitant valproate (Chesney et al., 2020; Huestis et al., 2019).

Side effects

The adverse effect profile of non-THC phytocannabinoids is best characterized for CBD, based on extensive data from randomized clinical trials for epilepsy. Emerging data suggest that CBG is well tolerated in healthy adults (Cuttler et al., 2024), whereas CBC and THCV appear benign at studied doses but lack large-scale trials. In contrast, CBGA exhibits dose-dependent proconvulsant activity in animal models (Anderson et al., 2021).

Adverse effect profile of CBD

The safety profile of CBD is well defined, with adverse effects that are typically mild to moderate but can be serious, particularly concerning liver function.

- **Gastrointestinal (GI) effects:** GI-related symptoms are the most frequently reported adverse effects in up to 59.5% of patients in some analyses. Diarrhea is the single most common complaint, followed by nausea and vomiting (Madeo et al., 2023). These effects are often dose-related and may resolve with time or dose adjustment.

- **Central nervous system effects:** Somnolence or sedation is a common CNS effect, reported in 25%–32% of patients. This effect is clearly dose-dependent and is significantly more likely in patients concurrently taking other sedating medications, such as clobazam (Chesney et al., 2020).
- **Metabolic effects:** Paradoxically, and in contrast to THC's appetite-stimulating properties, decreased appetite is a frequent side effect of CBD, reported in 30.5% of patients (Odds Ratio: 3.56 vs. placebo). This can be associated with weight loss, which requires monitoring, particularly in pediatric populations (Chesney et al., 2020).
- **Hepatocellular injury:** The most serious potential adverse effect of CBD is dose-dependent, drug-induced liver injury, manifesting as elevated serum transaminases (Alanine Aminotransferase/Aspartate Aminotransferase, ALT/AST). The overall incidence is approximately 7.4%, but this risk is magnified significantly by concomitant use of valproate, where the incidence skyrockets to 21%. Compared to placebo, the odds ratio for developing abnormal liver function tests with CBD is 11.19, underscoring the need for baseline and follow-up liver function test (LFT) monitoring in at-risk patients (Chesney et al., 2020).
- **Serious adverse events (SAEs):** While the overall risk of any SAE is higher with CBD compared to placebo (OR 2.30), most of these events, such as infections or status epilepticus, are related to the severe underlying medical conditions of the trial participants rather than direct drug toxicity (Chesney et al., 2020).

Emerging safety data on minor cannabinoids

- **CBG:** The first human trial of single-dose CBG showed excellent tolerability. Fewer than 10% of participants reported mild fatigue at a 20 mg dose, with no other significant adverse events noted (Cuttler et al., 2024). Long-term safety data are not yet available.
- **CBC:** Human data is minimal. A pharmacokinetic study, where participants received up to 26.4 mg/day, reported only a few cases of mild GI discomfort (Peters et al., 2022).
- **THCV:** Human safety trial data is pending. Theoretical concerns include mild GI effects or headache. Importantly, animal models suggest THCV does not produce intoxication at doses up to 10 mg/kg, distinguishing it from THC (Abioye et al., 2020).
- **CBGA:** CBGA carries a unique safety concern. Preclinical models in mice have found that it can exacerbate seizures at specific doses, acting as a proconvulsant. This finding precludes its use in patients with epilepsy (Anderson et al., 2021) (Table 6.3).

Overall, CBD's adverse-event spectrum is well defined—dominated by GI symptoms, somnolence, and transaminase elevations—whereas minor cannabinoids require larger human trials to fully delineate their safety profiles.

Table 6.3 Common adverse effects of non-THC phytocannabinoids.

Cannabinoid	Adverse effect	Frequency/ severity	Clinical comment	Key reference(s)
CBD	Gastrointestinal (diarrhea, nausea)	Up to 59.5% (GI overall)	Most common AE	(Madeo et al., 2023)
	Somnolence/ sedation	25%–32%	Dose-dependent;↑ with clobazam	(Chesney et al., 2020)
	Decreased appetite/ weight loss	30.5% (OR 3.56)	More common than placebo	(Chesney et al., 2020)
	Hepatocellular injury (ALT/ AST ↑)	7.4% overall; 21% with valproate	OR 11.19 vs placebo	(Chesney et al., 2020)
	Serious adverse events	OR 2.30 vs placebo	Mostly infection, status epilepticus	(Chesney et al., 2020)
CBG	Mild fatigue	<10%	No significant AEs at 20 mg single dose.	(Cuttler et al., 2024)
CBC	Mild GI discomfort	Few cases in PK study	Tolerated up to 26.4 mg/day	(Peters et al., 2022)
THCV	Mild GI/ headache (theoretical)	Limited human data	No intoxication at ≤ 10 mg/ kg in animals; human trials pending	(Abioye et al., 2020) (review)
CBGA	Seizure exacerbation	Dose-dependent in mice	Proconvulsant at certain doses	(Anderson et al., 2021)

Drug interactions

The most clinically relevant drug-drug interactions with non-THC phytocannabinoids are pharmacokinetic, resulting from the suppression of the cytochrome P450 (CYP) enzyme system in the liver. CBD is a powerful, dose-dependent inhibitor of certain essential CYP enzymes. This inhibition may interfere with the metabolism of simultaneously administered medications, leading to elevated plasma levels and potential toxicity. The interaction potential of minor cannabinoids is inadequately defined; hence, a prudent perspective regards all cannabinoids as possible CYP450 inhibitors. The dose-dependent interactions of CBD are particularly significant at dosages of 10 mg/kg/day or greater; still, caution is essential in any polypharmacy context, even at reduced doses.

Inhibition of cytochrome P450 2C19

The most significant metabolic interaction of CBD is the suppression of CYP2C19. This produces clinically significant effects with certain common medications, which are discussed as follows.

Clobazam: This benzodiazepine exhibits a typical interaction. The suppression of CYP2C19 by CBD results in a two- to three-fold increase in the levels of N-desmethylclobazam, the active metabolite. This directly correlates with an increased incidence of sedation, potentially affecting up to 77% of patients on this combination. The typical protocol involves a preemptive reduction of clobazam dosage by 25%–50% upon the introduction of CBD, along with vigilant observation for drowsiness (Lucas et al., 2018; Bansal et al., 2023).

Omeprazole: This prevalent proton-pump inhibitor can see a 207% increase in total systemic exposure due to CBD. Whereas this may enhance its acid-suppressing benefits, it simultaneously elevates the possibility of long-term side effects, requiring monitoring and perhaps dose modification (Bansal et al., 2023).

Inhibition of CYP2C9

The suppression of CYP2C9 by CBD presents a significant risk for interactions, especially with medications that possess a narrow therapeutic index. The interaction with warfarin is of paramount significance due to the potential for hemorrhage. CBD may elevate the international normalized ratio (INR) by 1.5 to 4. Management necessitates meticulous attention, generally entailing a proactive reduction of warfarin dosage by a minimum of 30% and weekly INR assessments until a stable level is attained (Bansal et al., 2023).

Losartan: Exposure to this antihypertensive may be elevated by 77%, thus amplifying its hypotensive effects while also heightening the likelihood of adverse reactions. It is advisable to monitor blood pressure and renal function (Bansal et al., 2023).

Inhibition of CYP3A4 and additional enzymes

Although less powerful than its impact on CYP2C19, CBD's suppression of CYP3A4 remains therapeutically significant.

Tacrolimus: The concentrations of this immunosuppressant may elevate two- to three-fold, posing a considerable risk of nephrotoxicity. This requires therapeutic drug monitoring (TDM) every 2–3 days during the introduction of CBD and subsequent dose modifications as necessary (Rong et al., 2018).

Caffeine (CYP1A2): CBD may inhibit CYP1A2, resulting in a 39% increase in caffeine exposure. Although not hazardous, this may result in jitteriness or insomnia; hence, patients should be counseled to restrict their caffeine consumption (Bansal et al., 2023).

Pharmacodynamic interactions

Separate from metabolic inhibition, pharmacodynamic interactions occur when cannabinoids have additive effects with other drugs. CNS depressants: Cannabinoids like CBD and CBG can cause sedation. When combined with other CNS depressants, such as benzodiazepines or opioids, this effect is synergistic, increasing the risk of profound drowsiness and respiratory depression. The management strategy involves starting with the lowest effective doses of both agents and thoroughly counseling the patient on the risks (Iffland and Grotenhermen, 2017) (Table 6.4).

Dose-dependent inhibition has been observed at CBD doses ≥ 10 mg/kg/day; lower doses may pose less risk but should still be monitored in polypharmacy settings.

Table 6.4 Clinically significant drug interactions with non-THC cannabinoids.

Interacting drug/class	Mechanism (CYP enzyme)	Clinical consequence	Management recommendation	Key reference(s)
Clobazam	Potent CYP2C19 inhibition	N-desmethylclobazam ↑ 2-3-fold; sedation incidence ↑ to 77%	Preemptive 25%–50% clobazam dose reduction	(Lucas et al., 2018; Bansal et al., 2023)
Warfarin	CYP2C9 inhibition	INR ↑ 1.5–4-fold; bleeding risk	Weekly INR; ↓ warfarin dose ≥ 30%	(Bansal et al., 2023)
Omeprazole	CYP2C19 inhibition	207% ↑ exposure	Monitor for dyspepsia and headache; consider dose ↓	(Bansal et al., 2023)
Losartan	CYP2C9 inhibition	77% ↑ exposure	Monitor BP, renal function; adjust dose	(Bansal et al., 2023)
Tacrolimus	CYP3A4 inhibition	Levels ↑ 2-3-fold; nephrotoxicity	TDM q 2–3 days; dose ↓ as needed	(Rong et al., 2018) (review)
Midazolam (probe)	CYP3A4 inhibition	56% ↑ exposure	Consider 25% dose ↓; monitor sedation	(Bansal et al., 2023)
Caffeine	CYP1A2 inhibition	39% ↑ exposure	Limit caffeine intake; monitor for jitteriness	(Bansal et al., 2023)
Benzodiazepines/ Opioids	Pharmacodynamic (additive CNS depression)	Synergistic sedation, respiratory depression	Start with the lowest effective doses; counsel on drowsiness	(Iffland and Grotenhermen, 2017)

Recommended doses for different conditions

Non-THC phytocannabinoid dosing is variable and depends on the specific cannabinoid, the clinical indication, the formulation, and the individual patient. A general clinical guideline is to "start low and go slow" to optimize efficacy while minimizing adverse effects (Allan et al., 2018). The formally established and standardized dosing protocols exist only for CBD's FDA-approved indications.

FDA-approved dosing protocols for CBD

These protocols are for oral CBD formulation and are based on extensive data from randomized controlled trials for refractory epilepsy (Table 6.5).

Dravet syndrome and Lennox–Gastaut syndrome: This protocol is generally started with a dose of 2.5 mg/kg, administered twice daily (BID). After one week, this is increased to a maintenance dose of 5 mg/kg BID. For further seizure control, the dose can be increased to a maximum of 10 mg/kg BID, for a total daily dose of 20 mg/kg.

TSC: TSC dosing starts at 2.5 mg/kg BID but follows a more gradual titration schedule. It is usually increased by 2.5 mg/kg BID every week until a higher maintenance dose of 12.5 mg/kg BID (25 mg/kg/day) is reached.

An important point to consider for all oral CBD dosing is how it interacts with food. A high-fat meal can increase the area under the curve (AUC), or total drug exposure, by four to five times, significantly enhancing bioavailability (Millar et al., 2018).

Investigational dosing for minor cannabinoids

Dosing recommendations for other minor cannabinoids are not yet established and are based on limited or preclinical data. The "start low, go slow" principle is especially critical for these compounds (Table 6.6).

Table 6.5 FDA-approved dosing for cannabidiol (Epidiolex) for refractory epilepsy.

Indication	Starting Dose	Titration Schedule	Maintenance Dose	Maximum Dose	Special Considerations
Dravet and Lennox–Gastaut	2.5 mg/kg BID	↑ to 5 mg/kg BID after 1 week	10 mg/kg BID (20 mg/kg/day)	20 mg/kg/day	Administer with food (↑ AUC ≈ 4–5 × with high-fat meal)
Tuberous sclerosis complex	2.5 mg/kg BID	↑ by 2.5 mg/kg BID weekly to 12.5 mg/kg BID	12.5 mg/kg BID (25 mg/kg/day)	25 mg/kg/day	Target plasma: 100–300 ng/mL

Table 6.6 Investigational dosing for other cannabinoids.

Canna-binoid	Investiga-tional dose (study)	Duration/ route	Evidence level	Key reference(s)	Notes/ corrections
CBG	20 mg (single dose)	Acute, oral	Human RCT	(Cuttler et al., 2024)	Well tolerated; no significant adverse events (AEs) reported
THCV	3–12.5 mg/kg/ day (rodent)	≥ 13 weeks, oral	Preclinical	(Wargent et al., 2013; Abioye et al., 2020) (review)	Human equivalent dose unknown; no intoxication observed in animals
CBC	Up to 26.4 mg/day	Variable, oral (co-formulated with CBD/ THC)	Human PK study	(Peters et al., 2022)	Demonstrated higher absorption vs. CBD/THC; safety data is limited
CBGA	1–100 mg/kg (rodent)	Acute to subacute, oral	Preclinical	(Anderson et al., 2021)	Anticonvulsant at low doses; proconvulsant at high doses—use caution

Note: *THCV and CBGA dosing are derived from animal studies; clinical dosing has not been established. All investigational cannabinoids should follow the "start low, go slow" principle until human data is available.*

CBG: The first human RCT for anxiety demonstrated that a single oral dose of 20 mg was well tolerated and effective (Cuttler et al., 2024). Chronic dosing schedules have not yet been determined.

The existing dosing information for THCV and CBGA is solely based on rat trials, with the corresponding human dose remaining undetermined. In these animal models, THCV has been reviewed at dosages of up to 12.5 mg/kg/day, whereas CBGA has been investigated across a broader range of 1–100 mg/kg (Wargent et al., 2013; Anderson et al., 2021). Because CBGA has the unusual ability to be both anticonvulsant and proconvulsant, its effects are difficult to predict, creating a clear need for caution.

The existing data regarding human tolerability for cannabichromene is still preliminary. One pharmacokinetic study found that doses up to 26.4 mg/day were safe. However, this finding has a significant caveat: CBC was not tested in isolation but as part of a combination product (Peters et al., 2022).

Dosage modifications for certain populations and scenarios

Dosage modifications are essential for patients with hepatic impairment or those on specific interfering drugs. Baseline LFTs are advised for all individuals commencing prolonged cannabis therapy (Table 6.7).

Hepatic Impairment: A dose adjustment is generally not required for patients with mild hepatic impairment (Child–Pugh A). For those with moderate hepatic impairment (Child–Pugh B), the starting dosage of CBD should be reduced to 50% of the standard dose, accompanied by monthly liver function test monitoring. CBD is contraindicated for those with severe hepatic impairment (Child–Pugh C).

Drug Interactions: When CBD is administered alongside CYP2C19 substrates, such as clobazam, a proactive dose decrease of 25%–50% for clobazam is advised to reduce the risk of severe sedation. When coadministered with valproate, CBD dosages should preferably remain under 20 mg/kg/day, and LFTs should be assessed at baseline and subsequently at 1, 3, and 6 months.

Variability in pharmacokinetics and therapeutic drug monitoring

The bioavailability of orally ingested cannabis is highly variable, influenced by factors such as first-pass metabolism and concurrent meal intake. This pharmacokinetic heterogeneity highlights the necessity for personalized dosage adjustment. TDM

Table 6.7 Special population dosing adjustments.

Population	Cannabinoid	Adjustment	Monitoring
Hepatic impairment			
Child–Pugh A	All	No adjustment	Baseline LFTs, then routine follow-up
Child–Pugh B	CBD	Start at 50% usual dose	LFTs monthly
Child–Pugh C	CBD	Contraindicated	–
Drug Interactions			
With CYP2C19 substrates	CBD	Consider CBD or substrate dose reduction.	Drug levels as indicated
With valproate	CBD	Avoid CBD doses ≥ 20 mg/kg/day.	LFTs at baseline, 1, 3, and 6 months
With clobazam	CBD	Reduce clobazam 25%–50% pre-emptively	Monitor sedation and, if available, N-desmethylclobazam levels

Note: *Baseline liver-function tests (LFTs) are recommended for all patients initiating long-term cannabinoid therapy, with more frequent monitoring in those with hepatic impairment or interacting medications.*

can, in certain instances, serve as an effective instrument. In the context of epilepsy, plasma concentrations of CBD ranging from 100 to 300 ng/mL have been correlated with optimal seizure reduction (Landmark et al., 2025). Non-THC phytocannabinoid dosing is variable and depends on the specific cannabinoid, the clinical indication, the formulation, and the individual patient. The formally established and standardized dosing protocols exist only for CBD's FDA-approved indications.

Acknowledgments

The author gratefully acknowledges Rhea Doshi for her diligent and valuable contributions to this manuscript's research and editorial review. Rhea Doshi provided significant assistance in research and editing during the preparation of this manuscript.

References

Abioye, A., Ayodele, O., Marinkovic, A., Patidar, R., Akinwekomi, A., Sanyaolu, A., 2020. Δ^9-Tetrahydrocannabivarin (THCV): a commentary on potential therapeutic benefit for the management of obesity and diabetes. J. Cannabis Res. 2522-5782. 2 (1). doi:10.1186/s42238-020-0016-7.

Allan, G.M., Ramji, J., Perry, D., Ton, J., Beahm, N.P., Crisp, N., Dockrill, B., Dubin, R.E., Findlay, T., Kirkwood, J., Fleming, M., Makus, K., Zhu, X., Korownyk, C., Kolber, M.R., McCormack, J., Nickel, S., Noël, G., Lindblad, A.J., 2018. Simplified guideline for prescribing medical cannabinoids in primary care. Can. Fam. Physician 0008350X. 64 (2), 111. Available from: http://www.cfp.ca/content/cfp/64/2/111.full.pdf.

Anderson, L.L., Heblinski, M., Absalom, N.L., Hawkins, N.A., Bowen, M.T., Benson, M.J., Zhang, F., Bahceci, D., Doohan, P.T., Chebib, M., McGregor, I.S., Kearney, J.A., Arnold, J.C., 2021. Cannabigerolic acid, a major biosynthetic precursor molecule in cannabis, exhibits divergent effects on seizures in mouse models of epilepsy. Br. J. Pharmacol. 14765381. 178 (24), 4826–4841. Available from: http://onlinelibrary.wiley.com/journal/10.1111/(ISSN)1476-5381. doi:10.1111/bph.15661.

Bansal, S., Zamarripa, C.A., Spindle, T.R., Weerts, E.M., Thummel, K.E., Vandrey, R., Paine, M.F., Unadkat, J.D., 2023. Evaluation of cytochrome P450–mediated cannabinoid-drug interactions in healthy adult participants. Clin. Pharmacol. Ther. 15326535. 114 (3), 693–703. Available from: http://onlinelibrary.wiley.com/journal/10.1002/(ISSN)1532-6535. doi:10.1002/cpt.2973.

Cascio, M.G., Zamberletti, E., Marini, P., Parolaro, D., Pertwee, R.G., 2015. The phytocannabinoid, Δ^9-tetrahydrocannabivarin, can act through 5-HT1A receptors to produce antipsychotic effects. Br. J. Pharmacol. 14765381. 172 (5), 1305–1318. Available from: http://onlinelibrary.wiley.com/journal/10.1111/%28ISSN%291476-5381. doi:10.1111/bph.13000.

Chesney, E., Oliver, D., Green, A., Sovi, S., Wilson, J., Englund, A., Freeman, T.P., McGuire, P., 2020. Adverse effects of cannabidiol: a systematic review and meta-analysis of randomized clinical trials. Neuropsychopharmacology 1740634X. 45 (11), 1799–1806. Available from: http://www.nature.com/npp/index.html. doi:10.1038/s41386-020-0667-2.

Cuttler, C., Stueber, A., Cooper, Z.D., Russo, E., 2024. Acute effects of cannabigerol on anxiety, stress, and mood: a double-blind, placebo-controlled, crossover, field trial. Sci. Rep. 20452322. 14 (1). Available from: https://www.nature.com/srep/. doi:10.1038/s41598-024-66879-0.

Devinsky, O., Cross, J.H., Laux, L., Marsh, E., Miller, I., Nabbout, R., Scheffer, I.E., Thiele, E.A., Wright, S., 2017. Trial of cannabidiol for drug-resistant seizures in the dravet syndrome. N. Engl. J. Med. 15334406. 376 (21), 2011–2020. Available from: http://www.nejm.org/doi/pdf/10.1056/NEJMoa1611618. doi:10.1056/NEJMoa1611618.

Devinsky, O., Patel, A.D., Cross, J.H., Villanueva, V., Wirrell, E.C., Privitera, M., Greenwood, S.M., Roberts, C., Checketts, D., VanLandingham, K.E., Zuberi, S.M., 2018. Effect of cannabidiol on drop seizures in the lennox–gastaut syndrome. N. Engl. J. Med. 15334406. 378 (20), 1888–1897. Available from: http://www.nejm.org/medical-index. doi:10.1056/NEJMoa1714631.

Ferber, S.G., Namdar, D., Hen-Shoval, D., Eger, G., Koltai, H., Shoval, G., Shbiro, L., Weller, A., 2020. The "entourage effect": terpenes coupled with cannabinoids for the treatment of mood disorders and anxiety disorders. Curr. Neuropharmacol. 18756190. 18 (2), 87–96. Available from: http://www.eurekaselect.com/603/journal/current-neuropharmacology. doi:10.2174/1570159X17666190903103923.

Huestis, M.A., Solimini, R., Pichini, S., Pacifici, R., Carlier, J., Busardò, F.P., 2019. Cannabidiol adverse effects and toxicity. Curr. Neuropharmacol. 18756190. 17 (10), 974–989. Available from: http://www.eurekaselect.com/CDN/download.php?param=Sk9VjUk5vBTFaMvQbkVOpL0NhOLz7E3LpzEwFLzA4wOElFDL1nBkrZnxb8YXqBwbdGljcYXRcpb2n4vczGRmqfHwkzODpk5OeDU1CNDcizM2uExMqjhhnMDlamMWlRiZbjU0wMzVliZjdMzNqQTcVYTcVYt&key=VWlxoRWl7uZEEUtQcVNEwc2RdZZkEVJQyTMytRDMc1Rj1c1NbDV2ALTQj3MzrItMejM1dNjQazMjwNERm2VEaeATcVYDTcVY. doi:10.2174/1570159X17666190603171901.

Iffland, K., Grotenhermen, F., 2017. An update on safety and side effects of cannabidiol: a review of clinical data and relevant animal studies. Cannabis Cannabinoid Res. 2378-8763. 2 (1), 139–154. doi:10.1089/can.2016.0034.

Landmark, C.J., Sætre, J., Gottås, A., Wolden, M., McQuade, T.A.P., Kjeldsen, S.F., Våtevik, A., Sætre, E., Svendsen, T., Burns, M.L., Øiestad, E.L., Johannessen, S.I., 2025. Pharmacokinetic variability and use of therapeutic drug monitoring of cannabidiol in patients with refractory epilepsy. Epilepsia 15281167. 66 (5), 1477–1491. Available from: http://onlinelibrary.wiley.com/journal/10.1111/(ISSN)1528-1167. doi:10.1111/epi.18284.

Laprairie, R.B., Bagher, A.M., Kelly, M.E.M., Denovan-Wright, E.M., 2015. Cannabidiol is a negative allosteric modulator of the cannabinoid CB 1 receptor. Br. J. Pharmacol. 0007-1188. 172 (20), 4790–4805. doi:10.1111/bph.13250.

Lucas, C.J., Galettis, P., Schneider, J., 2018. The pharmacokinetics and the pharmacodynamics of cannabinoids. Br. J. Clin. Pharmacol. 13652125. 84 (11), 2477–2482. Available from: http://onlinelibrary.wiley.com/journal/10.1111/(ISSN)1365-2125. doi:10.1111/bcp.13710.

Madeo, G., Kapoor, A., Giorgetti, R., Busardò, F.P., Carlier, J., 2023. Update on cannabidiol clinical toxicity and adverse effects: a systematic review. Curr. Neuropharmacol. 18756190. 21 (11), 2323–2342. Available from: http://www.eurekaselect.com/603/journal/current-neuropharmacology. doi:10.2174/1570159X21666230322143401.

Millar, S.A., Stone, N.L., Yates, A.S., O'Sullivan, S.E., 2018. A systematic review on the pharmacokinetics of cannabidiol in humans. Front. Pharmacol. 16639812. 9. Available from: http://www.frontiersin.org/Pharmacology. doi:10.3389/fphar.2018.01365.

Nachnani, R., Raup-Konsavage, W.M., Vrana, K.E., 2021. The pharmacological case for cannabigerol. J. Pharmacol. Exp. Ther. 15210103. 376 (2), 204–212. Available from: https://jpet.aspetjournals.org/content/376/2/204. doi:10.1124/jpet.120.000340.

Navarro, G., Varani, K., Reyes-Resina, I., de Medina, V.S., Rivas-Santisteban, R., Sánchez-Carnerero Callado, C., Vincenzi, F., Casano, S., Ferreiro-Vera, C., Canela, E.I., Borea, P.A., Nadal, X., Franco, R., 2018. Cannabigerol action at cannabinoid CB1 and CB2 receptors and at CB1–CB2 heteroreceptor complexes. Front. Pharmacol. 1663-9812. 9. doi:10.3389/fphar.2018.00632.

Peters, E.N., MacNair, L., Mosesova, I., Christians, U., Sempio, C., Klawitter, J., Land, M.H., Ware, M.A., Turcotte, C., Bonn-Miller, M.O., 2022. Pharmacokinetics of cannabichromene in a medical cannabis product also containing cannabidiol and Δ9-tetrahydrocannabinol:

a pilot study. Eur. J. Clin. Pharmacol. 14321041. 78 (2), 259–265. Available from: https://link.springer.com/journal/volumesAndIssues/228. doi:10.1007/s00228-021-03232-8.
Rong, C., Carmona, N.E., Lee, Y.L., Ragguett, R.M., Pan, Z., Rosenblat, J.D., Subramaniapillai, M., Shekotikhina, M., Almatham, F., Alageel, A., Mansur, R., Ho, R.C., McIntyre, R.S., 2018. Drug-drug interactions as a result of co-administering Δ9-THC and CBD with other psychotropic agents. Expert Opin. Drug Saf. 1744764X. 17 (1), 51–54. doi:10.1080/14740338.2017.1397128.
Russo EB. Taming THC: potential cannabis synergy and phytocannabinoid-terpenoid entourage effects. Br J Pharmacol. 2011;163(7):1344–1364. doi: 10.1111/j.1476-5381.2011.01238.x. PMID: 21749363
Russo, E., Guy, G.W., 2006. A tale of two cannabinoids: the therapeutic rationale for combining tetrahydrocannabinol and cannabidiol. Med. Hypotheses 03069877. 66 (2), 234–246. Available from: http://intl.elsevierhealth.com/journals/mehy/. doi:10.1016/j.mehy.2005.08.026.
Sepulveda, D.E., Vrana, K.E., Kellogg, J.J., Bisanz, J.E., Desai, D., Graziane, N.M., Raup-Konsavage, W.M., 2024. The potential of cannabichromene (CBC) as a therapeutic agent. J. Pharmacol. Exp. Ther. 15210103. 391 (2), 206–213. Available from: https://jpet.aspetjournals.org/content/391/2/206. doi:10.1124/jpet.124.002166.
Thiele, E.A., Bebin, E.M., Bhathal, H., Jansen, F.E., Kotulska, K., Lawson, J.A., O'Callaghan, F.J., Wong, M., Sahebkar, F., Checketts, D., Knappertz, V., 2021. Add-on cannabidiol treatment for drug-resistant seizures in tuberous sclerosis complex: a placebo-controlled randomized clinical trial. JAMA Neurol. 21686157. 78 (3), 285–292. Available from: http://archneur.jamanetwork.com/issues.aspx. doi:10.1001/jamaneurol.2020.4607.
Udoh, M., Santiago, M., Devenish, S., McGregor, I.S., Connor, M., 2019. Cannabichromene is a cannabinoid CB2 receptor agonist. Br. J. Pharmacol. 14765381. 176 (23), 4537–4547. Available from: http://onlinelibrary.wiley.com/journal/10.1111/(ISSN)1476-5381. doi:10.1111/bph.14815.
Valdeolivas, S., Navarrete, C., Cantarero, I., Bellido, M.L., Muñoz, E., Sagredo, O., 2015. Neuroprotective properties of cannabigerol in Huntington's disease: studies in R6/2 mice and 3-nitropropionate-lesioned mice. Neurotherapeutics 18787479. 12 (1), 185–199. Available from: http://www.springerlink.com/content/1933-7213/. doi:10.1007/s13311-014-0304-z.
Wargent, E.T., Zaibi, M.S., Silvestri, C., Hislop, D.C., Stocker, C.J., Stott, C.G., Guy, G.W., Duncan, M., Di Marzo, V., Cawthorne, M.A., 2013. The cannabinoid Δ9-tetrahydrocannabivarin (THCV) ameliorates insulin sensitivity in two mouse models of obesity. Nutr. Diabetes. 2044-4052. 3 (5), e68. doi:10.1038/nutd.2013.9.
Zagzoog, A., Mohamed, K.A., Kim, H.J.J., Kim, E.D., Frank, C.S., Black, T., Jadhav, P.D., Holbrook, L.A., Laprairie, R.B., 2020. In vitro and in vivo pharmacological activity of minor cannabinoids isolated from Cannabis sativa. Sci. Rep. 20452322. 10 (1). Available from: https://www.nature.com/srep/index.html. doi:10.1038/s41598-020-77175-y.

CHAPTER

Dronabinol

7

Ahmed Abdalla, Preethash Sundar, and Jamal Hasoon

Department of Anesthesiology, Critical Care, and Pain Medicine, The University of Texas Health Science Center at Houston, Houston, TX, United States

Introduction

Pain management continues to advance in pursuit of therapies that not only relieve discomfort but also address the multifaceted needs of individuals living with chronic illness. Among these approaches, cannabinoids have attracted increasing clinical attention for their potential to manage difficult symptoms such as persistent chronic pain (Fine and Rosenfeld, 2013; Mallick-Searle and St. Marie, 2019). A key compound in this domain is delta-9-tetrahydrocannabinol (THC), the primary psychoactive agent in *Cannabis sativa.* Although the use of cannabis remains influenced by legal and societal factors, growing interest in its therapeutic potential has driven the development of pharmacologically standardized cannabinoid medications (Whiting et al., 2015).

Dronabinol, a synthetic formulation of THC, represents a notable advancement in this area. It offers a regulated and reproducible alternative to plant-derived cannabis products (O'Donnell et al., 2023; National Academies of Sciences, Engineering, and Medicine, Health and Medicine Division, 2017). Its clinical utility was first recognized in 1985, when the US Food and Drug Administration (FDA) approved it for the treatment of chemotherapy-induced nausea and vomiting (CINV) in patients unresponsive to traditional antiemetics (O'Donnell et al., 2023; May and Glode, 2016). This approval signaled an essential shift in how cannabinoid compounds were viewed within the medical community. The FDA later approved dronabinol for the treatment of anorexia associated with weight loss in individuals with acquired immunodeficiency syndrome (AIDS) (O'Donnell et al., 2023; Mylan Pharmaceuticals Inc, 2022), recognizing its appetite-stimulating properties as beneficial for a population at high risk for severe nutritional deficiencies (O'Donnell et al., 2023; Mylan Pharmaceuticals Inc, 2022; Badowski and Perez, 2016).

Although these are the only conditions currently approved by the FDA, dronabinol's pharmacologic properties, particularly its interaction with the endocannabinoid system (ECS) involved in pain regulation, have generated increasing interest in its off-label use for managing chronic pain (Ueberall et al., 2022; Johnson et al., 2025). For clinicians involved in palliative care or the treatment of complex pain conditions,

Cannabinoids for Chronic Pain. https://doi.org/10.1016/B978-0-443-36773-1.00006-4

familiarity with dronabinol's mechanisms and clinical profile is essential for delivering well-rounded symptom management.

Mechanism of action

Dronabinol produces its therapeutic and psychoactive effects by acting on the body's ECS, a complex regulatory network that plays a key role in maintaining physiological balance (Sheikh and Dua, 2023; Mechoulam and Parker, 2013). The ECS comprises cannabinoid receptors, endogenous ligands, and the enzymes responsible for their synthesis and degradation (Di Marzo and Petrosino, 2007; Kilaru and Chapma, 2020). Two primary receptors have been identified: CB1 and CB2 (Di Marzo and Petrosino, 2007). CB1 receptors are among the most abundant G protein-coupled receptors in the central nervous system (CNS), densely expressed in areas involved in pain perception (periaqueductal gray, rostral ventromedial medulla, spinal cord dorsal horn), memory, motor control, and reward pathways (Sheikh and Dua, 2023; Woodhams et al., 2015). When activated, these receptors tend to suppress the release of neurotransmitters, producing an inhibitory effect on neural activity. In contrast, CB2 receptors are found mainly outside the CNS, especially on immune cells, where they help regulate inflammation and immune function (Sheikh and Dua, 2023; Turcotte et al., 2016).

Dronabinol functions as a partial agonist at both CB1 and CB2 receptors, mimicking the actions of endogenous cannabinoids but with potentially different potency and duration (O'Donnell et al., 2023; Pertwee, 2008). Its psychoactive properties and ability to stimulate appetite and reduce nausea are mainly attributed to CB1 activation in the brain (O'Donnell et al., 2023; Sideris et al., 2024). By binding to presynaptic CB1 receptors, dronabinol can modulate the release of various neurotransmitters, including inhibiting the release of excitatory transmitters like glutamate and promoting the release of inhibitory transmitters like gamma-aminobutyric acid (GABA) in specific neuronal circuits (Manzanares et al., 2006; Schurman et al., 2020). This modulation of synaptic transmission is believed to be a key mechanism underlying its analgesic effects. Furthermore, activation of cannabinoid receptors, particularly CB1, within nociceptive pathways at peripheral, spinal, and supraspinal levels contributes to analgesia (Woodhams et al., 2015; Milligan et al., 2020; Guindon and Hohmann, 2009). Dronabinol helps reduce the transmission of pain signals and can influence how patients emotionally and mentally experience pain (Woodhams et al., 2015; Milligan et al., 2020). While CB2 receptor activation is less associated with central psychoactive effects, its role in modulating inflammatory responses suggests it may also contribute to analgesia, particularly in inflammatory conditions (Turcotte et al., 2016; Guindon and Hohmann, 2009).

Overall, dronabinol acts through multiple mechanisms within the ECS, altering neurotransmitter activity and directly impacting how pain is processed and perceived throughout the body.

Indications

Dronabinol has received FDA approval for two specific medical indications. Its initial approval in 1985 was for the treatment of CINV in cancer patients who had failed to respond adequately to conventional antiemetic therapies (O'Donnell et al., 2023; May and Glode, 2016; Mylan Pharmaceuticals Inc, 2022). It serves as an essential second-line or adjunctive agent in managing this side effect of cancer treatment. Subsequently, it was approved for treating anorexia associated with weight loss in patients with AIDS, leveraging its appetite-stimulating properties (O'Donnell et al., 2023; Mylan Pharmaceuticals Inc, 2022; Badowski and Perez, 2016). Outside of these approved indications, dronabinol is increasingly considered for off-label use in chronic pain management, although high-quality evidence supporting these uses varies (Ueberall et al., 2022; Johnson et al., 2025). Significant research interest has focused on its potential efficacy in chronic neuropathic pain, a condition often refractory to standard analgesics (Mücke et al., 2018; Lynch and Campbell, 2011). Systematic reviews and metaanalyses suggest that cannabinoids, including dronabinol and related compounds, may offer modest pain relief in neuropathic pain conditions compared to placebo. However, the effect size is often limited, and side effects need careful consideration (Johnson et al., 2025; Mücke et al., 2018; Busse et al., 2021). Evidence supporting its use in other conditions, such as fibromyalgia, is still in the early stages. Some preliminary studies show potential benefit, but the lack of large, well-controlled trials limits its use as a primary therapy (Mücke et al., 2018; Walitt et al., 2016). Similarly, while cannabinoids are investigated for cancer-related pain, often as adjuncts to opioids, the evidence specifically supporting dronabinol as a primary analgesic in this context remains limited (Johnson et al., 2025; Mücke et al., 2018). Observational studies and smaller trials have reported its use in mixed chronic pain populations, sometimes showing patient-reported improvements but often constrained by methodological limitations (Ueberall et al., 2022; Banerjee and McCormack, 2019). For clinicians considering dronabinol off-label for chronic pain, it is essential to weigh these limitations carefully and involve patients in shared decision-making to assess potential benefits versus risks (Johnson et al., 2025; Busse et al., 2021).

Contraindications

Absolute contraindications

Dronabinol is contraindicated in patients with known hypersensitivity to THC, THC derivatives, sesame oil, or dronabinol itself. In such individuals, use of the medication may result in adverse reactions, including hives, widespread rash, lip or throat swelling, oral lesions, skin burning, difficulty breathing, and, in severe cases, anaphylaxis (Mylan Pharmaceuticals Inc, 2022; Caicedo et al., 2025).

Relative contraindications

Dronabinol should be prescribed with caution in patients with a history of schizophrenia, psychosis, or depression, as it may worsen psychiatric symptoms. Patients should be screened for underlying mental health conditions before initiating therapy. In those with known psychiatric disorders, dronabinol is generally not recommended; however, if prescribed, close monitoring is necessary to detect any new or worsening symptoms. It is also not advised to prescribe dronabinol with other medications that can cause or exacerbate symptoms of the above-mentioned diseases (Mylan Pharmaceuticals Inc, 2022; Caicedo et al., 2025).

Dronabinol may occasionally cause hypotension, hypertension, tachycardia, or syncope. It is not recommended for patients with known cardiovascular disease, as they may be more susceptible to these hemodynamic effects. In addition, it should be avoided in combination with other medications that can cause blood pressure or heart rate fluctuations (Mylan Pharmaceuticals Inc, 2022; Caicedo et al., 2025).

Dronabinol also carries a potential for abuse. Patients should be screened for a history of substance or alcohol use disorder, as such individuals are considered contraindicated for treatment. Any signs of new or worsening substance misuse during therapy should be closely monitored (O'Donnell et al., 2023; Mylan Pharmaceuticals Inc, 2022).

For mothers actively breastfeeding, it is advised to change medications during breastfeeding or to withhold breastfeeding for up to 9 days after the last dose of dronabinol to minimize infant exposure (Mylan Pharmaceuticals Inc, 2022; Baker et al., 2018).

Side effects

Side effects associated with dronabinol can be categorized as common or severe and are often grouped by the organ systems they affect. Common side effects include weakness/fatigue and dry mouth, palpitations, vasodilation, and facial flushing in the cardiovascular system. Gastrointestinal complaints, such as abdominal pain, nausea, and vomiting, are also frequently reported. The CNS effects are varied and may include dizziness, euphoria, somnolence, anxiety, amnesia, paranoia, nervousness, ataxia, confusion, depersonalization, and hallucinations (O'Donnell et al., 2023; Badowski, 2017).

Serious adverse effects, while less common, may involve multiple systems. General symptoms include chills, headache, and malaise. Cardiovascular effects, such as hypotension and tachycardia, have been noted, while gastrointestinal side effects may include diarrhea, fecal incontinence, anorexia, and elevated hepatic enzymes. Neurological complications, like depression, nightmares, speech difficulties, tinnitus, and sensory disturbances such as visual changes, have also been documented. Respiratory symptoms can include cough, rhinitis, and sinusitis, while dermatologic manifestations may involve flushing and sweating (O'Donnell et al., 2023; Mylan Pharmaceuticals Inc, 2022).

Tolerance is another critical consideration. Research has shown that individuals who use cannabinoids chronically, primarily through smoking marijuana, may require progressively higher doses to achieve the same therapeutic effect (Ueberall et al., 2022; Baker et al., 2018). Although physical dependence on dronabinol is rare, it can occur, particularly in patients with a history of substance or alcohol use disorder. Withdrawal symptoms may develop 12–16 h after abrupt discontinuation. They can include hot flashes, sweating, rhinorrhea, loose stools, hiccups, anorexia, sleep disturbances, and mood changes, with severity often correlating with the degree of prior use (O'Donnell et al., 2023; Sangiamo et al., 2023).

Counseling should be given to each patient, explaining the above-mentioned side effects. Screening should be conducted for patients to identify those at risk for substance abuse and dependence. For patients who are at risk for substance abuse and dependence, other options should be considered and presented, or support should be provided to the patient to avoid serious cases of dependence, abuse, and withdrawal (Schimrigk et al., 2017; Schlienz et al., 2018).

Drug interactions

Dronabinol is a cannabinoid with various drug interactions. It has a significant interaction with CNS depressants such as benzodiazepines and opioids. Cannabinoids have a CNS depressive effect and can exacerbate the effects of other CNS depressants if taken together (Ueberall et al., 2022; Yang et al., 2023).

Dronabinol is metabolized in the liver via extensive first-pass metabolism through cytochrome P450. The isoenzymes CYP2C9 and CYP3A4 are chiefly responsible for its metabolism, with CYP2C9 playing a key role in forming its active metabolite, 11-hydroxy-delta-9-THC. When dronabinol is coadministered with inhibitors of CYP2C9 or CYP3A4, clinicians should closely monitor for increased drug levels and the potential for enhanced adverse effects (Dean and Kane, 2020).

Although dronabinol does not act through anticholinergic pathways, some side effects may resemble anticholinergic symptoms. It primarily exerts its effects through cannabinoid receptors CB1 and CB2. Therefore caution is advised when prescribing dronabinol alongside medications with anticholinergic properties, as overlapping side effects may occur (Ueberall et al., 2022; Legare et al., 2022).

Patients should also be counseled to avoid alcohol while taking dronabinol. Both substances have sedative properties and can produce compounded effects such as drowsiness, dizziness, confusion, tachycardia, cognitive impairment, and even syncope. Concurrent use increases the risk of adverse outcomes and should be discouraged (Sangiamo et al., 2023; Schlienz et al., 2018; Weinberger et al., 2021).

Recommended doses for different conditions

Dronabinol dosing varies depending on the condition being treated.

For adults who have anorexia associated with weight loss in AIDS, the starting dosage is 2.5 mg orally twice daily, advised to be taken 1 h before lunch and 1 h before dinner. Patients may be advised to take the evening dose at bedtime to reduce the risk of CNS side effects. Mild CNS symptoms often resolve within 1 to 3 days without dose adjustment. However, for those experiencing persistent or more severe side effects, the dose may be reduced to 2.5 mg once daily in the evening (O'Donnell et al., 2023).

For CINV, the initial recommended dose is 5 mg taken orally 1–3 h before chemotherapy, followed by one capsule every 2–4 h for a total of 4–6 doses in 24 h. In elderly patients, a reduced starting dose of 2.5 mg may be used to minimize CNS complications. Dosage should be kept consistent in relation to meal timing and chemotherapy timing (O'Donnell et al., 2023; Mylan Pharmaceuticals Inc, 2022).

In treating chronic pain, dronabinol should be started at low doses, typically 2.5 mg once or twice daily, with gradual titration based on symptom control and tolerability. Usually, the dosage is started at 2.5 mg once or twice daily and increased incrementally until the desired pain relief is achieved. Dosage should not routinely exceed 20 mg, as increased dosage leads to increased CNS adverse effects (Narang et al., 2008). In addition, patients should be carefully monitored during titration for drug-related side effects.

Individualized considerations

Several patient-specific factors must be considered when determining appropriate dosing. Age is essential; the older a patient is, the more susceptible they are to CNS-related adverse effects. Since dronabinol is a lipophilic compound that accumulates in adipose tissue, patients with higher body fat percentages may experience prolonged drug effects due to slower clearance (Mylan Pharmaceuticals Inc, 2022; Yang et al., 2023; Hansen et al., 2024). Patients with hepatic impairment are at risk for reduced metabolism of dronabinol and should be monitored carefully or considered for alternative therapies. Special attention should be given to patients taking additional medications such as sedatives, opioids, CYP450 inhibitors, and anticholinergic drugs, as dronabinol can exacerbate adverse effects and cause drug interactions with other treatments (Mylan Pharmaceuticals Inc, 2022; Dean and Kane, 2020).

A thorough history of cannabis use is also essential. Patients with prior cannabis exposure, particularly those with a history of misuse, may develop tolerance and dependence, potentially requiring altered dosing strategies or closer monitoring (Weinberger et al., 2021; Bedi et al., 2010; Leung et al., 2022). Appropriate assessment and individualized treatment planning are critical to ensure dronabinol's safe and effective use in diverse patient populations.

Titration strategy

Titration should be scheduled methodically. Initial dosing may begin with 2.5 mg once daily at bedtime or before dinner, continuing for 1–3 days to assess tolerability. Based on efficacy and response to the patient's symptoms and adverse effects, the dosage can be increased to two times a day and then slowly increased to 5 mg as needed. Each patient is unique, and based on the patient's symptoms, adverse effects, and tolerance, the dosage can safely be increased to as much as 20 mg daily. In select cases, total daily doses can be increased beyond 20 mg. However, such escalations should be approached with caution, as higher doses are associated with an increased risk of CNS-related adverse effects, including sedation, dizziness, and cognitive impairment. During titration, patients should be reassessed every 3–7 days for signs of cognitive changes, psychiatric symptoms, pain control, or improvements in nausea, vomiting, or appetite (Mylan Pharmaceuticals Inc, 2022; Ueberall et al., 2022; Narang et al., 2008).

Special populations

For specific special populations, studies have been conducted on the safety and efficacy of cannabinoids.

Dronabinol use in pregnancy should be avoided. While data are limited, studies have shown that cannabinoids can cross the placenta and may be present in the umbilical cord blood of mothers who use marijuana. Animal studies have demonstrated decreased weight gain in offspring exposed to cannabinoids during gestation (Day and Richardson, 1991; Perez-Reyes and Wall, 1982).

Caution should be exercised for elderly patients, given the increased risk of CNS-related adverse effects due to sedation, cognitive function impairment, and risk of falling. No difference in efficacy was found between younger patients and elderly patients (Mylan Pharmaceuticals Inc, 2022; Narang et al., 2008).

For patients with hepatic impairment, cautious use is advised due to extensive first-pass liver metabolism; alternative therapies may be preferred (Ueberall et al., 2022; Legare et al., 2022).

Regarding lactation, there is limited data on dronabinol in breast milk. However, breastfeeding is generally discouraged during treatment and for up to nine days after the final dose (Day and Richardson, 1991; Perez-Reyes and Wall, 1982).

While dronabinol may offer therapeutic benefits for select patients, its use in special populations requires careful consideration. The risks associated with pregnancy, lactation, advanced age, and hepatic impairment require an individualized risk-benefit assessment. Clinicians should remain vigilant in monitoring for adverse effects and consider alternative therapies when appropriate to ensure patient safety.

References

Badowski, M.E., 2017. A review of oral cannabinoids and medical marijuana for the treatment of chemotherapy-induced nausea and vomiting: a focus on pharmacokinetic variability and pharmacodynamics. Cancer Chemother. Pharmacol. 14320843. 80 (3), 441–449. Available from: https://link.springer.de/link/service/journals/00280/index.htm. doi:10.1007/s00280-017-3387-5.

Badowski, M.E., Perez, S.E., 2016. Clinical utility of dronabinol in the treatment of weight loss associated with HIV and AIDS. HIV/AIDS Res. Palliat. Care 11791373. 8, 37–45.

Available from: https://www.dovepress.com/getfile.php?fileID=28954. doi:10.2147/HIV.S81420.

Baker, T., Datta, P., Rewers-Felkins, K., Thompson, H., Kallem, R.R., Hale, T.W., 2018. Transfer of inhaled cannabis into human breast milk. Obstet. Gynecol. 1873233X. 131 (5), 783–788. Available from: http://journals.lww.com/greenjournal. doi:10.1097/AOG.0000000000002575.

Banerjee, S., McCormack, S., 2019. Medical Cannabis for the Treatment of Chronic Pain: A Review of Clinical Effectiveness and Guidelines (Ottawa (ON): Canadian Agency for Drugs and Technologies in Health). 31532599

Bedi, G., Foltin, R.W., Gunderson, E.W., Rabkin, J., Hart, C.L., Comer, S.D., Vosburg, S.K., Haney, M., 2010. Efficacy and tolerability of high-dose dronabinol maintenance in HIV-positive marijuana smokers: a controlled laboratory study. Psychopharmacology. 14322072. 212 (4), 675–686. doi:10.1007/s00213-010-1995-4.

Busse, J.W., Vankrunkelsven, P., Zeng, L., Heen, A.F., Merglen, A., Campbell, F., Petter, L., Aertgeerts, B., Buchbinder, R., Coen, M., Juurlink, D., Samer, C., Siemieniuk, R.A.C., Kumar, N., Cooper, L., Brown, J., Lytvyn, L., Zeraatkar, D., Wang, L., Guyatt, G.H., Vandvik, P.O., Agoritsas, T., 2021. Medical cannabis or cannabinoids for chronic pain: a clinical practice guideline. BMJ 17561833. 374. Available from: http://www.bmj.com/. doi:10.1136/bmj.n2040.

Caicedo, D.A., Pérez-Mañá, C., Farré, M., Papaseit, E., 2025. An overview of the potential for pharmacokinetic interactions between drugs and cannabis products in humans. Pharmaceutics. 1999-4923. 17 (3), 319. doi:10.3390/pharmaceutics17030319.

Day, N.L., Richardson, G.A., 1991. Prenatal marijuana use: epidemiology, methodologic issues, and infant outcome. Clin. Perinatol. 00955108. 18 (1), 77–91. doi:10.1016/s0095-5108(18)30535-9.

Dean, L., Kane, M., 2020. Dronabinol therapy and CYP2C9 genotype. Med. Genet. Summ. 323–330 https://www.ncbi.nlm.nih.gov/books/NBK564166/. 33211456.

Di Marzo, V., Petrosino, S., 2007. Endocannabinoids and the regulation of their levels in health and disease. Curr. Opin. Lipidol. 0957-9672. 18 (2), 129–140. doi:10.1097/mol.0b013e32803dbdec.

Fine, P.G., Rosenfeld, M.J., 2013. The endocannabinoid system, cannabinoids, and pain. Rambam Maimonides Med. J. 20769172. 4 (4). doi:10.5041/RMMJ.10129.

Guindon, J., Hohmann, A.G., 2009. The endocannabinoid system and pain. CNS Neurol. Disord. Drug Targets 18715273. 8 (6), 403–421. Available from: http://docstore.ingenta.com/cgi-bin/ds_deliver/1/u/d/ISIS/54495694.1/ben/cnsnddt/2009/00000008/00000006/art00002/72A48A3AB8A5489C1263948195692CEF096D14A5D1.pdf?link=http://www.ingentaconnect.com/error/delivery&format=pdf. doi:10.2174/187152709789824660.

Hansen, J.S., Boix, F., Hasselstrøm, J.B., Sørensen, L., Kjolby, M., Gustavsen, S., Hansen, R., Petersen, T., Sellebjerg, F., Kasch, H., Rasmussen, P.V., Finnerup, N.B., Sædder, E.A., Svendsen, K.B., 2024. Pharmacokinetics and pharmacodynamics of cannabis-based medicine in a patient population included in a randomized, placebo-controlled, clinical trial. Clin. Transl. Sci. 17528062. 17 (1). Available from: http://onlinelibrary.wiley.com/journal/10.1111/(ISSN)1752-8062. doi:10.1111/cts.13685.

Johnson, B.W., Strand, N.H., Raynak, J.C., Jara, C., Habtegiorgis, K., Hand, B.A., Hong, S., Maloney, J.A., 2025. Cannabinoids in chronic pain management: a review of the history, efficacy, applications, and risks. Biomedicines 22279059. 13 (3). Available from: https://www.mdpi.com/journal/biomedicines. doi:10.3390/biomedicines13030530.

Kilaru, A., Chapma, K.D., 2020. The endocannabinoid system. Essays Biochem. 00711365. 64 (3), 485–499. Available from: https://portlandpress.com/essaysbiochem/article/64/3/485/225762/The-endocannabinoid-system. doi:10.1042/EBC20190086.

Legare, C.A., Raup-Konsavage, W.M., Vrana, K.E., 2022. Therapeutic potential of cannabis, cannabidiol, and cannabinoid-based pharmaceuticals. Pharmacology 14230313. 107 (3), 131–149. Available from: https://www.karger.com/journals/pha/pha_jh.htm. doi:10.1159/000521683.

Leung, J., Chan, G., Stjepanović, D., Chung, J.Y.C., Hall, W., Hammond, D., 2022. Prevalence and self-reported reasons of cannabis use for medical purposes in USA and Canada. Psychopharmacology 14322072. 239 (5), 1509–1519. Available from: https://www.springer.com/journal/213. doi:10.1007/s00213-021-06047-8.

Lynch, M.E., Campbell, F., 2011. Cannabinoids for treatment of chronic non-cancer pain; a systematic review of randomized trials. Br. J. Clin. Pharmacol. 13652125. 72 (5), 735–744. doi:10.1111/j.1365-2125.2011.03970.x.

Mallick-Searle, T., St. Marie, B., 2019. Cannabinoids in pain treatment: an overview. Pain Manag. Nurs. 15249042. 20 (2), 107–112. doi:10.1016/j.pmn.2018.12.006.

Manzanares, J., Julian, M., Carrascosa, A., 2006. Role of the cannabinoid system in pain control and therapeutic implications for the management of acute and chronic pain episodes. Curr. Neuropharmacol. 1570159X. 4 (3), 239–257. doi:10.2174/157015906778019527.

May, M.B., Glode, A.E., 2016. Dronabinol for chemotherapy-induced nausea and vomiting unresponsive to antiemetics. Cancer Manag. Res. 11791322. 8, 49–55. Available from: https://www.dovepress.com/getfile.php?fileID=30339. doi:10.2147/CMAR.S81425.

Mechoulam, R., Parker, L.A., 2013. The endocannabinoid system and the brain. Annu. Rev. Psychol. 15452085. 64, 21–47. Available from: http://arjournals.annualreviews.org/loi/psych. doi:10.1146/annurev-psych-113011-143739.

Milligan, A.L., Szabo-Pardi, T.A., Burton, M.D., 2020. Cannabinoid receptor type 1 and its role as an analgesic: an opioid alternative? J. Dual Diagn. 15504271. 16 (1), 106–119. Available from: http://www.tandfonline.com/toc/wjdd20/current. doi:10.1080/15504263.2019.1668100.

Mücke, M., Phillips, T., Radbruch, L., Petzke, F., Häuser, W., 2018. Cannabis-based medicines for chronic neuropathic pain in adults. Cochrane Database Syst. Rev. 1469493X. 3. Available from: http://as.wiley.com/WileyCDA/Brand/id-6.html. doi:10.1002/14651858.CD012182.pub2.

Mylan Pharmaceuticals Inc, Dronabinol capsule [package insert]. U.S. National Library of Medicine, Morgantown, WV. Retrieved June 1, 2025, from. (2022), Available from: https://dailymed.nlm.nih.gov/dailymed/drugInfo.cfm?setid=fd8c5d57-0173-4f62-8ff0-d8147986a976.

Narang, S., Gibson, D., Wasan, A.D., Ross, E.L., Michna, E., Nedeljkovic, S.S., Jamison, R.N., 2008. Efficacy of dronabinol as an adjuvant treatment for chronic pain patients on opioid therapy. J. Pain. 15288447. 9 (3), 254–264. doi:10.1016/j.jpain.2007.10.018.

National Academies of Sciences, Engineering, and Medicine, Health and Medicine Division, 2017. Board on Population Health and Public Health Practice, Committee on the Health Effects of Marijuana: An Evidence Review and Research Agenda, the Health Effects of Cannabis and Cannabinoids: The Current State of Evidence and Recommendations for Research. National Academies Press (US).

O'Donnell, B., Meissner, H., Gupta, V., 2023. Dronabinol, StatPearls.

Perez-Reyes, M., Wall, M.E., 1982. Presence of Δ 9 -tetrahydrocannabinol in human milk. N. Engl. J. Med. 0028-4793. 307 (13), 819–820. doi:10.1056/NEJM198209233071311.

Pertwee, R.G., 2008. The diverse CB 1 and CB 2 receptor pharmacology of three plant cannabinoids: Δ 9-tetrahydrocannabinol, cannabidiol and Δ 9-tetrahydrocannabivarin. Br. J. Pharmacol. 14765381. 153 (2), 199–215. doi:10.1038/sj.bjp.0707442.

Sangiamo, D.T., Weingarten, M.J., Choi, C.Y., Das, A., Liang, N.C., 2023. Experience with dronabinol consumption facilitated a stimulant effect of alcohol and affected alcohol-related changes in frontal cortical endocannabinoid levels in male rats. Behav. Brain Res. 18727549. 452. Available from: https://www.elsevier.com/locate/bbr. doi:10.1016/j.bbr.2023.114587.

Schimrigk, S., Marziniak, M., Neubauer, C., Kugler, E.M., Werner, G., Abramov-Sommariva, D., 2017. Dronabinol is a safe long-term treatment option for neuropathic pain patients. Eur. Neurol. 14219913. 78 (5-6), 320–329. Available from: https://www.karger.com/journals/ene/ene_jh.htm. doi:10.1159/000481089.

Schlienz, N.J., Lee, D.C., Stitzer, M.L., Vandrey, R., 2018. The effect of high-dose dronabinol (oral THC) maintenance on cannabis self-administration. Drug Alcohol Depend. 18790046. 187, 254–260. Available from: https://www.elsevier.com/locate/drugalcdep. doi:10.1016/j.drugalcdep.2018.02.022.

Schurman, L.D., Lu, D., Kendall, D.A., Howlett, A.C., Lichtman, A.H., 2020. Molecular mechanism and cannabinoid pharmacology. Handb. Exp. Pharmacol. 258, 323–353. Available from: http://www.springer.com/series/164. doi:10.1007/164_2019_298.

Sheikh, N.K., Dua, A., 2023. Cannabinoids. StatPearls https://www.ncbi.nlm.nih.gov/books/NBK556062/. 32310522.

Sideris, A., Lauzadis, J., Kaczocha, M., 2024. The basic science of cannabinoids. Anesth. Analg. 15267598. 138 (1), 42–53. Available from: http://journals.lww.com/anesthesia-analgesia/toc/publishahead. doi:10.1213/ANE.0000000000006472.

Turcotte, C., Blanchet, M.R., Laviolette, M., Flamand, N., 2016. The CB2 receptor and its role as a regulator of inflammation. Cell. Mol. Life Sci. 14209071. 73 (23), 4449–4470. Available from: http://link.springer.de/link/service/journals/00018/index.htm. doi:10.1007/s00018-016-2300-4.

Ueberall, M.A., Horlemann, J., Schuermann, N., Kalaba, M., Ware, M.A., 2022. Effectiveness and tolerability of dronabinol use in patients with chronic pain: a retrospective analysis of 12-week open-label real-world data provided by the German Pain e-Registry. Pain Med. 1526-2375. 23 (8), 1409–1422. doi:10.1093/pm/pnac010.

Walitt, B., Klose, P., Fitzcharles, M.A., Phillips, T., Häuser, W., 2016. Cannabinoids for fibromyalgia. Cochrane Database Syst. Rev. 1469493X. 7. Available from: http://as.wiley.com/WileyCDA/Brand/id-6.html. doi:10.1002/14651858.CD011694.pub2.

Weinberger, A.H., Zhu, J., Levin, J., Moeller, S.J., McKee, S.A., Goodwin, R.D., 2021. Changes in alcohol use by cannabis use status among adolescents and young adults in the United States: emerging evidence for both substitution and complementarity. Alcohol. Clin. Exp. Res. 15300277. 45 (12), 2536–2545. Available from: http://onlinelibrary.wiley.com/journal/10.1111/(ISSN)1530-0277. doi:10.1111/acer.14737.

Whiting, P.F., Wolff, R.F., Deshpande, S., Di Nisio, M., Duffy, S., Hernandez, A.V., Keurentjes, J.C., Lang, S., Misso, K., Ryder, S., Schmidlkofer, S., Westwood, M., Kleijnen, J., 2015. Cannabinoids for medical use: a systematic review and meta-analysis. JAMA 15383598. 313 (24), 2456–2473. Available from: http://jama.jamanetwork.com/article.aspx?articleid=2338251. doi:10.1001/jama.2015.6358.

Woodhams, S.G., Sagar, D.R., Burston, J.J., Chapman, V., 2015. The role of the endocannabinoid system in pain. Handb. Exp. Pharmacol. 18650325. 227, 119–143. Available from: http://www.springer.com/series/164. doi:10.1007/978-3-662-46450-2_7.

Yang, C., Williams, M.G., Agnello, R.N., 2023. Dronabinol useful for chronic pain: a case report. Biomed. J. Sci. Tech. Res. 49 (2), 40376–40378.

CHAPTER

The role of nabilone in pain management

8

David Rivetti[a] and Trent Emerick[b]

[a]*Department of Physical Medicine and Rehabilitation, University of Pittsburgh Medical Center, Pittsburgh, PA, United States*

[b]*Division of Chronic Pain, Department of Anesthesiology and Perioperative Medicine, University of Pittsburgh Medical Center, Pittsburgh, PA, United States*

Introduction

Synthetic cannabis derivatives remain a popular option for both patients and clinicians for a variety of clinical conditions. Nabilone is a synthetic cannabinoid that has structural similarity to delta-9-tetrahydrocannabinol (THC), the primary psychoactive component of Cannabis sativa (Cesamet (nabilone), 2021). It was initially developed in the 1970s to manufacture orally active cannabinoids with antiemetic properties but with more predictable pharmacokinetics and fewer psychoactive effects than plant-based THC (Bathula and Maciver, 2024). Nabilone was first approved in Canada in 1982 and subsequently received approval from the US Food and Drug Administration (FDA) in 1985 for the treatment of nausea and vomiting associated with cancer chemotherapy in patients who have not responded adequately to conventional antiemetic therapies (Ware et al., 2008). Nabilone is classified as a Schedule II controlled substance in the United States due to its potential for abuse and psychological dependence, with similar effects on mood and cognition to those of cannabis (Cesamet (nabilone), 2021). It is not FDA-approved for pain, although it may be efficacious in the off-label management of various types of pain, especially neuropathic pain. Presently, nabilone is not considered a first-line therapy and is generally reserved for cases as an adjunctive treatment or cases where standard treatments have been unsuccessful or poorly tolerated (Price et al., 2022).

Mechanism of action

Nabilone is an orally active synthetic cannabinoid. It is a partial agonist of both cannabinoid receptor type 1 (CB1) and cannabinoid receptor type 2 (CB2) receptors in the central nervous system, with greater affinity for CB1 receptors. The FDA labeling notes that nabilone's antiemetic effect is thought to be mediated by interaction with

Cannabinoids for Chronic Pain. https://doi.org/10.1016/B978-0-443-36773-1.00009-X

the CB1 receptor system in neural tissues (Cesamet (nabilone), 2021). However, preclinical and pharmacological studies confirm that nabilone also has affinity for CB2 receptors (Ashton et al., 2008; Pertwee, 2010) with activation associated with immunomodulatory and antiinflammatory effects (Meanti et al., 2025). Peak plasma concentrations are achieved in approximately 2 hours, and the half-life of metabolites may last up to 35 hours (Cesamet (nabilone), 2021).

CB1 agonism broadly suppresses the release of several neurotransmitters, including glutamate, GABA, acetylcholine, norepinephrine, serotonin, and dopamine. This effect on neurotransmitter release contributes to nabilone's antiemetic and neuromodulatory properties, particularly in regions of the brainstem associated with emesis, such as the dorsal vagal complex (Szabo and Schlicker, 2005; Pergolizzi et al., 2017). In addition to its antiemetic effects, nabilone's interaction with the endocannabinoid system is thought to underlie its analgesic and potential anxiolytic properties (Pergolizzi et al., 2017; Davis et al., 2007; Davis, 2008).

Experimental pain models have demonstrated that the analgesic effects of nabilone are not mediated by the potentiation of descending inhibitory pain pathways, but rather by the direct modulation of synaptic transmission in pain circuits (Redmond et al., 2008). Clinical studies indicate that nabilone is most effective in conditions characterized by central sensitization and neuropathic pain, rather than acute nociceptive pain (Davis, 2008; Toth et al., 2012).

Indications

It should be noted that nabilone only has FDA approval for chemotherapy-induced nausea and vomiting, and other indications and usages are experimental and considered off-label. Multiple professional societies have also weighed in on the available evidence. The relatively new American Academy of Neurology guidelines indicate that nabilone is probably more effective than a placebo for improving pain in patients with painful diabetic polyneuropathy (Price et al., 2022). Nabilone may help reduce pain scores and improve quality of life measures for patients with diabetic peripheral neuropathy when combined with standard-of-care pain therapies, and additionally provide benefits in sleep and quality of life (Toth et al., 2012). However, its clinical effect size is modest, and heterogeneity among studies is high (Meng et al., 2017; Tsang and Giudice, 2016). In direct comparison studies, nabilone is less effective than weak opioids such as dihydrocodeine for neuropathic pain. It is associated with more frequent side effects, including dizziness, drowsiness, and dry mouth (Frank et al., 2008). For multiple sclerosis-induced neuropathic pain, nabilone combined with gabapentin has been effective and well tolerated, providing significant pain relief compared to placebo (Turcotte et al., 2015). The American Burn Association guidelines note that while cannabinoids, including nabilone, have potential analgesic effects, they are not recommended for routine use in burn pain management due to limited data (Romanowski et al., 2020).

Nabilone has been investigated for the management of fibromyalgia, particularly for associated symptoms, including pain and sleep disturbance. The relationship is characterized by limited and low-quality evidence, suggesting that nabilone may provide modest improvements in pain, sleep quality, and functional outcomes in fibromyalgia patients, but with a higher rate of adverse effects compared to placebo or amitriptyline (Walitt et al., 2016; Ware et al., 2010; Skrabek et al., 2008). Given the complexity in managing the spectrum of fibromyalgia symptoms, combination therapy is often necessary. Cannabinoids are considered potential alternative pharmacological agents in fibromyalgia treatment, most notably for patients with sleep disturbances (Calandre et al., 2015).

The American Society of Clinical Oncology (ASCO) 2016 guidelines recognized the effectiveness of nabilone as an adjuvant analgesic in managing chronic pain in cancer survivors (Paice et al., 2016). However, variability in formulations and limited high-quality data preclude strong recommendations. The ASCO 2024 guidelines state that cannabinoids, such as nabilone, may improve refractory nausea or vomiting for patients on nausea-provoking antineoplastic agents who are already taking standard-of-care antiemetics (Braun et al., 2024). There remains uncertainty surrounding the impact of cannabinoids on other supportive care outcomes, emphasizing the need for additional research in this space. The National Comprehensive Cancer Network (NCCN) also notes that data supporting the use of cannabinoids as adjuvant analgesics for cancer pain is extremely limited and conflicting (Swarm et al., 2019). While nabilone is FDA-approved for chemotherapy-induced nausea and vomiting, its use for cancer pain is not endorsed by NCCN guidelines (Swarm et al., 2019). The emerging systematic reviews, including a 2023 Cochrane review, found low-certainty evidence that nabilone is ineffective in reducing pain associated with chemotherapy or radiochemotherapy in head and neck cancer and non-small cell lung cancer patients, and that synthetic THC analogs are not superior to low-dose codeine for moderate-to-severe cancer pain (Häuser et al., 2023).

Contraindications

Nabilone is contraindicated in any patient who has a history of cannabinoid hypersensitivity. While other absolute contraindications have not been established, the FDA has identified populations at higher risk of adverse reactions to nabilone, including those with cardiovascular disease, existing or previous psychiatric disorders, a history of substance use, pediatric and geriatric populations, pregnant and nursing women, and those with hepatic or renal impairment (Cesamet (nabilone), 2021).

The FDA notes that, given variations in response and tolerance to nabilone, patients should refrain from driving, operating machinery, or engaging in hazardous activity while on the medication. In addition, it should not be taken with alcohol, sedatives, hypnotics, or other psychoactive substances due to concern for potentiation of CNS effects (Tsang and Giudice, 2016).

Side effects

The most common side effects of nabilone include drowsiness, vertigo, dry mouth, euphoria, ataxia, sleep disturbance, dysphoria, and headaches (Cesamet (nabilone), 2021). In placebo-controlled clinical trials, these occurred in a substantial proportion of patients, with drowsiness and vertigo each affecting over 50% of patients, and dry mouth affecting approximately 36% of those included (Cesamet (nabilone), 2021). A systematic review and metaanalysis of placebo-controlled clinical trials found that, while most adverse events were mild to moderate and transient in nature, drowsiness (OR: 7.25; 95% CI: 1.64–31.95), dizziness (OR: 21.14; 95% CI: 2.92–152.75), and dry mouth (OR: 17.23; 95% CI: 4.33–68.55) were significantly more common in patients treated with nabilone compared to placebo (Bajtel et al., 2022).

Serious psychiatric reactions, including hallucinations and psychosis, can occur and may persist for 48–72 hours after cessation of nabilone. The drug label emphasizes that nabilone should be used under close supervision, especially during initiation and dose adjustments, due to the unpredictability of CNS and psychiatric effects (Cesamet (nabilone), 2021).

A review of the safety and efficacy of nabilone for pain management, including cancer pain, chronic noncancer pain, neuropathic pain, fibromyalgia, and pain associated with spasticity, revealed that nabilone was rarely associated with severe drug reactions that necessitated discontinuation. In addition, the likelihood of abuse was considered low (Tsang and Giudice, 2016).

Drug interactions

Nabilone is associated with several clinically significant drug interactions. The most important are potentiated CNS depression and psychomotor impairment when combined with other CNS depressants, including barbiturates, gabapentinoids, benzodiazepines, alcohol, opioids, antihistamines, muscle relaxants, and other sedatives or psychoactive drugs. Psychomotor function is particularly impaired with the concurrent use of diazepam, and such combinations should be avoided or used with caution (Cesamet (nabilone), 2021).

Sympathomimetic agents (e.g., amphetamines, ephedrine, cocaine) can result in significant hypertension, tachycardia, and possibly increased cardiotoxicity when taken concurrently with nabilone. When used with anticholinergic agents (e.g., atropine, scopolamine), there is a risk of additive tachycardia and drowsiness. Similarly, tricyclic antidepressants (e.g., amitriptyline) may also increase the risk of tachycardia, hypertension, and drowsiness (Cesamet (nabilone), 2021). Rarely, interactions with disulfiram or fluoxetine have been associated with hypomanic reactions in case reports involving marijuana, and opioid receptor blockade with naltrexone may enhance the effects of oral THC (Lacoursiere and Swatek, 1983; Stoll et al., 1991; Haney et al., 2003).

In addition, nabilone has interactions with drugs metabolized by cytochrome P450 (CYP) isoenzymes, particularly CYP2C9 and CYP3A4 (Campos et al., 2024). Coadministration of nabilone with CYP3A4 inhibitors, such as ketoconazole, can increase plasma concentrations, potentially leading to enhanced effects and an increased risk of adverse reactions. Conversely, CYP3A4 inducers, like rifampin, can decrease nabilone plasma levels, potentially reducing its efficacy. Furthermore, nabilone may interact with anticonvulsant medications, altering their pharmacokinetics and potentially leading to therapeutic failures or increased side effects (Campos et al., 2024). Therefore it is essential for clinicians to obtain a thorough history from patients, including current medications and substance use, before prescribing nabilone.

Recommended doses for different conditions

Clinical evidence suggests that nabilone can lead to small but significant reductions in pain when used as adjunctive therapy. For example, a study on diabetic peripheral neuropathic pain demonstrated that flexible-dose nabilone (1–4 mg/day) significantly improved pain scores, sleep quality, and overall patient status compared to placebo (Toth et al., 2012). Additionally, nabilone has shown benefits in fibromyalgia, with improvements in pain relief and functional outcomes (Skrabek et al., 2008). The typical dosing for pain management ranges from 1 to 4 mg per day, adjusted based on patient response and tolerability (Toth et al., 2012; Tsang and Giudice, 2016).

Randomized controlled trials have shown that nabilone, dosed at 0.5–1 mg at bedtime and up to 1 mg twice daily, can improve sleep quality and reduce pain scores compared to placebo. Still, these effects are generally minor, and small sample sizes and short durations limit the studies (Ware et al., 2010; Skrabek et al., 2008). When used as an adjunct antiemetic for patients receiving chemotherapy, the usual adult dosage is 1 or 2 mg orally twice daily, with a maximum dose of 6 mg daily divided into three doses. The FDA recommends that the initial dose of nabilone should be administered 1–3 hours before chemotherapeutic agents (Cesamet (nabilone), 2021). Nabilone has a longer duration of action than dronabinol, offering dosing twice daily, while dronabinol is typically given every 4–6 hours. Thus nabilone may be a more useful medication for chronic therapy (Davis, 2008).

Conclusion

Nabilone is a synthetic THC derivative that is FDA-approved for nausea and vomiting related to chemotherapy. There is a growing body of off-label evidence supporting its use for various chronic pain states, and it may also help with other quality measures, such as sleep. Nabilone is generally considered a third-line agent in clinical practice guidelines despite its potential benefits, primarily due to its side effect profile and reported efficacy. Many studies involving nabilone are of lower-quality evidence, and additional research is needed on nabilone, including the differences in its use compared to other cannabinoid derivatives. However, there are some critical side effects to consider, including drowsiness, vertigo, and dry mouth. Lastly, it is essential to note that there are potential interactions, including worsening CNS depression, when combined with other substances such as opioids, muscle relaxants, and benzodiazepines.

References

Ashton, J.C., Wright, J.L., McPartland, J.M., Tyndall, J.D.A., 2008. Cannabinoid CB1 and CB2 receptor ligand specificity and the development of CB2-selective agonists. Curr. Med. Chem. 0929-8673. 15 (14), 1428–1443. Available from: http://www.ingentaconnect.com/content/ben/cmc/2008/00000015/00000014/art00007. doi:10.2174/092986708784567716.

Bajtel, Á., Kiss, T., Tóth, B., Kiss, S., Hegyi, P., Vörhendi, N., Csupor-Löffler, B., Gede, N., Hohmann, J., Csupor, D., 2022. The safety of dronabinol and nabilone: a systematic review and meta-analysis of clinical trials. Pharmaceuticals. 1424-8247. 15 (1), 100. doi:10.3390/ph15010100.

Bathula, P.P., Maciver, M.B., 2024. Cannabinoids in treating chemotherapy-induced nausea and vomiting, cancer-associated pain, and tumor growth. Int. J. Mol. Sci. 1422-0067. 25 (1), 74. doi:10.3390/ijms25010074.

Braun, I.M., Bohlke, K., Abrams, D.I., Anderson, H., Balneaves, L.G., Bar-Sela, G., Bowles, D.W., Chai, P.R., Damani, A., Gupta, A., Hallmeyer, S., Subbiah, I.M., Twelves, C., Wallace, M.S., Roeland, E.J., 2024. Cannabis and cannabinoids in adults with cancer: ASCO guideline. J. Clin. Oncol. 1527-7755. 42 (13), 1575–1593. Available from: https://ascopubs.org/loi/jco. doi:10.1200/JCO.23.02596.

Calandre, E.P., Rico-Villademoros, F., Slim, M., 2015. An update on pharmacotherapy for the treatment of fibromyalgia. Expert Opin. Pharmacother. 1744-7666. 16 (9), 1347–1368. Available from: http://www.tandfonline.com/loi/ieop20. doi:10.1517/14656566.2015.1047343.

Campos, M.G., China, M., Cláudio, M., Capinha, M., Torres, R., Oliveira, S., Fortuna, A., 2024. Drug–cannabinoid interactions in selected therapeutics for symptoms associated with epilepsy, autism spectrum disorder, cancer, multiple sclerosis, and pain. Pharmaceuticals. 1424-8247. 17 (5), 613. doi:10.3390/ph17050613.

Cesamet (nabilone), 2021. Prescribing information. Bausch Health Companies Inc. https://www.accessdata.fda.gov/drugsatfda_docs/label/2022/018677Orig1s017lbl.pdf. Accessed May 27, 2025.

Davis, M., Maida, V., Daeninck, P., Pergolizzi, J., 2007. The emerging role of cannabinoid neuromodulators in symptom management. Support. Care Cancer. 0941-4355. 15 (1), 63–71. doi:10.1007/s00520-006-0180-0.

Davis, M.P., 2008. Oral nabilone capsules in the treatment of chemotherapy-induced nausea and vomiting and pain. Expert Opin. Investig. Drugs. 1354-3784. 17 (1), 85–95. Available from: http://www.expertopin.com/doi/pdf/10.1517/13543784.17.1.85. doi:10.1517/13543784.17.1.85.

Frank, B., Serpell, M.G., Hughes, J., Matthews, J.N.S., Kapur, D., 2008. Comparison of analgesic effects and patient tolerability of nabilone and dihydrocodeine for chronic neuropathic pain: randomised, crossover, double blind study. BMJ. 0959-8146. 336 (7637), 199–201. doi:10.1136/bmj.39429.619653.80.

Haney, M., Bisaga, A., Foltin, R.W., 2003. Interaction between naltrexone and oral THC in heavy marijuana smokers. Psychopharmacology. 0033-3158. 166 (1), 77–85. Available from: http://link.springer.de/link/service/journals/00213/index.htm. doi:10.1007/s00213-002-1279-8.

Häuser, W., Welsch, P., Radbruch, L., Fisher, E., Bell, R.F., Moore, R.A., 2023. Cannabis-based medicines and medical cannabis for adults with cancer pain. Cochrane Database Syst. Rev. 1465-1858. 2023 (6). Available from: https://www.cochranelibrary.com/cdsr/table-of-contents. doi:10.1002/14651858.CD014915.pub2.

Lacoursiere, R.B., Swatek, R., 1983. Adverse interaction between disulfiram and marijuana: a case report. Am. J. Psychiatry. 0002-953X. 140 (2), 243–244. doi:10.1176/ajp.140.2.243.

Meanti, R., Bresciani, E., Rizzi, L., Molteni, L., Coco, S., Omeljaniuk, R.J., Torsello, A., 2025. Cannabinoid receptor 2 (CB2R) as potential target for the pharmacological treatment of neurodegenerative diseases. Biomed. Pharmacother. 1950-6007. 186, 118044. Available from: https://www.sciencedirect.com/science/journal/07533322. doi:10.1016/j.biopha.2025.118044.

Meng, H., Johnston, B., Englesakis, M., Moulin, D.E., Bhatia, A., 2017. Selective cannabinoids for chronic neuropathic pain: a systematic review and meta-analysis. Anesth. Analg. 1526-7598. 125 (5), 1638–1652. Available from: http://journals.lww.com/anesthesia-analgesia/toc/publishahead. doi:10.1213/ANE.0000000000002110.

Paice, J.A., Portenoy, R., Lacchetti, C., Campbell, T., Cheville, A., Citron, M., Constine, L.S., Cooper, A., Glare, P., Keefe, F., Koyyalagunta, L., Levy, M., Miaskowski, C., Otis-Green, S., Sloan, P., Bruera, E., 2016. Management of chronic pain in survivors of adult cancers: American society of clinical oncology clinical practice guideline. J. Clin. Oncol. 1527-7755. 34 (27), 3325–3345. Available from: http://jco.ascopubs.org/content/34/27/3325.full.pdf+html. doi:10.1200/JCO.2016.68.5206.

Pergolizzi, J.V., Taylor, R., LeQuang, J.A., Zampogna, G., Raffa, R.B., 2017. Concise review of the management of iatrogenic emesis using cannabinoids: emphasis on nabilone for chemotherapy-induced nausea and vomiting. Cancer Chemother. Pharmacol. 1432-0843. 79 (3), 467–477. Available from: http://link.springer.de/link/service/journals/00280/index.htm. doi:10.1007/s00280-017-3257-1.

Pertwee, R.G., 2010. Receptors and channels targeted by synthetic cannabinoid receptor agonists and antagonists. Curr. Med. Chem. 0929-8673. 17 (14), 1360–1381. Available from: http://www.benthamdirect.org/pages/b_getarticlebyissue.php. doi:10.2174/092986710790980050.

Price, R., Smith, D., Franklin, G., Gronseth, G., Pignone, M., David, W.S., Armon, C., Perkins, B.A., Bril, V., Rae-Grant, A., Halperin, J., Licking, N., O'Brien, M.D., Wessels, S.R., MacGregor, L.C., Fink, K., Harkless, L.B., Colbert, L., Callaghan, B.C., 2022. Oral and topical treatment of painful diabetic polyneuropathy: practice guideline update summary: report of the AAN guideline subcommittee. Neurology. 1526-632X. 98 (1), 31–43. Available from: http://www.neurology.org. doi:10.1212/WNL.0000000000013038.

Redmond, W.J., Goffaux, P., Potvin, S., Marchand, S., 2008. Analgesic and antihyperalgesic effects of nabilone on experimental heat pain. Curr. Med. Res. Opin. 0300-7995. 24 (4), 1017–1024. doi:10.1185/030079908X280635.

Romanowski, K.S., Carson, J., Pape, K., Bernal, E., Sharar, S., Wiechman, S., Carter, D., Liu, Y.M., Nitzschke, S., Bhalla, P., Litt, J., Przkora, R., Friedman, B., Popiak, S., Jeng, J., Ryan, C.M., Joe, V., 2020. American Burn Association guidelines on the management of acute pain in the adult burn patient: a review of the literature, a compilation of expert opinion, and next steps. J. Burn Care Res. 1559-0488. 41 (6), 1129–1151. Available from: http://journals.lww.com/burncareresearch. doi:10.1093/jbcr/iraa119.

Skrabek, R.Q., Galimova, L., Ethans, K., Perry, D., 2008. Nabilone for the treatment of pain in fibromyalgia. J. Pain. 1528-8447. 9 (2), 164–173. doi:10.1016/j.jpain.2007.09.002.

Stoll, A.L., Cole, J.O., Lukas, S.E., 1991. A case of mania as a result of fluoxetine-marijuana interaction [4]. J. Clin. Psychiatry. 52 (6), 280–281.

Swarm, R.A., Paice, J.A., Anghelescu, D.L., Are, M., Bruce, J.Y., Buga, S., Chwistek, M., Cleeland, C., Craig, D., Gafford, E., Greenlee, H., Hansen, E., Kamal, A.H., Kamdar, M.M., LeGrand, S., Sean Mackey, M.R.M., Moryl, N., Nabell, L.M., Nesbit, S., O'Connor, N., Rabow, M.W., Rickerson, E., Shatsky, R., Sindt, J., Urba, S.G., Youngwerth, J.M., Hammond, L.J., Gurski, L.A., 2019. Adult cancer pain, version 3.2019, NCCN clinical practice guidelines in oncology. J. Natl. Compr. Canc. Netw. 1540-1405. 17 (8), 977–1007. doi:10.6004/jnccn.2019.0038.

Szabo, B., Schlicker, E., 2005. Effects of cannabinoids on neurotransmission. In: Pertwee R. G. eds. Handbook of Experimental Pharmacology, vol. 168. Springer Science and Business Media, LLC, Germany, pp. 327–365. 18650325. Available from: http://www.springer.com/series/164. doi: 10.1007/3-540-26573-2_11

Toth, C., Mawani, S., Brady, S., Chan, C., Liu, C., Mehina, E., Garven, A., Bestard, J., Korngut, L., 2012. An enriched-enrolment, randomized withdrawal, flexible-dose, double-blind, placebo-controlled, parallel assignment efficacy study of nabilone as adjuvant in the treatment of diabetic peripheral neuropathic pain. Pain. 1872-6623. 153 (10), 2073–2082. doi:10.1016/j.pain.2012.06.024.

Tsang, C.C., Giudice, M.G., 2016. Nabilone for the management of pain. Pharmacotherapy. 1875-9114. 36 (3), 273–286. Available from: http://onlinelibrary.wiley.com/journal/10.1002/(ISSN)1875-9114/issues. doi:10.1002/phar.1709.

Turcotte, D., Doupe, M., Torabi, M., Gomori, A., Ethans, K., Esfahani, F., Galloway, K., Namaka, M., 2015. Nabilone as an adjunctive to gabapentin for multiple sclerosis-induced neuropathic pain: a randomized controlled trial. Pain Med. 1526-4637. 16 (1), 149–159. Available from: http://onlinelibrary.wiley.com/journal/10.1111/(ISSN)1526-4637. doi:10.1111/pme.12569.

Walitt, B., Klose, P., Fitzcharles, M.A., Phillips, T., Häuser, W., 2016. Cannabinoids for fibromyalgia. Cochrane Database Syst. Rev. 1469-493X. 2016 (7). Available from: http://as.wiley.com/WileyCDA/Brand/id-6.html. doi:10.1002/14651858.CD011694.pub2.

Ware, M.A., Daeninck, P., Maida, V., 2008. A review of nabilone in the treatment of chemotherapy-induced nausea and vomiting. Ther. Clin. Risk Manag. 1176-6336. 4 (1), 99–107.

Ware, M.A., Fitzcharles, M.A., Joseph, L., Shir, Y., 2010. The effects of nabilone on sleep in fibromyalgia: results of a randomized controlled trial. Anesth. Analg. 0003-2999. 110 (2), 604–610. Available from: http://journals.lww.com/anesthesia-analgesia/toc/publishahead. doi:10.1213/ANE.0b013e3181c76f70.

CHAPTER

Cannabidiol

9

Hemant Kalia

C.R.I.S.P Center for Research & Innovation in Spine & Pain, Rochester, NY, United States

Introduction

Cannabidiol (CBD) constitutes one of over 100 cannabinoids identified in cannabis plants, typically representing 0.1%–2.9% of the plant's dry weight. Unlike delta-9-tetrahydrocannabinol (THC), CBD produces no euphoric effects and exhibits minimal abuse potential, making it an attractive therapeutic option for patients and clinicians concerned about psychoactive side effects. The compound was first isolated in 1940, and its structure was elucidated in 1963, but clinical interest remained limited until recent decades (Britch et al., 2021; Iffland and Grotenhermen, 2017).

The therapeutic landscape for CBD changed dramatically with the 2018 approval of Epidiolex (pharmaceutical-grade CBD oral solution) by the US Food and Drug Administration (FDA) for treatment-resistant epilepsy syndromes. This milestone marked the first FDA-approved medication derived from marijuana and established CBD as a legitimate pharmaceutical agent. Subsequently, research into CBD's therapeutic potential has expanded rapidly, with clinical trials investigating its efficacy across diverse medical conditions, including anxiety disorders, chronic pain, psychosis, and neurodegenerative diseases (Iffland and Grotenhermen, 2017; O'Sullivan et al., 2023; Sholler et al., 2020).

The legal status of CBD has evolved significantly following the 2018 Farm Bill, which legalized hemp-derived CBD products containing less than 0.3% THC at the federal level. However, the regulatory landscape remains complex, with variations between pharmaceutical and over-the-counter CBD products raising important quality and safety considerations for healthcare providers (Anon, n.d.a).

Mechanism of action (Britch et al., 2021; Schouten et al., 2024; Martinez Naya et al., 2023; Chayasirisobhon, 2021; Martinez Naya et al., 2023)

CBD's therapeutic effects arise through interactions with multiple molecular targets rather than a single receptor system, distinguishing it from THC's primarily CB1-mediated actions. This polypharmacological profile contributes to CBD's diverse

Cannabinoids for Chronic Pain. https://doi.org/10.1016/B978-0-443-36773-1.00003-9

therapeutic potential while complicating precise mechanistic understanding (Britch et al., 2021). Table 9.1 summarizes the mechanisms of action of CBD at various molecular targets.

Cannabinoid receptor interactions

Despite being a cannabinoid, CBD exhibits weak direct affinity for both CB1 and CB2 receptors. At CB1 receptors, CBD functions as a negative allosteric modulator, reducing the binding and functional activity of THC and other CB1 agonists. This mechanism explains CBD's ability to modulate THC's psychoactive effects and contributes to its neuroprotective properties. At CB2 receptors, CBD acts as a low-affinity agonist, mediating antiinflammatory effects primarily in immune cells (Britch et al., 2021; Schouten et al., 2024).

Serotonergic system

CBD functions as an agonist at 5-HT1A (receptors), a mechanism underlying many of its anxiolytic and antidepressant effects. Multiple studies have demonstrated that

Table 9.1 Cannabidiol (CBD) molecular targets and mechanisms of action.

Mechanism of action at various molecular targets			
Receptor/Target	Interaction type	Therapeutic effect	Clinical relevance
CB1 receptors	Negative allosteric modulator	Modulates THC effects, neuroprotection	Reduces psychoactive effects
CB2 receptors	Low affinity agonist	Antiinflammatory effects	Immune system modulation
5-HT1A receptors	Agonist	Anxiolytic, antidepressant effects	Anxiety and mood disorders
TRPV1 channels	Agonist, then desensitization	Analgesic, antiinflammatory	Pain management
PPARγ receptors	Agonist	Antiinflammatory, neuroprotective	Metabolic and inflammatory conditions
Adenosine uptake	Inhibits ENT1 transporter	Antiinflammatory, cardiovascular protection	Multiple organ systems
GABA uptake	Inhibits uptake	Anxiolytic, anticonvulsant effects	Seizure control, anxiety
Cytochrome P450	Inhibits CYP2C9, CYP3A4, others	Drug–drug interactions	Affects medication metabolism

5-HT1A receptor antagonists block CBD-induced anxiolytic effects in preclinical models. This serotonergic activity also contributes to CBD's antinausea and neuroprotective properties (Britch et al., 2021; Schouten et al., 2024).

TRPV1 channels

CBD activates transient receptor potential vanilloid 1 (TRPV1) channels, leading to rapid desensitization that contributes to analgesic and antiinflammatory effects. This mechanism is particularly relevant for pain management applications, as TRPV1 channels are expressed throughout pain pathways (Britch et al., 2021).

Other molecular targets

CBD enhances adenosine signaling by inhibiting the equilibrative nucleoside transporter 1 (ENT1), leading to antiinflammatory and cardiovascular protective effects. The compound also activates peroxisome proliferator-activated receptor gamma (PPARγ), contributing to its antiinflammatory and neuroprotective properties. In addition, CBD inhibits gamma-aminobutyric acid (GABA) and anandamide reuptake, potentially contributing to its anxiolytic and anticonvulsant effects (Britch et al., 2021; Martinez Naya et al., 2023).

Indications (O'Sullivan et al., 2023; Sholler et al., 2020; Whiting et al., 2015; von Wrede et al., 2021; Shannon et al., 2019; Blessing et al., 2015; McGuire et al., 2018; Schubart et al., 2014; Bhaskar et al., 2021)

FDA-approved indications

The strongest clinical evidence for CBD exists in pediatric epilepsy syndromes. Epidiolex is FDA-approved for treating seizures associated with Lennox–Gastaut syndrome, Dravet syndrome, and tuberous sclerosis complex in patients aged 1 year and older. Pivotal clinical trials demonstrated significant seizure reduction, with responder rates (≥50% seizure reduction) ranging from 37% to 56%, depending on seizure type.

Evidence-based off-label uses

Anxiety disorders: Multiple randomized controlled trials support CBD's anxiolytic effects. Studies using 300–600 mg of CBD demonstrated significant anxiety reduction in both healthy volunteers and patients with social anxiety disorder. The evidence is robust for acute anxiety situations, such as public speaking tests.

Psychosis and schizophrenia: Clinical trials using 600–1000 mg/day of CBD have shown antipsychotic effects comparable to conventional antipsychotics but

Table 9.2 CBD dosing recommendations by medical condition.

CBD dosing recommendations			
Condition	Dose range	Administration	Evidence level
Epilepsy (L-G-D)	10–20 mg/kg/day	2x daily	Strong (FDA)
Epilepsy (TS)	25 mg/kg/day	2x daily	Strong (FDA)
Anxiety	300–600 mg/day	1–2x daily	Moderate
Chronic pain	40 mg/day CBD	2x daily	Limited
PTSD	1.5 mg–300 mg/day	Spray/oral	Limited
Sleep disorders	300 mg/day	1x daily	Limited
Schizophrenia	200–800 mg/day	2x daily	Limited

with superior tolerability. CBD may be particularly beneficial as adjunctive therapy in treatment-resistant cases.

Posttraumatic stress disorder (PTSD): Emerging evidence suggests CBD may reduce PTSD symptoms, with studies showing improvements in sleep quality, nightmares, and hyperarousal symptoms. Dosing has varied widely, from 1.5 mg (nasal spray) to 300 mg orally daily.

Substance use disorders: CBD has shown promise in reducing drug cravings and relapse behavior, particularly for opioid and stimulant use disorders. The evidence suggests that CBD may enhance the extinction of drug-seeking behaviors.

Some of the established dosing recommendations used in these studies are summarized in Table 9.2.

Investigational uses

The existing research is evaluating CBD for chronic pain management, inflammatory bowel disease, autism spectrum disorders, and neurodegenerative diseases. However, evidence for these applications remains preliminary and requires larger confirmatory trials.

Contraindications (Whiting et al., 2015; Eadie et al., 2024; De Genna et al., 2023; Anon, n.d.b)

Absolute contraindications

Hypersensitivity: Patients with known hypersensitivity to CBD or any formulation components should not receive CBD therapy. Severe allergic reactions, including angioedema, have been reported.

Severe hepatic impairment: CBD is contraindicated in patients with decompensated liver disease due to an increased risk of hepatotoxicity. The drug undergoes extensive hepatic metabolism and may accumulate to toxic levels in severe hepatic dysfunction.

Relative contraindications

Pregnancy and lactation: CBD crosses the placenta and has been associated with developmental abnormalities in animal studies. The FDA strongly advises against CBD use during pregnancy or breastfeeding due to potential fetal neurodevelopmental consequences.

Moderate hepatic impairment: Patients with Child–Pugh B hepatic impairment require dose reduction and enhanced monitoring. Consider alternative therapies when possible.

Concomitant hepatotoxic medications: Exercise caution when prescribing CBD with other potentially hepatotoxic drugs, particularly valproate, as additive liver toxicity may occur.

Side effects (Iffland and Grotenhermen, 2017; Chen et al., 2024; Balachandran et al., 2021; Chesney et al., 2020; Eadie et al., 2024; Huestis et al., 2019)

Common adverse effects

The most frequently reported side effects include somnolence (25% of patients), decreased appetite (19%), diarrhea (19%), and fatigue (13%). These effects are generally dose-dependent and may diminish with continued therapy (Chesney et al., 2020).

Hepatotoxicity

Liver injury represents the most serious adverse effect associated with CBD therapy. Elevated liver enzymes (Aspartate aminotransferase (ALT)/Alanine aminotransferase (AST) >3 times the upper limit of normal) occur in approximately 10% of patients, with 75% of cases occurring in patients concomitantly taking valproate. A systematic review found CBD associated with a 5.85-fold increased risk of liver enzyme elevation compared to placebo (Chen et al., 2024; Chesney et al., 2020; Eadie et al., 2024).

Hepatotoxicity can occur even in healthy volunteers, with 31% of subjects receiving 1500 mg/day developing drug-induced liver injury criteria in one study. Risk factors include high doses (≥1000 mg/day), concomitant hepatotoxic medications, and preexisting liver disease (Eadie et al., 2024; Chen et al., 2024).

Cardiovascular effects

High-dose CBD may cause hypotension and bradycardia. In animal studies, doses of 300 mg/kg caused cardiac failure. The clinical significance in humans at therapeutic doses appears limited, but monitoring is advisable in patients with cardiovascular disease (Huestis et al., 2019).

Table 9.3 Major drug interactions with cannabidiol (CBD).

CBD interactions with major drugs			
Drug/Class	Mechanism	Clinical effect	Management
Warfarin	CYP450 inhibition	Increased INR/ bleeding risk	Monitor INR closely, reduce warfarin dose
Clobazam	CYP2C19 inhibition	60% increase in clobazam levels	Dose reduction may be needed
Valproate	Additive hepatotoxicity	Increased liver enzyme elevation risk	Monitor liver function
Cyclosporine/ Tacrolimus	CYP3A4 inhibition	Threefold increase in drug levels	Monitor drug levels closely
Acetaminophen	Additive hepatotoxicity	Increased liver injury risk	Avoid high doses, monitor liver function
CNS Depressants	Pharmacodynamic interaction	Enhanced sedation	Monitor for excessive sedation
Metformin	Additive gastroenteral (GI) effects	Increased diarrhea risk	Monitor GI symptoms

Psychiatric effects

Rarely, CBD may cause behavioral changes, irritability, or suicidal ideation. These effects are most concerning in patients with preexisting psychiatric conditions and require careful monitoring.

Drug-related side effects

CBD's extensive drug interactions may manifest as increased side effects from co-administered medications rather than direct CBD toxicity. Examples include excessive sedation from enhanced benzodiazepine effects or bleeding complications from warfarin potentiation (Table 9.3).

Drug interactions

CBD's inhibition of cytochrome P450 enzymes creates significant potential for drug interactions. The compound primarily inhibits CYP2C9, CYP2C19, CYP3A4, and CYP2D6, affecting the metabolism of numerous medications (Britch et al., 2021; Doohan et al., 2021; Smith and Gruber, 2023).

High-risk interactions

Anticoagulants: CBD increases warfarin levels by approximately 30%, requiring INR monitoring and dose adjustments. Similar effects may occur with other anticoagulants metabolized by CYP450 enzymes (Balachandran et al., 2021).

Antiepileptic drugs: CBD significantly increases clobazam levels (60% increase) and its active metabolite, norclobazam (500% increase). Dose reductions of clobazam are frequently necessary. Interactions with other antiepileptics, including valproate, lamotrigine, and rufinamide, require monitoring (Britch et al., 2021; Balachandran et al., 2021).

Immunosuppressants: CBD increases cyclosporine and tacrolimus levels by up to threefold, necessitating dose adjustments and enhanced monitoring. These interactions carry significant clinical consequences, given the narrow therapeutic indices of these medications (Balachandran et al., 2021).

Moderate-risk interactions

Central nervous system (CNS) depressants: CBD enhances sedation from benzodiazepines, opioids, and other CNS depressants through pharmacodynamic interactions. Patients should be counseled about increased drowsiness and fall risk.

Statins: CBD may increase levels of atorvastatin and simvastatin but not pravastatin or rosuvastatin. Monitor for muscle-related side effects with affected statins (Anon, n.d.c).

Clinical management (Sholler et al., 2020; Whiting et al., 2015; Blessing et al., 2015; McGuire et al., 2018; Schubart et al., 2014; O'Sullivan et al., 2023; Smith and Gruber, 2023; Anon, n.d.d; Bhaskar et al., 2021).

Healthcare providers should review all medications before initiating CBD therapy. The "grapefruit rule" provides a valuable screening tool—medications with grapefruit warnings are likely to interact with CBD. Consider therapeutic drug monitoring for medications with narrow therapeutic indices.

Recommended doses for different conditions

Epilepsy (FDA-approved dosing)

Starting dose: 2.5 mg/kg twice daily (5 mg/kg/day) for one week.

Maintenance dose: Increase to 5 mg/kg twice daily (10 mg/kg/day) in Week 2.

Maximum dose: Up to 10 mg/kg twice daily (20 mg/kg/day) for Lennox–Gastaut and Dravet syndromes, or 12.5 mg/kg twice daily (25 mg/kg/day) for tuberous sclerosis complex.

Titration: Weekly increases of 2.5 mg/kg twice daily, as tolerated. More rapid titration (every other day) may be appropriate in select patients.

Off-label conditions

Anxiety Disorders: 300–600 mg/day as single or divided doses. Start with 300 mg/day and adjust based on response. Lower doses (25–50 mg/day) may be effective for mild anxiety.

Chronic pain: Begin with 5 mg twice daily, increasing by 10 mg every 2–3 days up to 40 mg/day. If the response is inadequate, consider adding low-dose THC (2.5 mg/day initially).

Psychosis/Schizophrenia: 600–800 mg/day, divided into 2–3 doses. Initiate at 200 mg/day and titrate weekly, as tolerated.

PTSD: Variable dosing ranges from 1.5 mg (nasal spray) to 300 mg/day orally. Start with low doses and titrate to effect.

General dosing principles

Start low, go slow: Begin with the lowest effective dose and titrate gradually to minimize side effects. Most patients achieve therapeutic benefits at lower doses than initially anticipated.

Individual variation: Dosing requirements vary significantly based on body weight, metabolism, condition severity, and individual sensitivity. Maintain detailed dosing records to optimize therapy.

Hepatic impairment: Reduce initial doses by 50% in moderate hepatic impairment, and avoid use in severe impairment.

Conclusion

CBD represents a paradigm shift in cannabinoid medicine, offering clinicians a non-intoxicating therapeutic option with demonstrated efficacy across multiple conditions. The compound's FDA approval for pediatric epilepsy has established its legitimacy as a pharmaceutical agent, while ongoing research continues to expand its therapeutic applications.

The evidence base for CBD is strongest in epilepsy treatment, where large randomized controlled trials demonstrate clear clinical benefit. For anxiety disorders, psychosis, and PTSD, moderate-quality evidence supports therapeutic use, though larger confirmatory trials are needed. Applications in chronic pain, substance use disorders, and other conditions show promise but require additional research.

Healthcare providers must carefully consider CBD's safety profile, particularly the hepatotoxicity risk and extensive drug interactions. Regular monitoring of liver function and careful medication review are essential components of CBD therapy. Contraindications during pregnancy and in cases of severe liver disease must be respected.

The regulatory landscape for CBD continues evolving, with important distinctions between pharmaceutical and over-the-counter products. Physicians should preferentially prescribe FDA-approved formulations when available and exercise caution with unregulated products due to quality and potency variability.

Future research priorities include establishing optimal dosing regimens, identifying biomarkers for treatment response, evaluating long-term safety, and expanding the evidence base for nonepilepsy applications. As our understanding of CBD's mechanisms and clinical applications continues to grow, this cannabinoid will likely play an increasingly important role in modern medical practice.

References

Balachandran, P., Elsohly, M., Hill, K.P., 2021. Cannabidiol interactions with medications, illicit substances, and alcohol: a comprehensive review. J. Gen. Intern. Med. 1525-1497. 36 (7), 2074–2084. Available from: http://www.springerlink.com/content/120414/. doi:10.1007/s11606-020-06504-8.

Bhaskar, A., Bell, A., Boivin, M., Briques, W., Brown, M., Clarke, H., Cyr, C., Eisenberg, E., de Oliveira Silva, R.F., Frohlich, E., Georgius, P., Hogg, M., Horsted, T.I., MacCallum, C.A., Müller-Vahl, K.R., O'Connell, C., Sealey, R., Seibolt, M., Sihota, A., Smith, B.K., Sulak, D., Vigano, A., Moulin, D.E., 2021. Consensus recommendations on dosing and administration of medical cannabis to treat chronic pain: results of a modified Delphi process. J. Cannabis Res. 2522-5782. 3 (1). Available from: https://jcannabisresearch.biomedcentral.com/. doi:10.1186/s42238-021-00073-1.

Blessing, E.M., Steenkamp, M.M., Manzanares, J., Marmar, C.R., 2015. Cannabidiol as a potential treatment for anxiety disorders. Neurotherapeutics. 1878-7479. 12 (4), 825–836. Available from: http://www.springerlink.com/content/1933-7213/. doi:10.1007/s13311-015-0387-1.

Britch, S.C., Babalonis, S., Walsh, S.L., 2021. Cannabidiol: pharmacology and therapeutic targets. Psychopharmacology. 1432-2072. 238 (1), 9–28. Available from: http://link.springer.de/link/service/journals/00213/index.htm. doi:10.1007/s00213-020-05712-8.

Chayasirisobhon, S., 2021. Mechanisms of action and pharmacokinetics of cannabis. Perm. J. 1552-5775. 25 (1), 1–3. doi:10.7812/tpp/19.200.

Chen, S., Li, Y., Li, X., Wu, Q., Puig, M., Moulin, F., Gingrich, J., Guo, L., 2024. Metabolism and liver toxicity of cannabidiol. J. Environ. Sci. Health C. 2689-6583. 42 (3), 238–254. doi:10.1080/26896583.2024.2366741.

Chesney, E., Oliver, D., Green, A., Sovi, S., Wilson, J., Englund, A., Freeman, T.P., McGuire, P., 2020. Adverse effects of cannabidiol: a systematic review and meta-analysis of randomized clinical trials. Neuropsychopharmacology. 0893-133X. 45 (11), 1799–1806. doi:10.1038/s41386-020-0667-2.

De Genna, N.M., Kennon-McGill, S., Goldschmidt, L., Richardson, G.A., Chang, J.C., 2023. Factors associated with ever using cannabidiol in a cohort of younger pregnant people. Neurotoxicol. Teratol. 1872-9738. 96. Available from: www.elsevier.com/locate/neutera. doi:10.1016/j.ntt.2023.107162.

Doohan, P.T., Oldfield, L.D., Arnold, J.C., Anderson, L.L., 2021. Cannabinoid interactions with cytochrome P450 drug metabolism: a full-spectrum characterization. AAPS J. 1550-7416. 23 (4). doi:10.1208/s12248-021-00616-7.

Eadie, L., Lo, L.A., Boivin, M., Deol, J.K., MacCallum, C.A., 2024. Clinical guidance for cannabidiol-associated hepatotoxicity: a narrative review. J. Gastroenterol. Hepatol. 1440-1746. 39 (12), 2522–2532. Available from: http://onlinelibrary.wiley.com/journal/10.1111/(ISSN)1440-1746. doi:10.1111/jgh.16730.

Huestis, M.A., Solimini, R., Pichini, S., Pacifici, R., Carlier, J., Busardò, F.P., 2019. Cannabidiol adverse effects and toxicity. Curr. Neuropharmacol. 1875-6190. 17 (10),

974–989. Available from: http://www.eurekaselect.com/CDN/download.php?param=Sk9VjUk5vBTFaMvQbkVOpL0NhOLz7E3LpzEwFLzA4wOElFDL1nBkrZnxb8YXqBwbdGljcYXRcpb2n4vczGRmqfHwkzODpk5OeDU1CNDcizM2uExMqjhhnMDlamMWlRiZbjU0wMzVliZjdMzNqQTcVYTcVYt&key=VWlxoRWl7uZEEUtQcVNEwc2RdZZkEVJQyTMytRDMc1Rj1c1NbDV2ALTQj3MzrItMejM1dNjQazMjwNERm2VEaeATcVYDTcVY. doi:10.2174/1570159X17666190603171901.

Iffland, K., Grotenhermen, F., 2017. An update on safety and side effects of cannabidiol: a review of clinical data and relevant animal studies. Cann. Cannabinoid Res. 2378-8763. 2 (1), 139–154. doi:10.1089/can.2016.0034.

Martinez Naya, N., Kelly, J., Corna, G., Golino, M., Abbate, A., Toldo, S., 2023. Molecular and cellular mechanisms of action of cannabidiol. Molecules. 1420-3049. 28 (16), 5980. doi:10.3390/molecules28165980.

McGuire, P., Robson, P., Cubala, W.J., Vasile, D., Morrison, P.D., Barron, R., Taylor, A., Wright, S., 2018. Cannabidiol (CBD) as an adjunctive therapy in schizophrenia: a multicenter randomized controlled trial. J. Psychiatry. 1535-7228. 175 (3), 225–231. Available from: https://ajp.psychiatryonline.org/doi/pdf/10.1176/appi.ajp.2017.17030325. doi:10.1176/appi.ajp.2017.17030325.

O'Sullivan, S.E., Jensen, S.S., Nikolajsen, G.N., Bruun, H.Z., Bhuller, R., Hoeng, J., 2023. The therapeutic potential of purified cannabidiol. J. Cannabis Res. 2522-5782. 5 (1). Available from: https://jcannabisresearch.biomedcentral.com/. doi:10.1186/s42238-023-00186-9.

Schouten, M., Dalle, S., Mantini, D., Koppo, K., 2024. Cannabidiol and brain function: current knowledge and future perspectives. Front. Pharmacol. 1663-9812. 14. doi:10.3389/fphar.2023.1328885.

Schubart, C.D., Sommer, I.E.C., Fusar-Poli, P., de Witte, L., Kahn, R.S., Boks, M.P.M., 2014. Cannabidiol as a potential treatment for psychosis. Eur. Neuropsychopharmacol. 0924-977X. 24 (1), 51–64. doi:10.1016/j.euroneuro.2013.11.002.

Shannon, S., Lewis, N., Lee, H., Hughes, S., 2019. Cannabidiol in anxiety and sleep: a large case series. Perm. J. 1552-5767. 23 (1). doi:10.7812/tpp/18-041.

Sholler, D.J., Schoene, L., Spindle, T.R., 2020. Therapeutic efficacy of cannabidiol (CBD): a review of the evidence from clinical trials and human laboratory studies. Curr. Addict. Rep. 2196-2952. 7 (3), 405–412. Available from: https://link.springer.com/journal/40429. doi:10.1007/s40429-020-00326-8.

Smith, R.T., Gruber, S.A., 2023. Contemplating cannabis? The complex relationship between cannabinoids and hepatic metabolism resulting in the potential for drug-drug interactions. Front. Psychiatry 1664-0640. 13. Available from: http://www.frontiersin.org/Psychiatry. doi:10.3389/fpsyt.2022.1055481.

von Wrede, R., Helmstaedter, C., Surges, R., 2021. Cannabidiol in the treatment of epilepsy. Clin. Drug Investig. 1179-1918. 41 (3), 211–220. Available from: http://rd.springer.com/journal/40261. doi:10.1007/s40261-021-01003-y.

Whiting, P.F., Wolff, R.F., Deshpande, S., Di Nisio, M., Duffy, S., Hernandez, A.V., Keurentjes, J.C., Lang, S., Misso, K., Ryder, S., Schmidlkofer, S., Westwood, M., Kleijnen, J., 2015. Cannabinoids for medical use: a systematic review and meta-analysis. JAMA. 1538-3598. 313 (24), 2456–2473. Available from: http://jama.jamanetwork.com/article.aspx?articleid=2338251. doi:10.1001/jama.2015.6358.

Websites

Anon, n.d.a. Available from: https://www.fda.gov/news-events/congressional-testimony/hemp-production-and-2018-farm-bill-07252019.

Anon, n.d.b. Available from: https://www.fda.gov/consumers/consumer-updates/what-you-should-know-about-using-cannabis-including-cbd-when-pregnant-or-breastfeeding.

Anon, n.d.c. Available from: https://doh.dc.gov/sites/default/files/dc/sites/doh/publication/attachments/MedicalCannabisAdverseEffectsandDrugInteractions_0.pdf
Anon, n.d.d. Available from: https://www.accessdata.fda.gov/drugsatfda_docs/label/2018/210365lbl.pdf.

CHAPTER

Nabiximols

10

Robert Wayde Gilmore[a], Alexander Bautista[a], and Alaa Abd-Elsayed[b]

[a]*Department of Anesthesiology and Perioperative Medicine, University of Louisville, Louisville, KY, United States*

[b]*Departments of Anesthesiology and Chronic Pain Physician, University of Wisconsin-Madison, Madison, WI, United States*

Introduction

Nabiximols (USAN), marketed as Sativex, is a standardized whole-plant extract containing delta-9-tetrahydrocannabinol (THC) and cannabidiol (CBD) in a 1:1 ratio, delivered as an oromucosal spray. Each spray delivers approximately 2.7 mg of THC and 2.5 mg of CBD, with a formulation meeting pharmaceutical standards for quality and consistency (European Medicines Agency [EMA], 2023; Electronic Medicines Compendium [eMC], 2025). Developed by GW Pharmaceuticals in the United Kingdom, nabiximols received its first marketing authorization in 2010 for the symptomatic relief of spasticity in multiple sclerosis (MS) patients who had not responded adequately to other medications (Medicines and Healthcare products Regulatory Agency [MHRA], 2010; United States Adopted Names Council, 2010).

Mechanism of action

Nabiximols exerts effects through modulation of the endocannabinoid system. THC is a partial agonist at CB1 and CB2 receptors, reducing neurotransmitter release via presynaptic inhibition. CBD is a negative allosteric modulator of CB1 and interacts with multiple molecular targets (e.g., TRPV1, GPR55, FAAH), contributing to potential analgesic, anti-inflammatory, and anti-spastic effects. The combined 1:1 THC:CBD ratio may permit symptom control with fewer psychoactive effects than THC-dominant preparations (Wade et al., 2004; Koppel et al., 2014).

Indications

Nabiximols is approved in several regions for MS-associated spasticity, typically as add-on therapy after an initial response test period (Electronic Medicines

Cannabinoids for Chronic Pain. https://doi.org/10.1016/B978-0-443-36773-1.00021-0

Compendium [eMC], 2025; Medicines and Healthcare products Regulatory Agency [MHRA], 2010; United States Adopted Names Council, 2010). Randomized controlled trials and systematic reviews demonstrate significant efficacy in reducing MS spasticity and related symptoms (Notcutt et al., 2012; D'hooghe et al., 2021; Bethoux, 2024; Lakhan and Rowland, 2009). Early trials suggested potential benefits for cancer-related pain; however, large Phase 3 randomized trials did not meet primary endpoints, while showing acceptable safety and some improvements in secondary outcomes in certain subgroups (Portenoy et al., 2012; Fallon et al., 2017). Its use has also been explored for neuropathic pain and other indications, with mixed results (Lynch et al., 2014).

Contraindications and precautions

Nabiximols is contraindicated in patients with hypersensitivity to cannabinoids or excipients. It is not recommended for individuals with a history of schizophrenia, psychosis, or severe psychiatric disorders due to potential exacerbation by THC. Use during pregnancy and lactation is not recommended, as safety data remain insufficient. Caution is advised for patients with significant cardiovascular disease (e.g., arrhythmias, unstable angina) due to risks of tachycardia and orthostatic hypotension. Nabiximols is not indicated for individuals under 18 years of age. Patients with mild hepatic impairment generally require no adjustment; however, use in moderate to severe hepatic impairment is not recommended due to limited data on accumulation with chronic dosing (Electronic Medicines Compendium [eMC], 2025; Medicines and Healthcare products Regulatory Agency [MHRA], 2010; United States Adopted Names Council, 2010).

Adverse effects

Across clinical trials and long-term extension studies, nabiximols is generally well tolerated. Common adverse effects include dizziness, fatigue, somnolence, and disorientation; discontinuation due to adverse events occurs in a minority of patients. Available data up to 12 months do not suggest clinically meaningful cognitive deterioration in MS populations using nabiximols, though monitoring is advisable (Notcutt et al., 2012; Bethoux, 2024; Lakhan and Rowland, 2009; Motaghi et al., 2023; Liao et al., 2022; Celius and Vila, 2018).

Drug interactions

CYP2C9 and CYP3A4 primarily metabolize THC; CBD interacts with CYP3A4 and CYP2C19, among other pathways. Clinically essential interactions are possible with potent inhibitors or inducers of these enzymes and with narrow-therapeutic-index drugs (e.g., warfarin, certain antiepileptics). CNS depressants (opioids, benzodiazepines, alcohol) may have additive effects with nabiximols. Careful monitoring and dose adjustments of concomitant medications may be required when initiating or titrating therapy (Stout and Cimino, 2014; Cox et al., 2019; Alsherbiny and Li, 2019).

Regulatory and market access

Following its initial authorization in 2010, nabiximols has been approved in multiple countries and regions for MS spasticity, with local requirements governing initiation, continuation criteria, and reimbursement. Official product information (e.g., EMA SmPC; UK Electronic Medicines Compendium) outlines indications, dosing, contraindications, and safety considerations, while national assessments (e.g., MHRA Public Assessment Reports) provide additional regulatory context (Electronic Medicines Compendium (eMC), 2025; Medicines and Healthcare products Regulatory Agency (MHRA), 2010; United States Adopted Names Council, 2010; Netherlands Medicines Evaluation Board, 2010).

Recommended doses for different conditions

Dosing should be individualized and titrated gradually to balance efficacy and tolerability. Each spray delivers approximately 2.7 mg of THC and 2.5 mg of CBD. The oromucosal route enables transmucosal absorption, with some gastrointestinal absorption also possible. Suggested ranges below are derived from pivotal studies, product information, and post-marketing experience (Electronic Medicines Compendium (eMC), 2025; Medicines and Healthcare products Regulatory Agency (MHRA), 2010; Notcutt et al., 2012; Portenoy et al., 2012).

Indication	Initiation & Titration	Typical maintenance	Maximum dose
MS spasticity	Start with 1 spray per day; increase by 1–2 sprays every 1–2 days following an initial response test period.	6–12 sprays/day (≈16–32 mg THC)	12 sprays/day (≈32 mg THC)
Neuropathic pain	Start low; titrate to 4–8 sprays per day based on response and tolerability	Individualized	Generally, ≤12 sprays/day
Cancer pain (adjunct to opioids)	Self-titrate during a 2-week open-label run-in; responders continue in the randomized phase	6–10 sprays/day	12 sprays/day (tolerability-based)

Continued

Indication	Initiation & Titration	Typical maintenance	Maximum dose
Older adults	Initiate at lower doses; titrate slowly	As tolerated, with close monitoring	Individualized

Each spray ≈ contains approximately 2.7 mg of THC and 2.5 mg of CBD. Source: SmPC/eMC and clinical trials (Electronic Medicines Compendium (eMC), 2025. Sativex oromucosal spray: summary of product characteristics. Available from: https://www.medicines.org.uk/emc/product/602/smpc. (Accessed 10 August 2025); Medicines and Healthcare products Regulatory Agency (MHRA), 2010. Sativex Oromucosal Spray—Public Assessment Report. PL 18024/0009. Available from: https://mhraproducts4853.blob.core.windows.net/docs/ece5648ffc9053af77f7aeb7f5d90c67ba36e4e0. (Accessed 10 August 2025); *Notcutt, W., Langford, R., Davies, P., Ratcliffe, S., Potts, R., 2012. A placebo-controlled, parallel-group, randomized withdrawal study of subjects with symptoms of spasticity due to multiple sclerosis who are receiving long-term Sativex (nabiximols). Mult. Scler. J. 1477-0970. 18 (2), 219–228. Available from http://msj.sagepub.com/. doi:10.1177/1352458511419700; Portenoy, R.K., Ganae-Motan, E.D., Allende, S., Yanagihara, R., Shaiova, L., Weinstein, S., McQuade, R., Wright, S., Fallon, M.T., 2012. Nabiximols for opioid-treated cancer patients with poorly-controlled chronic pain: a randomized, placebo-controlled, graded-dose trial. J. Pain. 1528-8447. 13 (5), 438–449. doi:10.1016/j.jpain.2012.01.003).*

Conclusion

Nabiximols represents a milestone in cannabinoid therapeutics, offering a standardized, well-tolerated option for MS-related spasticity and, in select patients, for other chronic pain conditions. While efficacy in cancer-related pain has been inconsistent across large, randomized trials, safety has generally been acceptable. Ongoing work seeks to optimize delivery, identify responsive subgroups, and refine indications.

References

Alsherbiny, M.A., Li, C.G., 2019. Medicinal cannabis—potential drug interactions. Medicines. 2305-6320. 6 (1), 3. doi:10.3390/medicines6010003.

Bethoux, F., 2024. A randomized, double-blind, placebo-controlled trial to evaluate the effect of nabiximols oromucosal spray on clinical measures of spasticity in patients with multiple sclerosis. Mult. Scler. Relat. Disord. 89, 105740.

Celius, E.G., Vila, C., 2018. The influence of THC:CBD oromucosal spray on driving, cognitive function, and postural stability. Ther. Adv. Drug Saf. 9 (3). doi:10.1177/2042098618756802.

Cox, E.J., Maharao, N., Patilea-Vrana, G., Unadkat, J.D., Rettie, A.E., McCune, J.S., 2019. A marijuana–drug interaction primer: precipitant drugs, pharmacology, and pharmacokinetics. J. Clin. Pharmacol. 59 (7), 923–934. doi:10.1002/jcph.1387.

D'hooghe, M., Willekens, B., Delvaux, V., 2021. Nabiximols in patients with resistant MS spasticity: the Belgian experience. BMC Neurol. 21 (1), 227. doi:10.1186/s12883-021-02246-0.

Electronic Medicines Compendium (eMC), 2025. Sativex oromucosal spray: summary of product characteristics. Available from: https://www.medicines.org.uk/emc/product/602/smpc. (Accessed 10 August 2025).

European Medicines Agency (EMA), 2023. Sativex: summary of product characteristics. Available from: https://www.ema.europa.eu. (Accessed 10 August 2025).

Fallon, M.T., Lux, E.A., McQuade, R., Rossetti, S., Sanchez, R., Sun, W., Wright, S., Lichtman, A.H., Kornyeyeva, E., 2017. Sativex oromucosal spray as adjunctive therapy in advanced cancer patients with chronic pain unalleviated by optimized opioid therapy: two double-blind, randomized, placebo-controlled phase 3 studies. Br. J. Pain 2049-4645. 11 (3), 119–133. Available from: http://bjp.sagepub.com/. doi:10.1177/2049463717710042.

Koppel, B.S., Brust, J.C.M., Fife, T., Bronstein, J., Youssof, S., Gronseth, G., Gloss, D., 2014. Systematic review: efficacy and safety of medical marijuana in selected neurologic disorders: report of the Guideline Development Subcommittee of the American Academy of Neurology. Neurology 1526-632X. 82 (17), 1556–1563. Available from: http://www.neurology.org. doi:10.1212/WNL.0000000000000363.

Lakhan, S.E., Rowland, M., 2009. Whole plant cannabis extracts in the treatment of spasticity in multiple sclerosis: a systematic review. BMC Neurol. 1471-2377. 9. doi:10.1186/1471-2377-9-59.

Liao, H., Fan, P., Ruan, H., Gong, N., Qiu, W., Zhang, Y., Li, X., Li, H., Zhang, M., 2022. Experiences of social isolation among patients with neuromyelitis optica spectrum disorder in China: a qualitative study. Mult. Scler. Relat. Disord. 2211-0348. 60, 103711. doi:10.1016/j.msard.2022.103711.

Lynch, M.E., Cesar-Rittenberg, P., Hohmann, A.G., 2014. A double-blind, placebo-controlled, crossover pilot trial with extension using an oral mucosal cannabinoid extract for treatment of chemotherapy-induced neuropathic pain. J. Pain Symptom Manage. 1873-6513. 47 (1), 166–173. doi:10.1016/j.jpainsymman.2013.02.018.

Medicines and Healthcare products Regulatory Agency (MHRA), 2010. Sativex Oromucosal Spray—Public Assessment Report. PL18024/0009. Available from: https://mhraproducts4853.blob.core.windows.net/docs/ece5648ffc9053af77f7aeb7f5d90c67ba36e4e0. (Accessed 10 August 2025).

Motaghi, E., Ghasemi-Pirbaluti, M., Rashidi, M., Alasvand, M., Di Ciano, P., Bozorgi, H., 2023. The effect of tetrahydrocannabinol:cannabidiol oromucosal spray on cognition: a systematic review. Eur. J. Clin. Pharmacol. 1432-1041. 79 (3), 371–381. Available from: https://www.springer.com/journal/228. doi:10.1007/s00228-023-03454-y.

Netherlands Medicines Evaluation Board, 2010. Public assessment report: sativex oromucosal spray—decentralised procedure. Available from: https://www.geneesmiddeleninformatiebank.nl/pars/h111277.pdf. (Accessed 10 August 2025).

Notcutt, W., Langford, R., Davies, P., Ratcliffe, S., Potts, R., 2012. A placebo-controlled, parallel-group, randomized withdrawal study of subjects with symptoms of spasticity due to multiple sclerosis who are receiving long-term Sativex (nabiximols). Mult. Scler. J. 1477-0970. 18 (2), 219–228. Available from: http://msj.sagepub.com/. doi:10.1177/1352458511419700.

Portenoy, R.K., Ganae-Motan, E.D., Allende, S., Yanagihara, R., Shaiova, L., Weinstein, S., McQuade, R., Wright, S., Fallon, M.T., 2012. Nabiximols for opioid-treated cancer patients with poorly-controlled chronic pain: a randomized, placebo-controlled, graded-dose trial. J. Pain. 1528-8447. 13 (5), 438–449. doi:10.1016/j.jpain.2012.01.003.

Stout, S.M., Cimino, N.M., 2014. Exogenous cannabinoids as substrates, inhibitors, and inducers of human drug metabolizing enzymes: a systematic review. Drug Metab. Rev. 1097-9883. 46 (1), 86–95. doi:10.3109/03602532.2013.849268.

United States Adopted Names Council, 2010. Statement on a nonproprietary name: nabiximols.

Wade, D.T., Makela, P., Robson, P., House, H., Bateman, C., 2004. Do cannabis-based medicinal extracts have general or specific effects on symptoms in multiple sclerosis? A double-blind, randomized, placebo-controlled study on 160 patients. Mult. Scler. 10 (4), 434–441. doi:10.1191/1352458504ms1082oa.

CHAPTER

Medical marijuana

11

Kenneth D. Candido[a,b]

[a]*University of Illinois College of Medicine, Chicago, IL, United States*

[b]*Pain Management Services, Thorek Memorial Hospital, Chicago, IL, United States*

Introduction

The first recorded use of cannabis for medicinal purposes dates back to ancient China, at least as early as 2727 BC (Li, 1973), with Emperor Shen Nung prescribing it for various conditions and ailments, including gout, rheumatism, malaria, and cognitive issues such as memory loss. Different cultures have utilized cannabis for its perceived therapeutic properties, with references to its use found in texts from Egypt, India, Greece, and Italy (Rome). In India, Hindu sects used marijuana for both spiritual reasons and stress relief, while Indian physicians employed it to control pain, gastrointestinal issues, and insomnia. By at least 1500 BC, the Egyptians described its use as a treatment for inflammatory conditions, while in Greece, cannabis was used to address inflammation and pain in the face and ear. The focus has shifted to understanding the chemical composition of cannabis, particularly tetrahydrocannabinol (THC), and its effects on the body. This research has led to the development of cannabis-containing medications not only for pain relief but also for treating nausea (e.g., Marinol) and stimulating appetite, particularly in cancer patients experiencing chemotherapy-induced nausea and vomiting (CINV).

The United States Food and Drug Administration (FDA) has approved low-dose THC (initial oral dose 0.04 mg/kg b.i.d.) for treating nausea and vomiting associated with cancer chemotherapy for nonresponders, for treating AIDS-associated anorexia and weight loss, and CBD (initial oral dose 2.5 mg/kg b.i.d.) to treat rare, severe forms of epilepsy (Madras, 2019). The marijuana plant contains over 500 distinct chemicals. It produces more than 100 different cannabinoids, including the two structurally distinct agents, Δ9-tetrahydrocannabinol (THC) and cannabidiol (CBD) (Fig. 11.1), which are the principal compounds used in medical marijuana. Over the past few decades, THC concentrations in retail marijuana have risen dramatically, while CBD levels have declined, with THC:CBD ratios now eight times greater than they were approximately 20 years ago. Smokable marijuana (~20% THC; ~ 0.9 mg/kg) delivers THC at approximately 20 times the FDA-approved initial dose of oral THC and is now obtainable at THC:CBD ratios varying from 1:1 to 80:1 (Madras, 2019).

Cannabinoids for Chronic Pain. https://doi.org/10.1016/B978-0-443-36773-1.00022-2

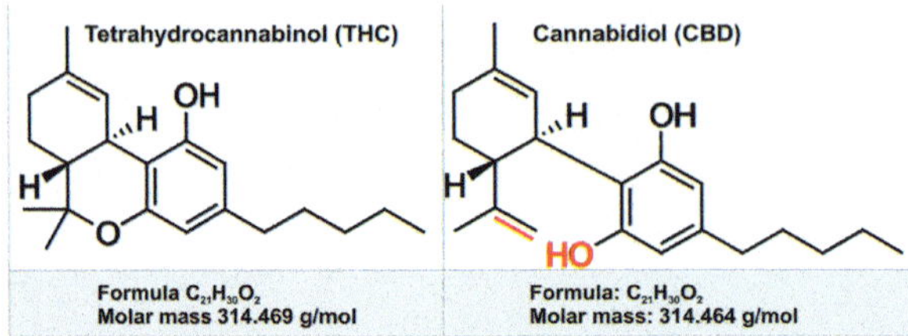

FIG. 11.1 Chemical structures of THC and CBD (Archie and Cucullo, 2019).

Limiting its therapeutic potential, THC in marijuana can acutely elicit psychosis, anxiolysis, intoxication, and cognitive impairment. With early initiation and prolonged use, marijuana is addictive and may therefore increase the risk for the individual user of developing schizophrenia and other psychoses, with higher use leading to higher risk (Madras, 2019). No comparable evidence implicates CBD in engendering euphoria, psychosis, cognitive impairment, anxiety, or addiction. High concentrations of THC and high ratios of THC:CBD in marijuana are associated with more robust euphoria, anxiety, and psychotic symptoms in otherwise normal people. Conversely, CBD mitigates the effects of THC by attenuating anxiety, cognitive deficits, or psychosis (Madras, 2019). CBD may blunt many of the adverse effects engendered by THC through mechanisms not yet elucidated.

The use of cannabis for medical purposes is approved in at least 40 U.S. states (as of March 2023), as well as in Washington, DC, and four out of five US territories (Stella et al., 2021). Its use relies upon understanding the respective cannabinoid receptors identified in humans, notably the two recognized receptors, CB1 and CB2, which are members of the seven-transmembrane G-protein-coupled receptors. One assessment suggests that cannabinoid and endocannabinoid compounds can activate a broader range of receptors. Endogenous cannabinoids are fatty acids with 20 carbon atoms (eicosanoids) capable of selectively binding to the CB1 and CB2 receptors; the latter is mainly present in immune system cells, including macrophages and other macrophage-derived cells. N-arachidonyl-ethanolamine, better known as anandamide (AEA), was the first endogenous molecule isolated and identified in 1992 in the pig brain, followed by the discovery in 1995 of 2-arachidonoylglycerol (2-AG) (Stella et al., 2021).

Cannabinoid-induced antinociception acts on various levels of the pain sensory pathway, activating the antinociceptive pathways mediated by cannabinoids. The activity of the CB1 receptor system increases in response to noxious stimuli. CB1 receptors have been identified on primary afferent neurons, including both C and Aβ/δ fibers, dorsal root ganglia (DRG), the spinal and terminal intrinsic neurons of the dorsal horn of the spinal cord, and neurons projecting to the brain (medulla, rostral ventromedial, periaqueductal gray matter, amygdala, and thalamus) (Stella et al., 2021). Activation of the CB1 receptor is involved in regulating pain, fear, sleep, appetite, memory, and motor responses, as well as producing psychoactive effects. At the spinal level, the analgesic effect of cannabinoids appears to result from the activation of the OP2 opioid receptor. The intrathecal administration of OP2 receptor antagonists cancels this analgesic effect. Periaqueductal gray matter is one area most strongly involved in mediating cannabinoid analgesia. This may be due, in part, to the direct inhibition of the release of gamma-aminobutyric acid in this brain area and the medulla, and, in part, to the inhibition of glutamate release. The hyperalgesia induced by the CB1 antagonist is attenuated by the administration of N-methyl-D-aspartate receptor antagonists (Stella et al., 2021). Activation of the CB2 receptor is involved in the upregulation or downregulation of the inflammatory response.

Pharmacologically active cannabinoids are classified as Class II according to the Biopharmaceutics Classification System for oral drugs, as they have very low water solubility and high lipophilicity; CBD (12.6 mg/L, logP 6.3, pKa 9.29) and THC (28.0 mg/L, logP 6.97, pKa 10.6) (Stella et al., 2021).

While the most common route of administration for cannabis-derived medications used for pain management is inhalation (smoking), the oral, oral transmucosal, rectal transmucosal, intranasal, topical, and transdermal routes have also been described. THC levels peak at 30 min and subside within 1–3.5 h when cannabis is smoked. When ingested orally or transmucosally, peak levels are not reached until 30 min to 2 h and do not subside for 5 to 8 h, due to the first-pass mechanism.

"Medicinal cannabis" refers to products derived from cannabis recommended to treat a medical condition (Weber and da Silva, 2022; Witek, 2021). "Medical marijuana" strains are derived by cross-breeding *Cannabis indica* and *Cannabis sativa* plant species to produce hybrid plants that vary in their contents of THC and CBD. Strains with THC content >20% are considered high THC, but strains need only have CBD >4% to be considered high CBD (Radhakrishnan et al., 2019). Regarding the therapeutic use of cannabinoids, the U.S. FDA has approved the following cannabis-derived drugs: dantrolene, CBD, and CBD. Dronabinol and nabilone, related to THC, are indicated for CINV. Dronabinol is also shown to stimulate appetite in states of exhaustion, such as AIDS. Finally, CBD is approved to treat two forms of pediatric epilepsy—the Dravet and Lennox–Gastaut syndromes—and seizures associated with tuberous sclerosis complex (Weber and da Silva, 2022). This chapter focuses on the analgesic benefits of marijuana and associated products for managing acute and chronic pain conditions and syndromes.

Prevalence of use and overall effectiveness of medical marijuana

Is medical marijuana (MM) or cannabis-based medication (CBM) effective for pain management? Over the past 20 years, there have been a mere 60 prospective studies assessing the actual benefits of CBM in chronic pain use, and some 72 systematic reviews since 1975. In a systematic review and metaanalysis published in the *British Medical Journal* in 2021, Wang et al. (2021) showed that the use of oral CBM for chronic pain is associated with about a 20% chance of reducing pain by ≥30% and can also improve physical function and sleep quality (Wang et al., 2021). These effects were observed regardless of the type of pain (neuropathic pain vs. nonneuropathic pain, tumor pain vs. nontumor pain) (Wang et al., 2021). These results would appear to be relatively modest. In addition, most systematic reviews indicate a relatively high prevalence of side effects and complications associated with CBM, which must be considered before making medical decisions to use this therapy in patients. Adverse effects were reported in most systematic reviews comparing cannabis with placebo (49/59, 83%) and in 20/24 (83%) of the reviews comparing cannabis to active drugs. Minor adverse effects (e.g., drowsiness, dizziness) were common and reported in over half of the reviews. Serious complications were not as common but were reported in 21/59 (36%) of reviews that reported adverse effects (Pratt et al., 2019). Who is typically prescribed medical cannabis? Studies show that identifying as Caucasian, being male, and earning >$60,000/year is associated with higher CBM use than not having those characteristics (Slawek et al., 2022). Some of the limitations in the use of CBM include the fact that, due to the Schedule I status of cannabis, medical cannabis is not covered by private insurance. Further, it must be paid for with cash or a debit card, which may pose a significant barrier to its use. Depending upon the state, medical cannabis products could cost over $150/month, and in Florida, it is somewhat prohibitive toward CBM use (Luque et al., 2021). The actual prevalence of CBM use is available through surveys only. In the U.S., national survey estimates of prescribed CBM were 1.1% among 12th graders and 17% among adults who reported past-year marijuana use (Park and Wu, 2017). Some of the difficulties arising from these types of surveys include the identification that many users of CBM may also be using marijuana recreationally. Medical-only (CBM) users appear to use it for evidence-based medical reasons and have a lower prevalence of substance use disorder than other marijuana users. In relation to nonmedical-only users ($n = 3339$), combined ($n = 362$), and medical-only ($n = 82$), CBM-only users had a higher prevalence of every medical condition examined. As compared to the combined use group, those using marijuana only for medical purposes were much less likely to have anxiety, alcohol, or nonmedical prescription opioid use disorders (Wall et al., 2019). This chapter takes a deeper look at several key studies and indications for the use of CBM, as well as risks and complications associated with that use.

Possible opioid-sparing benefits of CBM

The opioid epidemic in the United States and worldwide has stimulated great interest in identifying alternative analgesic pathways for managing both acute and chronic pain. Is MM one possible remedy to reduce the exposure of pain-suffering patients to the known deleterious effects of opioids, primarily the dreaded consequence of respiratory depression and death (Maher et al., 2019)? The emerging studies seem to suggest stable CBM use in patients previously prescribed opioids as primary analgesics for chronic pain management (Ross et al., 2021). When CBM is used by patients taking opioids, it does not significantly change the area under the curve of opioids or their metabolites, and there is a time delay to maximum serum concentration (Cmax) of opioids (Okusanya et al., 2020). It also has no significant effect on the pharmacokinetics of opioids. In one systematic review, 9 studies meeting criteria, involving 7222 participants, were included. There was a 64%–75% reduction in opioid dosage when used in combination with CBM. The use of CBM for opioid substitution was reported by 32%–59.3% of patients with noncancer chronic pain. One study reported a slight decrease in mean hospital admissions in the past calendar year ($P = .53$). It decreased mean emergency department visits in the past calendar year ($P = .39$) for patients who received CBM as an adjunct to opioids in the treatment of noncancer chronic pain compared to those who did not receive CBM (Okusanya et al., 2020). Overall, the review found a much higher reduction in opioid dosage, reduced emergency room visits, and hospital admissions for chronic noncancer pain by CBM users compared to people with no additional use of CBM (Okusanya et al., 2020). Another systematic review and metaanalysis included five randomized controlled trials (RCTs) in cancer patients and 12 observational studies (Noori et al., 2021). Regrettably, randomized trials provided high-certainty evidence that cannabis addition had little or no effect on pain relief (WMD − 0.18 cm; 95% CI − 0.38 to 0.02; on a 10 cm Visual Analog Scale [VAS] for pain) or sleep disturbance (WMD − 0.22 cm; 95% CI − 0.4 to − 0.06; on a 10 cm VAS for sleep disturbance; minimally important difference is 1 cm) among chronic cancer pain patients while increasing the incidence of nausea (Noori et al., 2021). A single-center, prospective study of chronic opioid use in chronic pain patients using CBM was published (Benedict et al., 2022). This is one of the very few attempts to determine whether concomitant use of CBM, prospectively undertaken in chronic opioid-using patients, has a real opioid-sparing benefit. There was a 67.1% average decrease in daily MME/patient from 49.9 to 16.4 MME at the first follow-up. There was a 73.3% decrease in MME at the second follow-up from 49.9 to 13.3 MME, with an ANOVA analysis denoting a significant difference of $P < 0.0001$ (Benedict et al., 2022). However, a large study of 371 patients from Portland, Oregon, found that concomitant opioid use for pain, coupled with CBM, found disturbing trends (Nugent et al., 2018). 18% of patients used CBM for pain control of the 371 assessed. No significant differences were detected in pain-related variables, depression, or anxiety between those who used medical cannabis and those who did not use it. CBM users had higher scores of risk for prescription opioid misuse (median = 17.0 vs. 11.5, $p < 0.001$), rates of hazardous alcohol use (25% vs. 16%, $p < 0.05$), and rates of nicotine use (42% vs. 26%, $p = 0.01$). Multivariable analyses indicated that medical cannabis use was significantly associated with risk of prescription opioid misuse ($\beta = 0.17, p = 0.001$), but not hazardous alcohol use (aOR = 1.96, 95% CI = 0.96–4.00, $p = 0.06$) or nicotine use (aOR = 1.61, 95% CI = 0.90–2.88, $p = 0.11$) (Nugent et al., 2018). These sobering results suggest that we do not have a consensus viewpoint as to whether or not opioid consumption can be reliably reduced when marijuana is consumed as an adjunct for pain management. However, unlike what is noted above about the majority of CBM users ("Introduction" section) (Caucasian; males; yearly income >$60,000), the results of the Portland group studied were in contrast to that finding. Participants who endorsed medical cannabis use were more likely to be male (81% vs. 48%), have an annual income below $30,000 (46% vs. 25%), and be less likely to be Caucasian

(70% vs. 86%) (all p-values < 0.05). Compared to national trends, the demographic differences could be responsible for the discrepancies in the results. Also, the most common pain diagnosis was chronic arthritis, which, by itself, begs the question as to why opioids were even being dispensed in the first place. One study from Michigan found that CBM use in chronic pain patients also consuming opioids for pain management was not associated with more aberrant drug-seeking or use or alcohol consumption when compared with a cohort of patients who use MM but not prescription pain medications. This further muddles the waters concerning whether or not CBM use can be considered a reliable analgesic and whether or not there is some reduced requirement for opioids in this patient population (Perron et al., 2015). In the largest review to date, conducted from 2006 to 2014 and concerning database reviews of 10% of U.S. pharmacy records concerning almost five million patients, it was identified that the legalization process of MM appears to have been associated somewhat with reduced opioid use (Shah et al., 2019). However, this period also corresponds to the nation's increasing vigilance and awareness of the exploding opioid epidemic and physicians' increasing reluctance to write opioid prescriptions. Therefore an extrapolation of the findings can be considered to be biased and not reflective of an association between the legalization process and actual reduced opioid requirements for chronic pain patients.

MM use in chronic cancer pain

Cancer-related pain appears to be an ideal situation in which MM could prove beneficial, but the evidence supporting its use is scarce. A 2025 publication assessing the efficacy of CBM in cancer, consisting of 13 systematic reviews and five additional primary studies (four RCTs and one cohort study), found that the certainty of evidence for most outcomes was low or very low (Braun et al., 2024). Despite this, data suggest that between 20% and >40% of adults with cancer report cannabis use (Braun et al., 2024). The authors concluded that, despite a lack of robust evidence supporting MM's use in cancer pain, anecdotal experiences among adults with cancer using cannabis and/or cannabinoids continue to promote strong incentives for use, even in the absence of objective evidence, creating ongoing clinical challenges and opportunities in the care of adults with cancer (Braun et al., 2024). Indeed, there is a high prevalence of cannabis use among cancer patients overall, regardless of whether they seek access to MM. One study of 267 patients found that over a quarter of respondents (26%) reported cannabis use in the past 30 days, and among those, 4.5% screened positive for cannabis use disorder. Participants who used cannabis most often consumed edibles (65%) or smoked cannabis (51%), and they were younger, more likely to be male, Black, and disabled, and had lower income and Medicaid insurance compared to participants who did not use cannabis. This demographic distinction differs from those who seek MM authorization (see above) (Azizoddin et al., 2023). A metaanalysis of the efficacy of CBM for managing cancer pain identified 2805 unique records, of which six randomized controlled trials were included in a systematic review (n = 1460 participants). Five studies were included in the meta-analysis (1442 participants). The primary outcome of interest was the absolute mean change from baseline to the end of treatment in average pain on a Numerical Rating Scale. Secondary outcomes included adverse effects and study dropouts. The authors concluded that the addition of MM to opioids was not effective in reducing cancer pain (Boland et al., 2020).

MM use in neuropathic pain

Neuropathic pain and the use of CBM were studied as part of a Cochrane database review, with results published in 2018 (Mücke et al., 2018). The authors selected randomized, double-blind controlled trials of medical cannabis, including plant-derived and synthetic cannabis-based medicines, compared against placebo or any other active treatment for conditions involving chronic neuropathic pain in adults. The studies included had a treatment duration of at least two weeks and a minimum of 10 participants per treatment arm. A total of 16 studies with 1750 participants were included, with study durations ranging from 2 to 26 weeks. The researchers found that cannabis-based medicines may increase the number of people achieving 50% or greater pain relief compared with placebo (21% vs. 17%; risk difference (RD) of 0.05 [95% confidence interval (CI) 0.00 to 0.09]; number needed to treat for an additional beneficial outcome (NNTB) 20 [95% CI 11 to 100]; 1001 participants, eight studies, low-quality evidence). They rated the evidence for improvement in the Patient Global Impression of Change with cannabis as being of very low quality (26% vs. 21%; RD of 0.09 [95% CI 0.01 to 0.17]; NNTB11 [95% CI 6 to 100]; 1092 participants, six studies). More participants withdrew from the studies due to adverse events with cannabis-based medicines (10% of participants) compared to placebo (5% of participants) (Boland et al., 2020). Another study found that the number needed to treat (NNT) for inhaled cannabis used for pain was 5.6. The authors concluded that this results in short-term reductions in chronic neuropathic pain for one in every five to six patients treated (NNT 5.6 with a Bayesian 95% credible interval ranging between 3.4 and 14). Based on the Initiative on Methods, Measurement, and Pain Assessment in Clinical Trials (IMMPACT) definition of at least moderate benefit, inhaled cannabis improved pain with an odds ratio of 3.2 (Andreae et al., 2015). One review found no long-term RCTs, comparing the use of MM for pain with alternative analgesics (Petzke et al., 2022). The authors referenced a 2021 statement by the International Association for the Study of Pain, which stated that the use of cannabinoids for the treatment of pain cannot, at this time, be endorsed due to a lack of evidence from high-quality research (Petzke et al., 2022). However, they indicated that if a clinician chooses to use marijuana products for chronic neuropathic pain as a third-line treatment option, the following dosages could be considered reliable: Recommended starting dosage for nabiximols: 2 puffs (5.4 mg THC and 5 mg CBD); increase by 1 or 2 puffs every second day; maximum dosage: 12 puffs (32.4 mg THC and 30 mg CBD) per day. Recommended starting dosage for nabilone: 2.5 mg (split into three doses over 24 h); increase by 0.8 mg every second day; dose range of 5–30 mg per day (Petzke et al., 2022). Adverse effects are not uncommon with MM used for neuropathic pain and contribute to reduced compliance with treatment and relatively high dropout rates (Bennici et al., 2021). Adverse effects with 6.7% THC exceeded those with 2.9% THC, with reports indicating that drug effects such as feeling high, impaired, stoned, sedated, changes in perceiving space, drunk, confused, and difficulty paying attention were more prevalent with the higher concentration (Bennici et al., 2021).

MM use in other pain states

Does MM serve useful purposes for pain management? The answer seems to be that the jury remains deadlocked as to whether or not there is sufficient benefit to its use to recommend it as anything beyond a tier-three type of intervention. One systematic review found that, from 1085 articles, 31 relevant systematic reviews were identified, including 23 for pain, 5 for spasticity, 6 for nausea and vomiting, and 12 for adverse events. Metaanalysis of 15 RCTs found that more patients taking cannabinoids attained at least a 30% pain reduction: risk ratio of 1.37 (95% CI 1.14 to 1.64), with an NNT of 11. Sensitivity analysis found study size and duration affected findings (subgroup differences, $P \leq .03$), with larger and longer RCTs finding no benefit (Allan et al., 2018). This large NNT is of some concern when considering the use of CBM. Another systematic review from a Canadian group found mixed evidence in favor of CBM use for disparate chronic pain states (Bell et al., 2024). Forty-seven studies relevant to pain management were reviewed, including 22 RCTs, 11 pre–post studies or uncontrolled trials, 11 cross-sectional or observational cohort studies, and 3 case series. Most studies (38/47) reported at least moderate benefits of CBM for chronic pain, seven were inconclusive or found insufficient evidence, and two reported mixed results (Bell et al., 2024). The authors recommend the use of CBM as monotherapy, replacement, or adjunct treatment in people living with chronic pain for the management of chronic pain, including central and/or peripheral neuropathic pain, to improve pain outcomes, with stronger evidence in favor of its use in HIV-related pain conditions (strong recommendation; moderate quality evidence) (Bell et al., 2024). Perhaps cannabis used moderately for chronic pain is superior in terms of analgesic benefit compared to heavier, daily use. One survey of 989 patients divided participants according to the frequency of daily use: heavy (≥5 times), moderate (3–4 times), or light (1–2 times). Heavy-use participants were younger, had lower income, less education, and were more likely to smoke cigarettes than light or moderate-use participants. Light-use participants were more likely to take concomitant pain medications (though not opioids or benzodiazepines). In pairwise analysis, light-use participants reported significantly lower pain severity scores than moderate or heavy-use participants (5.4 vs. 6.2 and 6.2, respectively, $p < 0.0001$) and lower pain interference scores than moderate or heavy-use participants (4.4 vs. 5.4 and 5.5, respectively, $p < 0.0001$). Light-use participants also reported lower positive effects (23.5 vs. 25.2, $p = 0.037$) and negative effects (24.7

vs. 26.8, $p = 0.011$) than heavy-use participants (Boehnke et al., 2020). Heavy-use participants preferred high THC: high CBD ratios (54.1% vs. 46.1% and 25% for moderate and light-use participants, respectively), while light-use participants preferred low THC: high CBD ratios (39.7% vs. 22% for moderate and 7.2% for heavy-use participants, respectively) (Boehnke et al., 2020). This would imply that heavy use of marijuana can be associated with a negative effect, similar to reports detailing higher opioid use in chronic pain. From the perspective of pain syndromes that possibly benefit from CBM or MM, studies seem to indicate that musculoskeletal pain (Mangual-Perez et al., 2022) can be positively affected, with weak evidence in its favor. Gynecological pain, particularly pain from endometriosis and chronic pelvic pain, is another area with scant research showing benefit, besides anecdotal evidence (Sinclair et al., 2023). Similarly, small studies offer hope that patients with rheumatological conditions could benefit from CBM. One survey of 319 patients found that the average monthly consumed dose of MC was 31, 35, 36, and 32 g, with mean pain level reductions of 77%, 82%, 83%, and 57%, and mean sleep quality improvements of 78%, 71%, 87%, and 76% among patients with fibromyalgia, mechanical, neuropathic, and inflammatory problems, respectively (Habib et al., 2021). In orthopedics, CBM use has a paucity of studies looking at safety and efficacy. Variability in the methodologies used in cannabis research makes it challenging to conclude dosing, routes, and frequency of administration. Most of the existing evidence suggests that medical cannabis use is effective. Still, this efficacy has been demonstrated only when there is no comparator or cannabis is compared with a placebo (Madden et al., 2019). For chronic low back pain (LBP), a systematic review of published studies found minimal high-quality evidence for CBM as a first-line treatment option, despite some favorable reviews (Lee et al., 2023). Twelve studies were included in the synthesis: one RCT, six observational studies (one prospective, four retrospective, and one crossover), and five case studies. All study results, except for the RCT, indicated a decrease in LBP levels or opioid use over time after medical cannabis use. The RCT reported no statistically significant difference in LBP between cannabis and placebo groups (Lee et al., 2023). Another review of 10 studies of LBP and CBM found a lack of evidence favoring CBM for chronic LBP (Kulesza et al., 2023).

Adverse effects of MM use

MM use, particularly in daily and heavier users, is associated with an unpleasant withdrawal syndrome upon abrupt cessation of use (Coughlin et al., 2021). A cohort of 527 adults with chronic pain seeking medical cannabis certification or recertification was recruited between February 2014 and June 2015 in one study. A three-class LCA model, including a mild (41%), moderate (34%), and severe (25%) symptom class, parsimoniously represented withdrawal symptoms experienced by people using medical cannabis. The stability of withdrawal symptoms using a three-class LTA at 12 and 24 months ranged from 0.58 to 0.87, with the most stability in the mild withdrawal class. Younger age predicted greater severity and worsening of withdrawal over time (Coughlin et al., 2021). While MM may be safer than opioids in terms of not engendering a risk of respiratory depression, there are nevertheless associated cardiac, pulmonary, and psychological risks linked to chronic, daily use, sometimes of a severe nature (Zeraatkar et al., 2022). A review identified 39 eligible studies that enrolled 12,143 adult patients with chronic pain. Very low-certainty evidence suggests that adverse events are common (prevalence: 26.0%; 95% CI 13.2% to 41.2%) among users of medical cannabis for chronic pain, particularly psychiatric adverse events (prevalence: 13.5%; 95% CI 2.6% to 30.6%) (Zeraatkar et al., 2022). Although males constitute a higher percentage of patients using MM, females appear to have a higher incidence of adverse events compared to males (Aviram et al., 2022). In 429 patients with noncancer-related pain (64% males), women reported more adverse events than men in the central nervous system (73% compared with 53%), gastrointestinal system (79% compared with 65%), musculoskeletal system (40% compared with 26%), and psychological AEs (54% compared with 39%). No significant differences were found in cardiovascular and ophthalmic AEs (Aviram et al., 2022). Sleep disturbances are also higher in females than males using CBM, for unknown reasons (Cranford et al., 2017).

Summary and conclusions

Both MM and CBM are possible options for the management of acute and, more commonly, chronic pain conditions, including both cancer-related and noncancer-related pain syndromes. The evidence in favor of their use is complicated by a lack of standardization in dosing and routes of administration, as well as by the heterogeneity of pain types being assessed. At best, their use can be considered a level or tier-three type of adjunctive therapy, with the benefits primarily being a lack of respiratory depression effects seen with chronic opioid consumption. It remains unclear whether the analgesic benefits are due to the marijuana itself or to the provision of a more apathetic mental state in the face of pain, owing to the known cognitive effects of marijuana. Study limitations notwithstanding, there does appear to be a role for medical marijuana in the armamentarium of the pain physician, with appropriate caveats and precautions regarding the possible high rates of side effects associated with their use being taken into consideration. Future studies will require a more rigorous approach to categorizing actual dosing and discrete pain conditions being studied before a more enthusiastic viewpoint can be adopted regarding medical marijuana use.

References

Allan, G.M., Finley, C.R., Ton, J., Perry, D., Ramji, J., Crawford, K., Lindblad, A.J., Korownyk, C., Kolber, M.R., 2018. Systematic review of systematic reviews for medical cannabinoids. Can. Fam. Physician 0008-350X. 64 (2), e78. Available from: http://www.cfp.ca/content/cfp/64/2/e78full.pdf.

Andreae, M.H., Carter, G.M., Shaparin, N., Suslov, K., Ellis, R.J., Ware, M.A., Abrams, D.I., Prasad, H., Wilsey, B., Indyk, D., Johnson, M., Sacks, H.S., 2015. Inhaled cannabis for chronic neuropathic pain: a meta-analysis of individual patient data. J. Pain. 1528-8447. 16 (12), 1221–1232. Available from: http://www.elsevier.com/inca/publications/store/6/2/3/1/4/7/index.htt. doi:10.1016/j.jpain.2015.07.009.

Archie, S.R., Cucullo, L., 2019. Harmful effects of smoking cannabis: a cerebrovascular and neurological perspective. Front. Pharmacol. 1663-9812. 10, 1481. doi:10.3389/fphar.2019.01481.

Aviram, J., Lewitus, G.M., Vysotski, Y., Berman, P., Shapira, A., Procaccia, S., Meiri, D., 2022. Sex differences in medical cannabis-related adverse effects. Pain. 0304-3959. 163 (5), 975–983. doi:10.1097/j.pain.0000000000002463.

Azizoddin, D.R., Cohn, A.M., Ulahannan, S.V., Henson, C.E., Alexander, A.C., Moore, K.N., Holman, L.L., Boozary, L.K., Sifat, M.S., Kendzor, D.E., 2023. Cannabis use among adults undergoing cancer treatment. Cancer. 0008-543X. 129 (21), 3498–3508. doi:10.1002/cncr.34922.

Bell, A.D., MacCallum, C., Margolese, S., Walsh, Z., Wright, P., Daeninck, P.J., Mandarino, E., Lacasse, G., Deol, J.K., de Freitas, L., St. Pierre, M., Belle-Isle, L., Gagnon, M., Bevan, S., Sanchez, T., Arlt, S., Monahan-Ellison, M., O'Hara, J., Boivin, M., Costiniuk, C., 2024. Clinical practice guidelines for cannabis and cannabinoid-based medicines in the management of chronic pain and co-occurring conditions. Cannabis Cannabinoid Res. 2578-5125. 9 (2), 669–687. doi:10.1089/can.2021.0156.

Benedict, G., Sabbagh, A., Conermann, T., 2022. Medical cannabis used as an alternative treatment for chronic pain demonstrates reduction in chronic opioid use—a prospective study. Pain Physician. 2150-1149. 25 (1), E113. Available from: http://www.painphysicianjournal.com/.

Bennici, A., Mannucci, C., Calapai, F., Cardia, L., Ammendolia, I., Gangemi, S., Calapai, G., Soler, D.G., 2021. Safety of medical cannabis in neuropathic chronic pain management. Molecules. 1420-3049. 26 (20), 6257. doi:10.3390/molecules26206257.

Boehnke, K.F., Scott, J.R., Litinas, E., Sisley, S., Williams, D.A., Clauw, D.J., 2020. High-frequency medical cannabis use is associated with worse pain among individuals with chronic pain. J. Pain. 1526-5900. 21 (5-6), 570–581. doi:10.1016/j.jpain.2019.09.006.

Boland, E.G., Bennett, M.I., Allgar, V., Boland, J.W., 2020. Cannabinoids for adult cancer-related pain: systematic review and meta-analysis. BMJ Support. Palliat. Care. 2045-4368. 10 (1), 14–24. Available from: http://spcare.bmj.com/. doi:10.1136/bmjspcare-2019-002032.

Braun, I.M., Bohlke, K., Abrams, D., Anderson, H., et al., 2024. Cannabis and cannabinoids in adults with cancer: ASCO guideline. J. Clin. Oncol. 42 (13), 1575–1593.

Coughlin, L.N., Ilgen, M.A., Jannausch, M., Walton, M.A., Bohnert, K.M., 2021. Progression of cannabis withdrawal symptoms in people using medical cannabis for chronic pain. Addiction. 0965-2140. 116 (8), 2067–2075. doi:10.1111/add.15370.

Cranford, J.A., Arnedt, J.T., Conroy, D.A., Bohnert, K.M., Bourque, C., Blow, F.C., Ilgen, M., 2017. Prevalence and correlates of sleep-related problems in adults receiving medical cannabis for chronic pain. Drug Alcohol Depend. 0376-8716. 180, 227–233. doi:10.1016/j.drugalcdep.2017.08.017.

Habib, G., Khazin, F., Artul, S., 2021. The effect of medical cannabis on pain level and quality of sleep among rheumatology clinic outpatients. Pain Res. Manag. 2021, 1756588.

Kulesza, B., Mazurek, M., Kurzepa, J., 2023. Can cannabidiol (CBD) help with low back pain? Ann. Agric. Environ. Med. 1232-1966. 30 (3), 549–554. doi:10.26444/aaem/168674.

Lee, C., Danielson, E.C., Beestrum, M., Eurich, D.T., Knapp, A., Jordan, N., 2023. Medical cannabis and its efficacy/effectiveness for the treatment of low-back pain: a systematic review. Curr. Pain Headache Rep. 1531-3433. 27 (12), 821–835. doi:10.1007/s11916-023-01189-0.

Li, H.L., 1973. An archaeological and historical account of cannabis in China. Econ. Bot. 1874-9364. 28 (4), 437–448. doi:10.1007/BF02862859.

Luque, J.S., Okere, A.N., Reyes-Ortiz, C.A., Williams, P.M., 2021. Mixed methods study of the potential therapeutic benefits from medical cannabis for patients in Florida. Complement. Ther. Med. 0965-2299. 57, 102669. doi:10.1016/j.ctim.2021.102669.

Madden, K., George, A., van der Hoek, N.J., Borim, F.M., Mammen, G., Bhandari, M., 2019. Cannabis for pain in orthopedics: a systematic review focusing on study methodology. Can. J. Surg. 0008-428X. 62 (6), 369–380. doi:10.1503/cjs.001018.

Madras, B.K., 2019. Tinkering with THC-to-CBD ratios in Marijuana. Neuropsychopharmacology. 0893-133X. 44 (1), 215–216. doi:10.1038/s41386-018-0217-3.

Maher, D.P., Carr, D.B., Hill, K., McGeeney, B., Weed, V., Jackson, W.C., DiBenedetto, D.J., Moriarty, E.M., Kulich, R.J., 2019. Cannabis for the treatment of chronic pain in the era of an opioid epidemic: a symposium-based review of sociomedical science. Pain Med. 1526-2375. 20 (11), 2311–2323. doi:10.1093/pm/pnx143.

Mangual-Perez, D., Tresgallo-Pares, R., Ramirez-Gonzalez, M., Torres-Lugo, N., et al., 2022. Patient experience and perspective on medical cannabis as an alternative for musculoskeletal pain management. JAAOS Glob. Res. Rev. 6, e2200055.

Mücke, M., Phillips, T., Radbruch, L., Petzke, F., Häuser, W., 2018. Cannabis-based medicines for chronic neuropathic pain in adults. Cochrane Database Syst. Rev. 1469-493X. 2018 (3). Available from: http://as.wiley.com/WileyCDA/Brand/id-6.html. doi:10.1002/14651858.CD012182.pub2.

Noori, A., Miroshnychenko, A., Shergill, Y., Ashoorion, V., Rehman, Y., Couban, R.J., Buckley, D.N., Thabane, L., Bhandari, M., Guyatt, G.H., Agoritsas, T., Busse, J.W., 2021. Opioid-sparing effects of medical cannabis or cannabinoids for chronic pain: a systematic review and meta-analysis of randomised and observational studies. BMJ Open. 2044-6055. 11 (7), e047717. doi:10.1136/bmjopen-2020-047717.

Nugent, S.M., Yarborough, B.J., Smith, N.X., Dobscha, S.K., Deyo, R.A., Green, C.A., Morasco, B.J., 2018. Patterns and correlates of medical cannabis use for pain among patients prescribed long-term opioid therapy. Gen. Hosp. Psychiatry 0163-8343. 50, 104–110. doi:10.1016/j.genhosppsych.2017.11.001.

Okusanya, B.O., Asaolu, I.O., Ehiri, J.E., Kimaru, L.J., Okechukwu, A., Rosales, C., 2020. Medical cannabis for the reduction of opioid dosage in the treatment of non-cancer chronic pain: a systematic review. Syst. Rev. 2046-4053. 9 (1), 167. doi:10.1186/s13643-020-01425-3.

Park, J.-Y., Wu, L.-T., 2017. Prevalence, reasons, perceived effects, and correlates of medical marijuana use: a review. Drug Alcohol Depend. 0376-8716. 177, 1–13. doi:10.1016/j.drugalcdep.2017.03.009.

Perron, B.E., Bohnert, K., Perone, A.K., Bonn-Miller, M.O., Ilgen, M., 2015. Use of prescription pain medications among medical cannabis patients: comparisons of pain levels, functioning, and patterns of alcohol and other drug use. J. Stud. Alcohol Drugs. 1937-1888. 76 (3), 406–413. doi:10.15288/jsad.2015.76.406.

Petzke, F., Tölle, T., Fitzcharles, M.A., Häuser, W., 2022. Cannabis-based medicines and medical cannabis for chronic neuropathic pain. CNS Drugs. 1179-1934. 36 (1), 31–44. Available from: http://rd.springer.com/journal/40263. doi:10.1007/s40263-021-00879-w.

Pratt, M., Stevens, A., Thuku, M., Butler, C., Skidmore, B., Wieland, L.S., Clemons, M., Kanji, S., Hutton, B., 2019. Benefits and harms of medical cannabis: a scoping review of systematic reviews. Syst. Rev. 2046-4053. 8 (1), 320. doi:10.1186/s13643-019-1243-x.

Radhakrishnan, R., Ranganathan, M., D'Souza, D.C., 2019. Medical marijuana: what physicians need to know. J. Clin. Psychiatry 1555-2101. 80 (5). Available from: https://www.psychiatrist.com/JCP/article/_layouts/ppp.psych.controls/BinaryViewer.ashx?Article=/JCP/article/Pages/2019/v80/18ac12537.aspx&Type=Article. doi:10.4088/JCP.18ac12537.

Ross, J., Slawek, D.E., Zhang, C., Starrels, J.L., Levin, F.R., Sohler, N.L., Minami, H., Arnsten, J.H., Cunningham, C.O., 2021. First-year trajectories of medical cannabis use among adults taking opioids for chronic pain: an observational cohort study. Pain Med. 1526-2375. 22 (12), 3080–3088. doi:10.1093/pm/pnab257.

Shah, A., Hayes, C.J., Lakkad, M., Martin, B.C., 2019. Impact of medical marijuana legalization on opioid use, chronic opioid use, and high-risk opioid use. J. Gen. Intern. Med. 0884-8734. 34 (8), 1419–1426. doi:10.1007/s11606-018-4782-2.

Sinclair, J., Abbott, J., Proudfoot, A., Armour, M., 2023. The place of cannabinoids in the treatment of gynecological pain. Drugs. 0012-6667. 83 (17), 1571–1579. doi:10.1007/s40265-023-01951-z.

Slawek, D.E., Curtis, S.A., Arnsten, J.H., Cunningham, C.O., 2022. Clinical approaches to cannabis: a narrative review. Med. Clin. North Am. 1557-9859. 106 (1), 131–152. Available from: http://www.medical.theclinics.com/. doi:10.1016/j.mcna.2021.08.004.

Stella, B., Baratta, F., Pepa, C.D., Arpicco, S., Gastaldi, D., Dosio, F., 2021. Cannabinoid formulations and delivery systems: current and future options to treat pain. Drugs. 0012-6667. 81 (13), 1513–1557. doi:10.1007/s40265-021-01579-x.

Wall, M.M., Liu, J., Hasin, D.S., Blanco, C., Olfson, M., 2019. Use of marijuana exclusively for medical purposes. Drug Alcohol Depend. 0376-8716. 195, 13–15. doi:10.1016/j.drugalcdep.2018.11.009.

Wang, L., Hong, P.J., May, C., Rehman, Y., Oparin, Y., Hong, C.J., Hong, B.Y., AminiLari, M., Gallo, L., Kaushal, A., Craigie, S., Couban, R.J., Kum, E., Shanthanna, H., Price, I., Upadhye, S., Ware, M.A., Campbell, F., Buchbinder, R., Agoritsas, T., Busse, J.W., 2021. Medical cannabis or cannabinoids for chronic non-cancer and cancer related pain: a systematic review and meta-analysis of randomised clinical trials. BMJ. 1756-1833. 374, n1034. doi:10.1136/bmj.n1034.

Weber, C.A.T., da Silva, A.G., 2022. Medical marijuana: what are we talking about? Braz. J. Psychiatry 1809-452X. 44 (6), 574–575. Available from: https://www.scielo.br/j/rbp/a/XSyzpXCWv5F9tWPtVGkVmrv/?format=pdf&lang=en. doi:10.47626/1516-4446-2022-2608.

Witek, T.J., 2021. Please don't call it medical marijuana unless it is; but it probably isn't. Can. J. Public Health. 0008-4263. 112 (1), 74–77. doi:10.17269/s41997-020-00333-2.

Zeraatkar, D., Cooper, M.A., Agarwal, A., Vernooij, R.W.M., Leung, G., Loniewski, K., Dookie, J.E., Ahmed, M.M., Hong, B.Y., Hong, C., Hong, P., Couban, R., Agoritsas, T., Busse, J.W., 2022. Long-term and serious harms of medical cannabis and cannabinoids for chronic pain: a systematic review of non-randomised studies. BMJ Open. 2044-6055. 12 (8), e054282. doi:10.1136/bmjopen-2021-054282.

CHAPTER

Uses of cannabinoids in pain management

12

Danielle Kohr and Michael Kugler
Department of Pain Medicine, Fachklinik Löwenstein, SLK Kliniken Heilbronn GmbH, Löwenstein, Germany

Introduction

Pain management remains one of the most challenging aspects of modern medicine, particularly in chronic conditions where conventional treatments may be insufficient or poorly tolerated. cannabinoids have garnered increasing attention as potential therapeutic agents for pain relief. While the scientific evidence regarding their efficacy is still evolving, both clinical studies and real-world data suggest that cannabis-based medicines may offer relief to specific patient populations, particularly those with neuropathic pain, spasticity, mixed pain syndromes, and cancer-related pain (McDonagh et al., 2022; BfArM, 2022; Maher et al., 2019). This chapter explores the indications for cannabinoids in pain therapy, the current state of scientific research, and real-world treatment outcomes.

Scientific evidence and limitations of existing research

The application of cannabis derivatives in pain therapy reflects a nuanced landscape shaped by varying degrees of clinical evidence and patient outcomes (McDonagh et al., 2022; Wang et al., 2021; Finn et al., 2021; Bilbao and Spanagel, 2022; Bell et al., 2024; Solmi et al., 2023; Russo, 2019; Bialas et al., 2022; Fisher et al., 2021).

One of the primary challenges in evaluating the efficacy of cannabis in pain therapy is the design of clinical trials. Most studies have compared cannabis-based treatments to a placebo rather than conventional analgesics, such as opioids, nonsteroidal antiinflammatory drugs (NSAIDs), or anticonvulsants (McDonagh et al., 2022; Wang et al., 2021; Barakji et al., 2023). As a result, there is a lack of comparative data to determine whether cannabis-based medicines are superior, equivalent, or inferior to standard treatments (Karst, 2024).

In addition, pain is a highly subjective symptom and cannot be measured with objective parameters like blood pressure or glucose levels. Unlike other medical conditions with clear biomarkers, pain perception varies significantly among individuals and is influenced by psychological, emotional, and contextual factors (Russo, 2019).

Cannabinoids for Chronic Pain. https://doi.org/10.1016/B978-0-443-36773-1.00010-6

Even within a single condition, such as fibromyalgia syndrome, pain mechanisms can be vastly different, making it difficult to standardize treatment responses across patient populations (Khurshid et al., 2021).

Studies investigating the effectiveness of cannabis-based treatments for pain management have generally shown an average symptom improvement of approximately 30% (Wang et al., 2021; Fisher et al., 2021). For comparison, placebo-controlled studies have reported up to 20% improvement in pain symptoms, emphasizing the complex nature of pain perception and the role of expectation in symptom relief (Vase and Wartolowska, 2019). Despite growing interest in cannabis-based therapies, current research remains limited by a notable scarcity of direct comparative studies with established analgesics such as opioids, anticonvulsants, or NSAIDs. Furthermore, most randomized controlled trials exhibit substantial methodological heterogeneity, particularly in relation to the specific cannabis formulations investigated (Karst, 2024).

Despite the large variability in the grade of scientific evidence and inconsistencies surrounding the effectiveness of cannabis in pain management (Fisher et al., 2021), robust observational data exist, and patient experiences, alongside real-world evidence, may hold substantial relevance given the complexities of pain assessment and should be duly considered in both clinical practice and research (BfArM, 2022; Russo, 2019; Lavie-Ajayi and Shvartzman, 2019).

Real-world data on cannabis use in pain therapy

Beyond clinical trials, real-world data provide additional insight into the use of cannabinoids in pain management (BfArM, 2022; Russo, 2019; Lavie-Ajayi and Shvartzman, 2019). One of the most significant sources of such data is the observational study conducted by the Federal Institute for Drugs and Medical Devices (BfArM) in Germany, which evaluated cannabis prescriptions in a large patient population (BfArM, 2022). This survey encompasses approximately 21,000 submitted datasets related to treatments involving cannabis-based medicines covered by statutory health insurance. The final report of this study largely aligns with the findings of existing scientific studies. Key insights from this real-world evaluation include:

- Chronic pain was the primary indication for cannabis-based therapy in approximately 75% of cases.
- Fifteen percent of patients were treated for cancer-related pain, often accompanied by additional symptoms such as nausea, vomiting, and cachexia.
- Spasticity associated with multiple sclerosis (MS) is another frequently documented indication.

Further demographic and clinical findings from the observational study:

- Gender distribution was nearly equal, and the average patient was around 45 years old.

- The average duration of symptoms before initiating cannabis therapy was approximately 8 years.
- Anesthesiologists were the most common prescribers of cannabis-based treatments, followed by general practitioners and neurologists.
- Synthetic tetrahydrocannabinol (THC) (e.g., dronabinol) was prescribed in approximately 60% of cases, while cannabis flowers and extracts each accounted for approximately 15% of prescriptions.
- THC and cannabidiol (CBD) were combined in approximately 10% of cases.
- The average daily dose of THC was 15 mg/day for extracts and synthetic derivatives. In contrast, unrealistically high daily doses of 250 mg were reported for cannabis flowers.

Patients reported positive treatment outcomes:

- Seventy-five percent of patients experienced symptom improvement, with 70% reporting a better quality of life overall.
- Adverse effects were common but not severe, with dizziness and fatigue being the most frequently reported side effects.
- One-third of patients discontinued treatment within one year, with the primary reasons being:
  - Lack of effectiveness (40%)
  - Side effects (25%)
  - Patient death (20%), particularly in palliative care settings

Interestingly, patients using cannabis flowers reported better therapeutic success, fewer side effects, and a lower discontinuation rate compared to those using synthetic THC or extracts. However, euphoric side effects were more commonly reported among users of cannabis flowers (BfArM, 2022).

Cannabinoids in pain management: THC and CBD effects

THC and cannabidiol have a wide range of therapeutic effects, some of which are similar and others that differ. Their diverse mechanisms of action can be strategically utilized to manage chronic pain and related symptoms (Finn et al., 2021).

THC and its role in pain relief

THC is primarily responsible for the analgesic effects in cannabis-based therapies. Its pain-relieving properties are mainly mediated through the activation of cannabinoid type 1 (CB1) receptors, which are predominantly located in the central nervous system (CNS) (Pagano et al., 2022). By modulating neurotransmitter release and neural excitability, THC influences the processing and perception of pain, particularly in neuropathic and chronic pain conditions (Anthony et al., 2020). In addition to its therapeutic potential, THC also possesses psychoactive properties, which are primarily attributable to its central CB1 receptor activity and are associated with the majority of CNS-related adverse effects (Finn et al., 2021; Anthony et al., 2020).

Furthermore, THC is known to stimulate appetite, which is particularly helpful for patients suffering from cachexia (weight loss) or reduced appetite due to cancer or other chronic illnesses (Bilbao and Spanagel, 2022). In addition, THC can help mitigate nausea and vomiting, making it particularly useful for cancer patients undergoing chemotherapy (Warr and Hesketh, 2020; Sukpiriyagul et al., 2023).

However, THC is a psychoactive compound, which means it affects the central nervous system and can alter consciousness and mood. These effects can include euphoria and undesirable side effects, such as dizziness, drowsiness, and cognitive disturbances like memory and concentration impairments (Finn et al., 2021; Anthony et al., 2020). These side effects should be carefully considered, especially in patients who are elderly or have cognitive impairments.

CBD and its therapeutic benefits

On the other hand, CBD has more subtle pain-relieving effects compared to THC. While it is only weakly analgesic on its own, CBD plays a significant role in modulating pain through its action on CB2 receptors, which are predominantly located in the immune system and peripheral tissues (Pagano et al., 2022; Mlost et al., 2020). CBD is not psychoactive, meaning it does not cause the mood-altering effects associated with THC (de Almeida and Devi, 2020). This makes it particularly suitable for patients who require pain relief without experiencing the cognitive and psychoactive side effects of THC (Pagano et al., 2022; de Almeida and Devi, 2020).

CBD also enhances muscle relaxation and has a calming effect, which is beneficial in managing anxiety, muscle tension, and spasticity (de Almeida and Devi, 2020). This makes it particularly useful for patients with MS, neuropathic pain, and conditions involving chronic muscle spasms.

Moreover, CBD has proven to be antiinflammatory, which is crucial for patients suffering from chronic inflammatory pain (de Almeida and Devi, 2020). By modulating immune response and reducing excessive inflammation, CBD can play an important role in conditions like rheumatoid arthritis, inflammatory bowel disease, and autoimmune disorders (Mlost et al., 2020). Its ability to reduce immune system overactivity can also aid in managing conditions like fibromyalgia, where immune dysregulation contributes to pain perception.

Combining THC and CBD for enhanced pain control

The combination of THC and CBD in a balanced ratio can produce a synergistic effect that enhances analgesia while minimizing adverse effects (Pagano et al., 2022; Anand et al., 2021). THC is primarily responsible for pain relief and muscle relaxation, whereas CBD can mitigate some of the psychoactive side effects of THC, such as anxiety, dizziness, and cognitive impairment. This combination approach has proven particularly beneficial in managing neuropathic pain, fibromyalgia, and spasticity, especially in neurological conditions like MS (Rog et al., 2007).

In neuropathic pain, both THC and CBD can help reduce the hyperactivity of nerve cells, which is a key feature in conditions like post-herpetic neuralgia, trigeminal neuralgia, and diabetic neuropathy (Seevathee et al., 2024). The combination of THC and CBD may help patients achieve a better quality of life by improving sleep (often disrupted in chronic pain conditions), enhancing mood, and alleviating stress while maintaining effective pain control (Sokolaj et al., 2024).

While both cannabinoids contribute meaningfully to pain regulation, they serve distinct roles. THC is especially effective in treating severe pain, such as that associated with spasticity or cancer. In contrast, CBD complements this effect by offering antiinflammatory benefits, promoting muscle relaxation, and reducing anxiety and cognitive side effects (Sokolaj et al., 2024). For patients with chronic or complex pain syndromes, a carefully balanced combination of THC and CBD offers a comprehensive approach that maximizes therapeutic benefits while minimizing side effects (Seevathee et al., 2024; Ueberall et al., 2022).

In clinical practice, the optimal use of THC and CBD should be tailored to the individual, considering the type and severity of pain, the patient's previous treatment responses, and their tolerance to psychoactive effects. Regular monitoring and dose adjustments based on clinical response are essential for safe and effective pain management.

Comparisons between different cannabis preparations

Various formulations of cannabis-based treatments have been investigated, including synthetic THC derivatives (such as dronabinol), cannabis extracts, and dried cannabis flowers. Scientific studies have found no substantial differences in efficacy among these formulations, suggesting that the choice of cannabis product may be influenced more by patient preference than by clear therapeutic advantages (Fisher et al., 2021; Hameed et al., 2023).

However, real-world data indicate that patients using cannabis flowers often report better symptom control, fewer side effects, and lower discontinuation rates compared to those using synthetic THC or extracts (BfArM, 2022). Despite this, cannabis flowers are also more frequently associated with euphoric effects, which could impact their acceptability in clinical settings.

Indications for cannabis in pain management

Cannabinoids exert their analgesic effects primarily through the endocannabinoid system, a complex network of receptors and signaling molecules that modulate pain perception, inflammation, and neurological function (Mlost et al., 2020). While the use of cannabis in pain therapy has been historically controversial, growing clinical experience has led to an increasing acceptance of its potential benefits, particularly in cases where standard pain medication fails or causes significant side effects (BfArM, 2022; Russo, 2019; Karst, 2024; Lavie-Ajayi and Shvartzman, 2019).

Based on the available evidence, cannabis-based therapy is recommended in two primary approaches:

- Monotherapy: Cannabis is used as the sole treatment for pain management.
- Combination therapy (add-on therapy): Cannabis is combined with conventional analgesics, such as opioids or anticonvulsants, to enhance pain relief.

Expert opinion generally favors a combination therapy approach, where cannabis-based treatments are used as an adjunct rather than a standalone therapy (Ueberall et al., 2022). However, clinical evidence does not strongly support the superiority of combination therapy over monotherapy.

Given the available evidence, recommendations suggest that cannabinoids should be considered when conventional analgesics fail or produce intolerable side effects (Bell et al., 2024).

The main indications for cannabis-based medications in pain therapy include the following.

Neuropathic pain

Neuropathic pain arises from damage or dysfunction of the nervous system and is notoriously difficult to treat with conventional analgesics. Cannabinoids have been studied in various neuropathic pain conditions (Bell et al., 2024; Ueberall et al., 2022; Häuser et al., 2018), including:

- Peripheral neuropathies (e.g., mono- and polyneuropathies, diabetic neuropathy, postherpetic neuralgia)
- Central neuropathies (e.g., spinal cord injury, stroke-related pain, MS)
- Trigeminal neuralgia
- Phantom limb pain

Patients with neuropathic pain often report significant symptom relief with cannabis-based therapies, particularly when standard treatments, such as gabapentinoids, antidepressants, or opioids, provide insufficient relief or cause intolerable side effects.

Painful spasticity (MS, thalamic pain, and related conditions)

Spasticity-related pain is another well-recognized indication for cannabinoids, particularly in MS and conditions affecting the central nervous system (Fisher et al., 2021; Häuser et al., 2018; Filippini et al., 2019; Boehnke et al., 2024; Pantoja-Ruiz et al., 2022). In MS patients, cannabis-based medicines, such as a THC/CBD combination, have been approved in several countries for their ability to reduce muscle spasms, pain, and stiffness (Filippini et al., 2019).

Mixed pain syndromes (a combination of neuropathic and nociceptive pain)

Certain chronic pain conditions involve a combination of neuropathic and nociceptive components (Bell et al., 2024; Bialas et al., 2022; Khurshid et al., 2021). Cannabinoids have been explored in conditions such as:

- Chronic back pain
- Myofascial pain syndrome
- Fibromyalgia syndrome
- Complex regional pain syndrome

Although the effectiveness of cannabis in mixed pain syndromes remains debated, some patients report symptom relief, particularly when it is used as part of a multimodal pain management approach.

Palliative and cancer pain management

In oncology and palliative care, cannabinoids are frequently considered as an adjunct to traditional analgesics (Wang et al., 2021; Häuser et al., 2018; Byars et al., 2019; Chung et al., 2020; Paice et al., 2016). Their potential benefits extend beyond pain relief to include:

- Reduction of chemotherapy-induced nausea and vomiting (Warr and Hesketh, 2020; Sukpiriyagul et al., 2023; Badowski, 2017; Grimison et al., 2020)
- Improved appetite and weight maintenance (treatment of cachexia) (Wang et al., 2019)
- Mood stabilization and reduction of anxiety or depression (Byars et al., 2019)

Cannabinoids are particularly relevant for cancer patients who suffer from opioid-resistant pain, experience significant opioid-related side effects (e.g., nausea, constipation), or require additional symptom control.

Opioid sparing

A key advantage of cannabis-based therapy lies in its potential to reduce or even replace opioid use in patients with chronic pain (Ueberall et al., 2022; Rosic et al., 2021; Boehnke et al., 2016). This opioid-sparing effect is especially relevant in the context of the ongoing opioid crisis, offering a safer alternative with a lower risk of dependence, tolerance, and overdose (Boehnke et al., 2024). Clinical experience and emerging data indicate that many patients can significantly reduce or discontinue opioid medications after incorporating cannabis into their pain management plans (Boehnke et al., 2016, 2024). The coadministration of cannabinoids and opioids may also help prevent the development of opioid tolerance, thereby reducing the need for dose escalation (Boehnke et al., 2016, 2024). In addition, cannabis has been shown to ease opioid withdrawal symptoms, further supporting its role in facilitating opioid reduction strategies (Barlowe et al., 2019; Vigil et al., 2017). By enhancing safety and reducing the risks associated with long-term opioid therapy, cannabis represents a valuable component of comprehensive pain care.

Additional indications with limited evidence

While pain relief remains the primary therapeutic goal, cannabinoids have been explored in other conditions with varying levels of evidence (Bell et al., 2024; Boehnke et al., 2024; Shustorovich et al., 2024). Some additional indications include:

- Sleep disorders related to chronic pain (Bell et al., 2024)
- Reduction of opioid consumption in patients with long-term opioid use (Vigil et al., 2017)
- Visceral pain (Solmi et al., 2023; Häuser et al., 2018; Bort et al., 2024), particularly in conditions such as Crohn's disease and ulcerative colitis
- Rheumatoid arthritis and other inflammatory joint disorders (Bryk and Starowicz, 2021; Sánchez-Flórez et al., 2022)
- Chronic pelvic pain syndromes, such as endometriosis-related pain or interstitial cystitis (Sinclair et al., 2021)
- Headaches (Bell et al., 2024):
  - Cannabis-based therapy is primarily considered for secondary headaches, particularly those of traumatic origin, such as traumatic brain injury or intracranial hemorrhages, where it is often used as monotherapy.
  - Primary headaches (e.g., migraines, tension-type headaches, cluster headaches): cannabinoids are generally not effective as a standalone treatment. Instead, they may be used in combination with standard therapies, primarily to address associated symptoms such as anxiety and sleep disturbances, which are commonly seen in migraine patients.

Given the available evidence, current recommendations suggest that cannabinoids should be considered when conventional analgesics fail or produce intolerable side effects. While monotherapy with cannabis is possible, an add-on approach in combination with standard pain treatments is generally favored by experts, especially to mitigate side effects or in particular indications where the CBD effect is beneficial.

Despite ongoing debates, cannabinoids continue to play an increasingly important role in pain management. With further research and optimized treatment protocols, their use may become more refined, offering a practical option for patients with complex and treatment-resistant pain conditions.

AI disclosure

During the preparation of this work, the author(s) used ChatGPT to enhance readability, grammar, and clarity. After using this tool/service, the author(s) reviewed and edited the content as needed and take(s) full responsibility for the content of the publication.

References

Anand, U., Pacchetti, B., Anand, P., Sodergren, M.H., 2021. Cannabis-based medicines and pain: a review of potential synergistic and entourage effects. Pain Manag. 1758-1877. 11 (4), 395–403. Available from: http://www.futuremedicine.com/loi/pmt. doi:10.2217/pmt-2020-0110.

Anthony, A.T., Rahmat, S., Sangle, P., Sandhu, O., Khan, S., 2020. Cannabinoid receptors and their relationship with chronic pain: a narrative review. Cureus. 2168-8184. 12 (9). doi:10.7759/cureus.10436.

Badowski, M.E., 2017. A review of oral cannabinoids and medical marijuana for the treatment of chemotherapy-induced nausea and vomiting: a focus on pharmacokinetic variability and pharmacodynamics. Cancer Chemother. Pharmacol. 1432-0843. 80 (3), 441–449. Available from: http://www.link.springer.de/link/service/journals/00280/index.htm. doi:10.1007/s00280-017-3387-5.

Barakji, J., Korang, S.K., Feinberg, J., Maagaard, M., Mathiesen, O., Gluud, C., Jakobsen, J.C., 2023. Cannabinoids versus placebo for pain: a systematic review with meta-analysis and trial sequential analysis. PLoS One. 1932-6203. 18 (1). Available from: https://journals.plos.org/plosone/article/file?id=10.1371/journal.pone.0267420&type=printable. doi:10.1371/journal.pone.0267420.

Barlowe, T.S., Koliani-Pace, J.L., Smith, K.D., Gordon, S.R., Gardner, T.B., 2019. Effects of medical cannabis on use of opioids and hospital visits by patients with painful chronic pancreatitis. Clin. Gastroenterol. Hepatol. 1542-7714. 17 (12), 2608. Available from: http://www.elsevier.com/inca/publications/store/6/7/2/7/4/3/index.htt. doi:10.1016/j.cgh.2019.01.018.

Bell, A.D., MacCallum, C., Margolese, S., Walsh, Z., Wright, P., Daeninck, P.J., Mandarino, E., Lacasse, G., Deol, J.K., de Freitas, L., Pierre, M.S., Belle-Isle, L., Gagnon, M., Bevan, S., Sanchez, T., Arlt, S., Monahan-Ellison, M., O'Hara, J., Boivin, M., Costiniuk, C., 2024. Clinical practice guidelines for cannabis and cannabinoid-based medicines in the management of chronic pain and co-occurring conditions. Cannabis Cannabinoid Res. 2378-8763. 9 (2), 669–687. Available from: http://www.liebertpub.com/overview/cannabis-and-cannabinoid-research/633/. doi:10.1089/can.2021.0156.

BfArM, 2022. Abschlussbericht Der Begleiterhebung Nach § 31 Absatz 6 Des Fünften Buches Sozialgesetzbuch Zur Verschreibung Und Anwendung von Cannabisarzneimitteln. Available from: https://www.bfarm.de/SharedDocs/Downloads/DE/Bundesopiumstelle/Cannabis/Abschlussbericht_Begleiterhebung.pdf (Accessed 1 June 2025).

Bialas, P., Fitzcharles, M.A., Klose, P., Häuser, W., 2022. Long-term observational studies with cannabis-based medicines for chronic non-cancer pain: a systematic review and meta-analysis of effectiveness and safety. Eur. J. Pain. 1532-2149. 26 (6), 1221–1233. Available from: http://onlinelibrary.wiley.com/journal/10.1002/(ISSN)1532-2149. doi:10.1002/ejp.1957.

Bilbao, A., Spanagel, R., 2022. Medical cannabinoids: a pharmacology-based systematic review and meta-analysis for all relevant medical indications. BMC Med. 1741-7015. 20 (1). doi:10.1186/s12916-022-02459-1.

Boehnke, K.F., Litinas, E., Clauw, D.J., 2016. Medical cannabis use is associated with decreased opiate medication use in a retrospective cross-sectional survey of patients with chronic pain. J. Pain. 1528-8447. 17 (6), 739–744. Available from: http://www.elsevier.com/inca/publications/store/6/2/3/1/4/7/index.htt. doi:10.1016/j.jpain.2016.03.002.

Boehnke, K.F., Wu, C.L., Clauw, D.J., 2024. Thoughtfully integrating cannabis products into chronic pain treatment. Anesth. Analg. 1526-7598. 138 (1), 5–15. Available from: http://journals.lww.com/anesthesia-analgesia/toc/publishahead. doi:10.1213/ANE.0000000000005904.

Bort, M., Olchowy, C., Olchowy, A., Nawrot-Hadzik, I., Smardz, J., Wieckiewicz, M., 2024. Efficiency of cannabis and cannabidiol in managing chronic pain syndromes: a comprehensive narrative review. Dent. Med. Probl. 2300-9020. 61 (5), 765–782. Available from: https://dmp.umw.edu.pl/pdf/2024/61/5/765.pdf. doi:10.17219/dmp/193020.

Bryk, M., Starowicz, K., 2021. Cannabinoid-based therapy as a future for joint degeneration. Focus on the role of CB2 receptor in the arthritis progression and pain: an updated review. Pharmacol. Rep. 1734-1140. 73 (3), 681–699. doi:10.1007/s43440-021-00270-y.

Byars, T., Theisen, E., Bolton, D.L., 2019. Using cannabis to treat cancer-related pain. Semin. Oncol. Nurs. 0749-2081. 35 (3), 300–309. Available from: http://www.us.elsevierhealth.com/product.jsp?isbn=07492081. doi:10.1016/j.soncn.2019.04.012.

Chung, M., Kim, H.K., Abdi, S., 2020. Update on cannabis and cannabinoids for cancer pain. Curr. Opin. Anaesthesiol. 1473-6500. 33 (6), 825–831. Available from: http://journals.lww.com/co-anesthesiology/pages/default.aspx. doi:10.1097/ACO.0000000000000934.

de Almeida, D.L., Devi, L.A., 2020. Diversity of molecular targets and signaling pathways for CBD. Pharmacol. Res. Perspect. 2052-1707. 8 (6). Available from: http://onlinelibrary.wiley.com/journal/10.1002/(ISSN)2052-1707. doi:10.1002/prp2.682.

Filippini, G., Tj, L., Dwan, K., D'Amico, R., Borrelli, F., AA, I., Minozzi, S., 2019. Cannabis and cannabinoids for people with multiple sclerosis. Cochrane Database Syst. Rev. doi:10.1002/14651858.CD013444.

Finn, D.P., Haroutounian, S., Hohmann, A.G., Krane, E., Soliman, N., Rice, A.S.C., 2021. Cannabinoids, the endocannabinoid system, and pain: a review of preclinical studies. Pain. 1872-6623. 162, S5. Available from: https://journals.lww.com/pain/Pages/currenttoc.aspx. doi:10.1097/j.pain.0000000000002268.

Fisher, E., Moore, R.A., Fogarty, A.E., Finn, D.P., Finnerup, N.B., Gilron, I., Haroutounian, S., Krane, E., Rice, A.S.C., Rowbotham, M., Wallace, M., Eccleston, C., 2021. Cannabinoids, cannabis, and cannabis-based medicine for pain management: a systematic review of randomised controlled trials. Pain. 1872-6623. 162, S45. Available from: https://journals.lww.com/pain/Pages/currenttoc.aspx. doi:10.1097/j.pain.0000000000001929.

Grimison, P., Mersiades, A., Kirby, A., Lintzeris, N., Morton, R., Haber, P., Olver, I., Walsh, A., McGregor, I., Cheung, Y., Tognela, A., Hahn, C., Briscoe, K., Aghmesheh, M., Fox, P., Abdi, E., Clarke, S., Della-Fiorentina, S., Shannon, J., Gedye, C., Begbie, S., Simes, J., Stockler, M., 2020. Oral THC:CBD cannabis extract for refractory chemotherapy-induced nausea and vomiting: a randomised, placebo-controlled, phase II crossover trial. Ann. Oncol. 0923-7534. 31 (11), 1553–1560. doi:10.1016/j.annonc.2020.07.020.

Hameed, M., Prasad, S., Jain, E., Dogrul, B.N., Al-Oleimat, A., Pokhrel, B., Chowdhury, S., Edzel Lorraine, C., Mitra, S., Quinonez, J., Ruxmohan, S., Stein, J., 2023. Medical cannabis for chronic nonmalignant pain management. Curr. Pain Headache Rep. 1531-3433. 27 (4), 57–63. doi:10.1007/s11916-023-01101-w.

Häuser, W., Finn, D.P., Kalso, E., Krcevski-Skvarc, N., Kress, H.G., Morlion, B., Perrot, S., Schäfer, M., Wells, C., Brill, S., 2018. European Pain Federation (EFIC) position paper on appropriate use of cannabis-based medicines and medical cannabis for chronic pain management. Eur. J. Pain. 1532-2149. 22 (9), 1547–1564. Available from: http://onlinelibrary.wiley.com/journal/10.1002/(ISSN)1532-2149. doi:10.1002/ejp.1297.

Karst, M., 2024. Overview: chronic pain and cannabis-based medicines. Pharmacopsychiatry. 0176-3679. 57 (03), 152–159. doi:10.1055/a-2231-6630.

Khurshid, H., Qureshi, I.A., Jahan, N., Went, T.R., Sultan, W., Sapkota, A., Alfonso, M., 2021. A systematic review of fibromyalgia and recent advancements in treatment: is medicinal cannabis a new hope? Cureus. 2168-8184. 13 (8). doi:10.7759/cureus.17332.

Lavie-Ajayi, M., Shvartzman, P., 2019. Restored self: a phenomenological study of pain relief by cannabis. Pain Med. 1526-2375. 20 (11), 2086–2093. doi:10.1093/pm/pny176.

Maher, D.P., Carr, D.B., Hill, K., McGeeney, B., Weed, V., Jackson, W.C., Dibenedetto, D.J., Moriarty, E.M., Kulich, R.J., 2019. Cannabis for the treatment of chronic pain in the era of an opioid epidemic: a symposium-based review of sociomedical science. Pain Med. 1526-4637. 20 (11), 2311–2323. Available from: http://oxfordjournals.org/en/press/announcements-from-oup/pain-medicine.html. doi:10.1093/pm/pnx143.

McDonagh, M.S., Morasco, B.J., Wagner, J., Ahmed, A.Y., Fu, R., Kansagara, D., Chou, R., 2022. Cannabis-based products for chronic pain. Ann. Intern. Med. 0003-4819. 175 (8), 1143–1153. doi:10.7326/m21-4520.

Mlost, J., Bryk, M., Starowicz, K., 2020. Cannabidiol for pain treatment: focus on pharmacology and mechanism of action. Int. J. Mol. Sci. 1422-0067. 21 (22), 1–22. Available from: https://www.mdpi.com/1422-0067/21/22/8870/pdf. doi:10.3390/ijms21228870.

Pagano, C., Navarra, G., Coppola, L., Avilia, G., Bifulco, M., Laezza, C., 2022. Cannabinoids: therapeutic use in clinical practice. Int. J. Mol. Sci. 1422-0067. 23 (6), 3344. doi:10.3390/ijms23063344.

Paice, J.A., Portenoy, R., Lacchetti, C., Campbell, T., Cheville, A., Citron, M., Constine, L.S., Cooper, A., Glare, P., Keefe, F., Koyyalagunta, L., Levy, M., Miaskowski, C., Otis-Green, S., Sloan, P., Bruera, E., 2016. Management of chronic pain in survivors of adult cancers: american Society of Clinical Oncology clinical practice guideline. J. Clin. Oncol. 1527-7755. 34 (27), 3325–3345. Available from: http://jco.ascopubs.org/content/34/27/3325.full.pdf+html. doi:10.1200/JCO.2016.68.5206.

Pantoja-Ruiz, C., Restrepo-Jimenez, P., Castañeda-Cardona, C., Ferreirós, A., Rosselli, D., 2022. Cannabis and pain: a scoping review. Braz. J. Anesthesiol. 0104-0014. 72 (1), 142–151. doi:10.1016/j.bjane.2021.06.018.

Rog, D., Nurmikko, T., Young, C., 2007. Oromucosal Δ9-tetrahydrocannabinol/cannabidiol for neuropathic pain associated with multiple sclerosis: an uncontrolled, open-label, 2-year extension trial. Clin. Ther. 0149-2918. 29 (9), 2068–2079. doi:10.1016/j.clinthera.2007.09.013.

Rosic, T., Kapoor, R., Panesar, B., Naji, L., Chai, D.B., Sanger, N., Marsh, D.C., Worster, A., Thabane, L., Samaan, Z., 2021. The association between cannabis use and outcome in pharmacological treatment for opioid use disorder. Harm Reduct. J. 1477-7517. 18 (1). doi:10.1186/s12954-021-00468-6.

Russo, E.B., 2019. Cannabis and pain. Pain Med. 1526-4637. 20 (11), 2083–2085. Available from: http://oxfordjournals.org/en/press/announcements-from-oup/pain-medicine.html. doi:10.1093/pm/pnz227.

Sánchez-Flórez, J.C., Seija-Butnaru, D., Valero, E.G., Acosta, C.D.P.A., Amaya, S., 2022. Pain management strategies in rheumatoid arthritis: a narrative review. J. Pain Palliat. Care Pharmacother. 1536-0539. 35 (4), 291–299. Available from: https://www.tandfonline.com/loi/ippc20. doi:10.1080/15360288.2021.1973647.

Seevathee, K., Kessomboon, P., Manimmanakorn, N., Luangphimai, S., Thaneerat, T., Wanaratna, K., Plengphanich, S., Thaenkham, T., Sena, W., 2024. Efficacy and safety of transdermal medical cannabis (THC:CBD:CBN formula) to treat painful diabetic peripheral neuropathy of lower extremities. Med. Cannabis Cannabinoids. 2504-3889. 8 (1), 1–14. Available from: https://www.karger.com/Journal/Home/275986. doi:10.1159/000542511.

Shustorovich, A., Corroon, J., Wallace, M.S., Sexton, M., 2024. Biphasic effects of cannabis and cannabinoid therapy on pain severity, anxiety, and sleep disturbance: a scoping review. Pain Med. 1526-4637. 25 (6), 387–399. Available from: http://oxfordjournals.org/en/press/announcements-from-oup/pain-medicine.html. doi:10.1093/pm/pnae004.

Sinclair, J., Collett, L., Abbott, J., Pate, D.W., Sarris, J., Armour, M., 2021. Effects of cannabis ingestion on endometriosis-associated pelvic pain and related symptoms. PLoS One. 1932-6203. 16 (10). Available from: https://journals.plos.org/plosone/article/file?id=10.1371/journal.pone.0258940&type=printable. doi:10.1371/journal.pone.0258940.

Sokolaj, E., Assareh, N., Anderson, K., Aubrey, K.R., Vaughan, C.W., 2024. Cannabis constituents for chronic neuropathic pain; reconciling the clinical and animal evidence. J. Neurochem. 1471-4159. 168 (11), 3685–3698. Available from: http://onlinelibrary.wiley.com/journal/10.1111/(ISSN)1471-4159. doi:10.1111/jnc.15964.

Solmi, M., De Toffol, M., Kim, J.Y., Choi, M.J., Stubbs, B., Thompson, T., Firth, J., Miola, A., Croatto, G., Baggio, F., Michelon, S., Ballan, L., Gerdle, B., Monaco, F., Simonato, P., Scocco, P., Ricca, V., Castellini, G., Fornaro, M., Murru, A., Vieta, E., Fusar-Poli, P., Barbui, C., Ioannidis, J.P.A., Carvalho, A.F., Radua, J., Correll, C.U., Cortese, S., Murray, R.M., Castle, D., Shin, J.I., Dragioti, E., 2023. Balancing risks and benefits of cannabis use: umbrella review of meta-analyses of randomised controlled trials and observational studies. BMJ. 1756-1833. 382, e072348. doi:10.1136/bmj-2022-072348.

Sukpiriyagul, A., Chartchaiyarerk, R., Tabtipwon, P., Smanchat, B., Prommas, S., Bhamarapravatana, K., Suwannarurk, K., 2023. Oral tetrahydrocannabinol (THC): cannabinoid (CBD) cannabis extract adjuvant for reducing chemotherapy-induced nausea and vomiting (CINV): a randomized, double-blinded, placebo-controlled, crossover trial. Int. J. Womens Health. 1179-1411. 15, 1345–1352. Available from: https://www.dovepress.com/getfile.php?fileID=92011. doi:10.2147/IJWH.S401938.

Ueberall, M.A., Essner, U., Silván, C.V., Mueller-Schwefe, G.H.H., 2022. Comparison of the effectiveness and tolerability of nabiximols (THC:CBD) oromucosal spray versus oral dronabinol (THC) as add-on treatment for severe neuropathic pain in real-world clinical practice: retrospective analysis of the German pain e-registry. J. Pain Res. 1178-7090. 15, 267–286. Available from: https://www.dovepress.com/comparison-of-the-effectiveness-and-tolerability-of-nabiximols-thccbd–peer-reviewed-fulltext-article-JPR. doi:10.2147/JPR.S340968.

Vase, L., Wartolowska, K., 2019. Pain, placebo, and test of treatment efficacy: a narrative review. Br. J. Anaesth. 0007-0912. 123 (2), e254. doi:10.1016/j.bja.2019.01.040.

Vigil, J.M., Stith, S.S., Adams, I.M., Reeve, A.P., 2017. Associations between medical cannabis and prescription opioid use in chronic pain patients: a preliminary cohort study. PLoS One. 1932-6203. 12 (11). Available from: http://journals.plos.org/plosone/article/file?id=10.1371/journal.pone.0187795&type=printable. doi:10.1371/journal.pone.0187795.

Wang, J., Wang, Y., Tong, M., Pan, H., Li, D., 2019. Medical cannabinoids for cancer cachexia: a systematic review and meta-analysis. Biomed. Res. Int. 2314-6133. 2019, 1–6. doi:10.1155/2019/2864384.

Wang, L., Hong, P.J., May, C., Rehman, Y., Oparin, Y., Hong, C.J., Hong, B.Y., Aminilari, M., Gallo, L., Kaushal, A., Craigie, S., Couban, R.J., Kum, E., Shanthanna, H., Price, I., Upadhye, S., Ware, M.A., Campbell, F., Buchbinder, R., Agoritsas, T., Busse, J.W., 2021. Medical cannabis or cannabinoids for chronic non-cancer and cancer related pain: a systematic review and meta-analysis of randomised clinical trials. BMJ. 1756-1833. 374. Available from: http://www.bmj.com/. doi:10.1136/bmj.n1034.

Warr, D., Hesketh, P., 2020. Cannabinoids as antiemetics: everything that's old is new again. Ann. Oncol. 0923-7534. 31 (11), 1425–1426. doi:10.1016/j.annonc.2020.08.2104.

CHAPTER

13

Proposed recommendations for the use of cannabinoids for chronic pain management

Danielle Kohr and Michael Kugler

Department of Pain Medicine, Fachklinik Löwenstein, SLK Kliniken Heilbronn GmbH, Löwenstein, Germany

Introduction

The use of cannabis in clinical practice has undergone significant transformation, particularly following its legalization for medical purposes. This shift, combined with evolving treatment guidelines and expanded insurance coverage in some countries, has positioned cannabis as an increasingly viable option for the management of chronic pain and other serious medical conditions.

The following recommendations are grounded in the existing scientific literature, extensive data from our specialized pain center, and findings from the large observational study conducted by the Federal Institute for Drugs and Medical Devices (BfArM) in Germany (*Abschlussbericht Der Begleiterhebung Nach § 31 Absatz 6 Des Fünften Buches Sozialgesetzbuch Zur Verschreibung Und Anwendung von Cannabisarzneimitteln*, 2022). These recommendations are further bolstered by over a decade of personal clinical experience, involving more 5000 patients treated with cannabis-based oils. Many of these patients were longitudinally monitored, providing valuable insights into the long-term effectiveness, tolerability, and practical considerations of cannabinoid-based therapy.

Indications for the use of cannabis-based medicines

Indications for the use of cannabis-based medicines include all pain patients and conditions that would generally qualify for combination therapy involving nonopioid analgesics, opioids, anticonvulsants, and antidepressants. Cannabis is rarely used as a first-line agent. As a guiding principle, these conventional treatment options should

Cannabinoids for Chronic Pain. https://doi.org/10.1016/B978-0-443-36773-1.00008-8

have been thoroughly explored, and the patient must be considered pharmacologically treatment-refractory before cannabinoid therapy is initiated (Bell et al., 2024; MacCallum and Russo, 2018; Häuser et al., 2018). Assessing and documenting the response—or lack thereof—is essential and recommended for standard, approved medications.

This applies particularly to neuropathic pain syndromes—such as phantom limb pain, mononeuropathies (e.g., intercostal neuralgia or postthoracotomy syndrome), trigeminal neuralgia, postherpetic neuralgia, neuropathic pain following surgical procedures, polyneuropathy, or central poststroke pain (Häuser et al., 2018). In patients who are resistant to conventional analgesics and coanalgesics, cannabis represents a viable therapeutic option.

Cannabis-based medicines are also considered appropriate for the treatment of mixed pain syndromes and musculoskeletal conditions, particularly when standard therapies have been ineffective or poorly tolerated (Bell et al., 2024; Bialas et al., 2022; Khurshid et al., 2021). This includes complex disorders such as persistent spinal pain syndrome (PSPS Type 1 and Type 2) and fibromyalgia, where both nociceptive and neuropathic components are present. Patients with these conditions often experience widespread pain, fatigue, sleep disturbances, and mood disorders, all of which may be positively influenced by cannabinoid therapy (Bell et al., 2024; Bourke et al., 2022).

Another specific indication is spastic or cramping pain associated with neurological conditions (Häuser et al., 2018; Boehnke et al., 2024; Fisher et al., 2021) such as multiple sclerosis and poststroke spasticity. In these conditions, cannabinoids—particularly those in the form of balanced tetrahydrocannabinol (THC)/cannabidiol (CBD) formulations—have shown efficacy in reducing muscle tone, alleviating spasms, and relieving associated pain (Fisher et al., 2021).

Furthermore, cannabis can serve as a supportive ("roborant") medication to counteract anorexia, cachexia, and nausea or vomiting (Badowski, 2017; Warr and Hesketh, 2020)—particularly when these symptoms are present in the context of cancer-related pain (Sukpiriyagul et al., 2023; Grimison et al., 2020).

Cannabinoid-based therapies may offer significant benefits for patients suffering from pain-related sleep disturbances, posttraumatic stress disorder (PTSD), anxiety, and depression—particularly when conventional treatments have proven insufficient or poorly tolerated (Bialas et al., 2022; Legare et al., 2022; Poli et al., 2018; Ware et al., 2010). Clinical evidence and patient-reported outcomes suggest that cannabis-based medicines can improve sleep quality by reducing pain intensity and nocturnal awakenings (Bell et al., 2024; Boehnke et al., 2024). In patients with PTSD, cannabinoids—especially those with balanced THC/CBD ratios—have demonstrated the potential to alleviate nightmares, hyperarousal, and intrusive symptoms (Jetly et al., 2015; Betthauser et al., 2015). Furthermore, anxiety and depressive symptoms, which often coexist with chronic pain or trauma-related disorders, may be positively influenced by carefully titrated cannabinoid treatments.

Primary headache disorders, such as migraine, tension-type headache, or cluster headache, are generally not suitable for monotherapy with cannabis. However, adjunctive use in combination with standard treatments—such as calcitonin gene-related peptide (CGRP) antibody therapy for migraine—may be beneficial (Bell et al., 2024; Häuser et al., 2018). In contrast, secondary headache disorders, such as those following traumatic brain injury, may justify a therapeutic trial with cannabis-based treatments.

Other pain syndromes with poor response to guideline-based pharmacologic therapy—including abdominal and pelvic pain (e.g., Crohn's disease, endometriosis, interstitial cystitis, coccygodynia)—may also benefit from cannabinoid therapy (Häuser et al., 2018; Sinclair et al., 2021; Bort et al., 2024; Solmi et al., 2023).

Another indication of full-spectrum cannabis is its potential therapeutic application in epilepsy (Solmi et al., 2023). There have been case reports of increased seizure susceptibility with THC, but for certain types of epilepsy, CBD has been shown to reduce seizure frequency (Laux et al., 2019; Szaflarski et al., 2018). Therefore the addition of CBD in full-spectrum cannabis may help mitigate this risk and provide therapeutic benefits in such cases.

Finally, patients with rheumatoid arthritis and other inflammatory conditions, such as Crohn's disease or ulcerative colitis, may respond positively to cannabis treatment (Bell et al., 2024; Bort et al., 2024). These effects may include symptom relief, potential enhancement of antiinflammatory actions (Bryk and Starowicz, 2021) from baseline immunosuppressive therapy, and a reduced need for coanalgesics and their associated side effects.

Cannabis and its active compounds

Cannabis plants contain a variety of active compounds known as cannabinoids. In medical applications, two primary cannabinoids derived from the plant are commonly used:

- THC—sometimes referred to as dronabinol; however, it is essential to note that dronabinol is the synthetic form of THC
- CBD

Both THC and CBD contribute significantly to the regulation of pain, but their roles differ (Anthony et al., 2020). While THC is primarily effective in managing severe pain, especially in the context of spasticity or cancer pain, CBD offers a complementary role in reducing inflammation, improving muscle relaxation, and minimizing anxiety or cognitive side effects (Bryk and Starowicz, 2021; Finn et al., 2021). For patients with chronic or complex pain syndromes, combining THC and CBD offers a balanced approach that can maximize therapeutic effects while minimizing undesirable side effects (Pagano et al., 2022; de Almeida and Devi, 2020).

In clinical practice, the optimal use of THC and CBD should be guided by individual patient needs, the nature of the pain being treated, and the patient's tolerance to psychoactive effects. The dosage is initiated at a low level and titrated slowly, adhering to the "start low, go slow, and stay low" approach, adjusted to the patient's therapeutic response and tolerability. Regular monitoring and adjustment of dosages based on patient response are critical to ensuring the best outcomes in pain management.

THC and CBD have varied and sometimes distinct effects, which make them suitable for different therapeutic purposes.

- THC is primarily responsible for the analgesic effects. It acts as a mild muscle relaxant, stimulates appetite, and can alleviate nausea and vomiting. However, it is a psychoactive substance, meaning it influences the central nervous system. THC affects the brain, alters consciousness and mood, and can induce central nervous system side effects such as dizziness, drowsiness, and cognitive disturbances like memory and concentration difficulties (Warr and Hesketh, 2020; Anthony et al., 2020; Finn et al., 2021; Bilbao and Spanagel, 2022).
- CBD has a mild analgesic effect, but it is generally weaker in pain relief compared to THC. However, its muscle-relaxing and calming effects are somewhat more potent than THC's. Unlike THC, CBD is nonpsychoactive and can reduce the CNS side effects of THC, helping to alleviate unwanted effects. In addition, CBD can stimulate appetite, alleviate nausea and vomiting, has anxiolytic (anxiety-reducing) properties, and possesses antiinflammatory effects. It can also help modulate immune system overactivity and has a neuroprotective effect, potentially protecting nerve cells (Bryk and Starowicz, 2021; Mlost et al., 2020; Anand et al., 2021; Schubart et al., 2011).

Medical cannabis preparations typically contain either THC alone or a combination of THC and CBD. They are available in a wide range of formulations and routes of administration, including smoking, vaping, oral capsules, oils, and oromucosal sprays. Both active compounds can be derived from the female cannabis plant or synthetically produced. Oily solutions offer the advantage of longer duration of action, ease of application, and availability (Russo, 2019). The therapeutic efficacy in pain management is primarily attributed to THC-containing preparations, which are considered to provide the predominant analgesic effect (Bell et al., 2024). While CBD contributes minimally to pain relief, its combined use with THC is typically indicated when CNS side effects associated with THC arise, as CBD can help mitigate these effects. In addition, the presence of epilepsy is another indication for the use of a combination formulation containing both THC and CBD (Laux et al., 2019; Szaflarski et al., 2018).

The recommendation to initiate treatment with THC is often based on its potent analgesic properties. THC is commonly recommended for effective relief of conditions such as severe chronic pain or cancer-related discomfort (Bell et al., 2024). In cases of spasticity, a combination of THC and CBD is often preferred, as THC helps reduce muscle spasms and pain. At the same time, CBD enhances the effects by promoting muscle relaxation and reducing inflammation. This combination offers a more comprehensive approach to managing spasticity, addressing both pain and underlying muscle tension, and leading to better therapeutic outcomes compared to THC alone (Bell et al., 2024).

If CNS side effects, such as dizziness, cognitive impairment, or anxiety, arise during THC therapy, it is advisable to introduce CBD as a complementary treatment. As a nonpsychoactive compound, CBD can help mitigate the adverse effects of THC, particularly those related to mood disturbances, memory issues, and other cognitive disruptions (Schubart et al., 2011; Freeman et al., 2019).

The combined use of THC and CBD also offers a more holistic treatment approach, targeting pain and spasticity while also providing additional benefits, such as reducing inflammation and improving mood stability and sleep (Bell et al., 2024).

In addition to THC and CBD, cannabis plants contain a variety of other compounds, including terpenes. Terpenes are secondary plant metabolites that are also extracted for use in full-spectrum medical cannabis products. However, the composition of these substances in cannabis extracts is often inconsistent, making their medically meaningful application difficult in most cases.

Although terpenes can interact with THC and CBD, their concentrations in full-spectrum cannabis oils are typically low. Moreover, the exact composition and quantity of individual terpenes are usually unknown. Both the type and amount of terpenes can vary significantly, mainly depending on the cannabis strain as well as the methods of cultivation, drying, and processing (Anand et al., 2021; Russo, 2011).

Cannabis can also be administered via inhalation, which requires a specialized vaporizer. Due to its high heat stability, the active compounds in cannabis are only released at elevated temperatures. For inhalation, dried cannabis flowers are typically utilized (Ware et al., 2010).

Inhaled medical cannabis offers a rapid onset of action, but the effects are relatively short-lived. As a result, repeated administration (5–6 times per day) is required, which can be cumbersome and impractical. In addition, the availability of medical cannabis for inhalation remains limited at present.

The rapid onset of action associated with inhaled cannabis often leads to central nervous system side effects, such as concentration difficulties, drowsiness, and dizziness (Legare et al., 2022). However, compared to other forms of cannabis administration, such as oils, tinctures, capsules, or oral sprays, inhaled cannabis generally appears to be better tolerated in terms of side effects. Nevertheless, inhalation is often less suitable for long-term treatment (Russo, 2019).

It is important to note that smoking cannabis flowers is not recommended in medical treatments. Smoking can produce harmful combustion byproducts, and it isn't easy to control the dosage and therapeutic effects effectively (MacCallum and Russo, 2018; Häuser et al., 2018; Carter et al., 2004).

Before initiating treatment with medical cannabis, the standard treatments recommended in clinical guidelines should always be exhausted (Bell et al., 2024; Häuser et al., 2018). These treatments typically include—depending on the condition—nonopioid analgesics, weak and strong opioids, antidepressants as adjunctive therapy (e.g., for burning neuropathic pain), anticonvulsants for shooting nerve pain, or medications for nausea and vomiting.

The use of cannabis is not an alternative to these treatments; it should only be considered when other therapies have proven insufficient or as a substitute when certain medications cannot be used.

Adverse events

The various cannabinoids can interact with a wide range of medications, including antihypertensives, anticoagulants, antibiotics, pain relievers such as aspirin and ibuprofen, sedatives, and antidepressants. These interactions can lead to undesirable side effects, as well as potential amplification or reduction of the impact of both the cannabinoids and the other medications (Bell et al., 2024; MacCallum and Russo, 2018; Häuser et al., 2018; Carter et al., 2004).

Side effects of medications occur with varying frequencies: very common, i.e., in more than 1 in 10 patients (>10%); common, in up to 1 in 10 patients (1–10%); occasional, in up to 1 in 100 patients (0.1–1%); rare, in up to 1 in 1000 patients (0.01–0.1%); very rare, in up to 1 in 10,000 patients (<0.01%).

Despite all precautions, complications can arise. Preexisting conditions, comorbidities, and individual characteristics may influence the frequency of side effects.

Medical cannabis is generally considered to be better tolerated than other medications. However, as with all drugs, adverse effects are possible (Bell et al., 2024). Below, the most well-known side effects are listed, though other side effects, possibly yet to be discovered, cannot be ruled out.

General side effects

- Very common: Dry mouth and fatigue.

CNS side effects

- As medical cannabis affects the CNS, the following side effects are frequently reported:
  - Dizziness, balance issues, and impaired coordination lead to an increased risk of falls.
  - Drowsiness, sleepiness, concentration and memory disturbances, confusion, disorientation, and memory impairment (e.g., amnesia).
  - Changes in perception, such as the sensation of being detached from one's body, thoughts, or surroundings (depersonalization, dissociation).
  - Sensory distortions (e.g., hallucinations) and paranoid reactions (e.g., excessive mistrust, delusions of persecution).
  - Mood swings, such as euphoria or depression, may occur. Occasionally, suicidal thoughts may arise.
  - Nervousness, anxiety, headaches, and blurred vision.

- Occasionally, cannabis may impair the function of muscles involved in speech, leading to speech difficulties such as slurred or indistinct speech.

Effects on heart function and circulation

- Occasionally, an irregular or rapid heartbeat may occur, perceived as palpitations or tachycardia (Holt et al., 2024). This also carries a risk of heart attack. In patients with preexisting cardiac conditions, a cardiological evaluation should be conducted to assess contraindications for cannabis therapy (see contraindications) (Richards et al., 2019).
- Blood pressure may occasionally become too low or too high, and blood vessels may dilate, leading to facial flushing, for example (Goyal et al., 2017).

Effects on the digestive system

- Common side effects include diarrhea, increased appetite, and weight gain. Occasional side effects include abdominal pain, nausea, and vomiting.

Allergic reactions

- Cannabis-based medicines may, in some cases, trigger allergic reactions (e.g., itching, rash, facial redness, watery eyes, respiratory difficulties). Severe reactions, such as circulatory shock, are rare.

Additional notes

- Side effects are more likely to occur at the beginning of treatment and are typically transient. Permanent side effects are infrequent. The medication should be discontinued if side effects, such as diarrhea or itching, persist.
- Given that cannabis medications are used in low doses tailored to the individual's needs and symptoms, the risk of developing dependency with correct medical usage is minimal. Although there is an inherent risk of addiction, it is very low due to the slow onset of effects, the form of administration, and close medical supervision.

When should medical cannabis not be administered?

Medical cannabis should not be prescribed in the following circumstances:

- In patients with coronary artery disease or other severe cardiovascular conditions
- In individuals with schizophrenia or other severe psychiatric disorders, such as psychosis or depression

- In patients with a known allergy to cannabis or any other components of cannabis-based medications, caution should be exercised
- During pregnancy
- While breastfeeding

Clinical recommendations and dosing strategies (adults)

The best way to reduce potential adverse effects associated with THC use is through a safe, low, and gradual initiation and titration process. We recommend the following initiation plan:

- Synthetic THC per day:
  - Week 1:2 × 2.5 mg
  - Week 2:2 × 5 mg
  - Week 3:3 × 5 mg
- Full-Spectrum Cannabis (THC: CBD 10:10):
  - Week 1:2 × 0.25 mL (2.5 mg THC + 2.5 mg CBD)
  - Week 2:2 × 0.5 mL (5 mg THC + 5 mg CBD)

By the end of the dosing phase, an approximately 80% improvement in symptoms should be achieved, accompanied by acceptable tolerability, primarily regarding central nervous system effects. If, after approximately 4 weeks, symptom improvement is less than 50%, the treatment will likely be insufficient in the long term. Typically, patients are initiated on cannabis-based medications as part of an inpatient multimodal pain management approach, and it is understandable that, under daily life conditions, an initial improvement may eventually stabilize at placebo-level effectiveness.

Increasing the dose beyond this point rarely results in substantial improvement, but higher doses can be attempted if no side effects are present. For synthetic THC, the dose can be increased up to 3 × 7.5 mg, and for full-spectrum cannabis, up to 2 × 1.0 mL (10 mg THC + 10 mg CBD).

In rare and particularly severe cases, clinically relevant symptom improvement has been observed only at daily THC doses exceeding 30 mg. However, elevated dosages are associated with a substantially increased risk of adverse effects on the central nervous system.

Patients, particularly those with comorbidities, can be started on synthetic THC as part of a hospital-based multimodal pain management regimen. During inpatient treatment, cannabis dosage should not only be carefully titrated, but the withdrawal of potentially addictive medications—such as opioids and pregabalin—should also be initiated, as dependency significantly reduces the likelihood of therapeutic success. This observation aligns with our clinical experience in combination therapies involving neurostimulation procedures and concurrent opioid use. Before initiating a neurostimulation trial, any ongoing opioid therapy should be gradually tapered and ultimately discontinued.

It is advisable to begin treatment with synthetic THC solutions, as it appears logical to initially attempt monotherapy. In addition, synthetic THC solutions are cost-effective and readily accessible. Interestingly, based on our clinical experience, patients with multiple sclerosis often respond better to pure THC-containing preparations, despite literature indicating that full-spectrum cannabis products are generally more effective and better tolerated. It is our view that a transition to full-spectrum cannabis should be considered only if CNS side effects occur or in the presence of epilepsy. In cases where the therapeutic response is insufficient or the effect diminishes during outpatient treatment, patients could opt to trial full-spectrum cannabis, if desired.

In our experience, the synthetic THC solution shows moderate effectiveness in inflammatory systemic diseases like rheumatoid arthritis. However, its effectiveness is generally limited. Full-spectrum cannabis products should be considered in cases where the synthetic THC solution fails to provide sufficient relief.

In geriatric patients (>70 years), a slower titration and a maximum dose of 3 × 2.5 mg are generally more effective than higher doses. This is likely because lower doses produce fewer adverse effects, enabling longer treatment durations and enhancing therapeutic efficacy. Interestingly, higher doses may sometimes exacerbate pain or fail to provide the same level of pain relief as lower doses.

Suppose a patient shows improvement after 4 weeks but experiences CNS side effects following the above titration scheme. In that case, reducing the morning and afternoon doses is advisable while potentially increasing the evening dose. For example, the regimen could be adjusted to 2.5 mg in the morning, 2.5 mg at midday, and 7.5 mg in the evening. In the case of full-spectrum cannabis products, the morning dose can be divided into two separate administrations—one in the morning and one at midday—while maintaining the evening dose (0.25 mL in the morning, 0.25 mL at midday, and 0.5 mL in the evening). This adjustment is expected to reduce daytime side effects and improve nocturnal sleep quality.

In patients with fibromyalgia, pain frequently diminishes during warmer months, and, in such cases, patients may reduce their daily dose, with some even opting to pause therapy altogether.

Combination therapies with additional analgesics and coanalgesics may be necessary in specific clinical situations. This is particularly true for patients suffering from severe cancer pain, who typically require a combination of opioids and cannabis-based preparations. Likewise, in neuropathic pain conditions such as polyneuropathy or thalamic pain, combination therapy is often essential. Positive outcomes have been observed with the combination of the anticonvulsant gabapentin and cannabis preparations, where both agents, administered at lower doses, provide effective relief. However, for patients experiencing severe cancer pain, high-dose combination therapy with a focus on opioids is required.

In cases of primary headache disorders, combination therapy may also be effective; however, conventional pain management must be established before the introduction of cannabis-based treatments. In patients with migraine, for instance, the combination of CGRP antibodies and cannabis may result in near-complete cessation of attacks.

Should other adverse events occur, such as heart palpitations, diarrhea, or dry mouth, which are not dose-dependent, cannabis therapy should be discontinued. Diarrhea represents a challenging side effect, as it may not manifest until weeks or months into therapy, and it often coincides with significant analgesic effects.

Concluding remarks

Cannabis-based medicines have emerged as a valuable addition to the therapeutic arsenal in modern pain medicine. Their potential to address a wide range of complex and often treatment-resistant pain conditions—ranging from neuropathic and musculoskeletal pain to spasticity and pain-related sleep disturbances—makes them an essential option within individualized, multimodal pain management strategies. While further high-quality studies are warranted, the available evidence, registry data, and extensive clinical experience support the judicious use of cannabinoids, particularly when standard therapies have been exhausted or are poorly tolerated. A responsible, well-monitored approach that considers indication, product selection, dosage, and patient-specific factors is essential to ensure safety and therapeutic success. As the field continues to evolve, cannabinoid therapy is likely to play an increasingly defined role in evidence-based pain care.

AI disclosure

During the preparation of this work, the author(s) used ChatGPT to improve readability, grammar, and clarity of language. After using this tool/service, the author(s) reviewed and edited the content as needed and take(s) full responsibility for the content of the publication.

References

Abschlussbericht Der Begleiterhebung Nach § 31 Absatz 6 Des Fünften Buches Sozialgesetzbuch Zur Verschreibung Und Anwendung von Cannabisarzneimitteln, 2022. Available from: https://www.bfarm.de/SharedDocs/Downloads/DE/Bundesopiumstelle/Cannabis/Abschlussbericht_Begleiterhebung.pdf.

Anand, U., Pacchetti, B., Anand, P., Sodergren, M.H., 2021. Cannabis-based medicines and pain: a review of potential synergistic and entourage effects. Pain Manag. 17581877. 11 (4), 395–403. Available from: http://www.futuremedicine.com/loi/pmt. doi:10.2217/pmt-2020-0110.

Anthony, A.T., Rahmat, S., Sangle, P., Sandhu, O., Khan, S., 2020. Cannabinoid receptors and their relationship with chronic pain: a narrative review. Cureus. 2168-8184. 12 (9). doi:10.7759/cureus.10436.

Badowski, M.E., 2017. A review of oral cannabinoids and medical marijuana for the treatment of chemotherapy-induced nausea and vomiting: a focus on pharmacokinetic variability and pharmacodynamics. Cancer Chemother. Pharmacol. 14320843. 80 (3), 441–449. Available from: https://link.springer.de/link/service/journals/00280/index.htm. doi:10.1007/s00280-017-3387-5.

Bell, A.D., MacCallum, C., Margolese, S., Walsh, Z., Wright, P., Daeninck, P.J., Mandarino, E., Lacasse, G., Deol, J.K., de Freitas, L., Pierre, M.S., Belle-Isle, L., Gagnon, M., Bevan, S., Sanchez, T., Arlt, S., Monahan-Ellison, M., O'Hara, J., Boivin, M., Costiniuk, C., 2024. Clinical practice guidelines for cannabis and cannabinoid-based medicines in the management of chronic pain and co-occurring conditions. Cannabis Cannabinoid Res. 23788763. 9 (2), 669–687. Available from: http://www.liebertpub.com/overview/cannabis-and-cannabinoid-research/633/. doi:10.1089/can.2021.0156.

Betthauser, K., Pilz, J., Vollmer, L.E., 2015. Use and effects of cannabinoids in military veterans with posttraumatic stress disorder. Am. J. Health Syst. Pharm. 15352900. 72 (15), 1279–1284. Available from: http://www.ajhp.org/content/72/15.toc. doi:10.2146/ajhp140523.

Bialas, P., Fitzcharles, M.A., Klose, P., Häuser, W., 2022. Long-term observational studies with cannabis-based medicines for chronic non-cancer pain: a systematic review and meta-analysis of effectiveness and safety. Eur. J. Pain. 15322149. 26 (6), 1221–1233. Available from: http://onlinelibrary.wiley.com/journal/10.1002/(ISSN)1532-2149. doi:10.1002/ejp.1957.

Bilbao, A., Spanagel, R., 2022. Medical cannabinoids: a pharmacology-based systematic review and meta-analysis for all relevant medical indications. BMC Med. 1741-7015. 20 (1). doi:10.1186/s12916-022-02459-1.

Boehnke, K.F., Wu, C.L., Clauw, D.J., 2024. Thoughtfully integrating cannabis products into chronic pain treatment. Anesth. Analg. 15267598. 138 (1), 5–15. Available from: http://journals.lww.com/anesthesia-analgesia/toc/publishahead. doi:10.1213/ANE.0000000000005904.

Bort, M., Olchowy, C., Olchowy, A., Nawrot-Hadzik, I., Smardz, J., Wieckiewicz, M., 2024. Efficiency of cannabis and cannabidiol in managing chronic pain syndromes: a comprehensive narrative review. Dent. Med. Probl. 23009020. 61 (5), 765–782. Available from: https://dmp.umw.edu.pl/pdf/2024/61/5/765.pdf. doi:10.17219/dmp/193020.

Bourke, S.L., Schlag, A.K., O'Sullivan, S.E., Nutt, D.J., Finn, D.P., 2022. Cannabinoids and the endocannabinoid system in fibromyalgia: a review of preclinical and clinical research. Pharmacol. Ther. 1879016X. 240. Available from: www.elsevier.com/locate/pharmthera. doi:10.1016/j.pharmthera.2022.108216.

Bryk, M., Starowicz, K., 2021. Cannabinoid-based therapy as a future for joint degeneration. Focus on the role of CB2 receptor in the arthritis progression and pain: an updated review. Pharmacol. Rep. 1734-1140. 73 (3), 681–699. doi:10.1007/s43440-021-00270-y.

Carter, G.T., Weydt, P., Kyashna-Tocha, M., Abrams, D.I., Francisco, S., 2004. Medicinal Cannabis: Rational Guidelines for Dosing. The Cyber Anthropology Institute, Seattle WA, USA.

de Almeida, D.L., Devi, L.A., 2020. Diversity of molecular targets and signaling pathways for CBD. Pharmacol. Res. Perspect. 20521707. 8 (6). Available from: http://onlinelibrary.wiley.com/journal/10.1002/(ISSN)2052-1707. doi:10.1002/prp2.682.

Finn, D.P., Haroutounian, S., Hohmann, A.G., Krane, E., Soliman, N., Rice, A.S.C., 2021. Cannabinoids, the endocannabinoid system, and pain: a review of preclinical studies. Pain. 18726623. 162, S5. Available from: https://journals.lww.com/pain/Pages/currenttoc.aspx. doi:10.1097/j.pain.0000000000002268.

Fisher, E., Moore, R.A., Fogarty, A.E., Finn, D.P., Finnerup, N.B., Gilron, I., Haroutounian, S., Krane, E., Rice, A.S.C., Rowbotham, M., Wallace, M., Eccleston, C., 2021. Cannabinoids, cannabis, and cannabis-based medicine for pain management: a systematic review of randomised controlled trials. Pain. 18726623. 162, S45. Available from: https://journals.lww.com/pain/Pages/currenttoc.aspx. doi:10.1097/j.pain.0000000000001929.

Freeman, A.M., Petrilli, K., Lees, R., 2019. How Does cannabidiol (CBD) influence the acute effects of delta-9-tetrahydrocannabinol (THC) in humans? A systematic review. Neurosci. Biobehav. Rev. 107, 696–712. doi:10.1016/j.neubiorev.2019.09.036.

Goyal, H., Awad, H.H., Ghali, J.K., 2017. Role of cannabis in cardiovascular disorders. J. Thorac. Dis. 20776624. 9 (7), 2079–2092. Available from: http://jtd.amegroups.com/article/download/14437/pdf. doi:10.21037/jtd.2017.06.104.

Grimison, P., Mersiades, A., Kirby, A., Lintzeris, N., Morton, R., Haber, P., Olver, I., Walsh, A., McGregor, I., Cheung, Y., Tognela, A., Hahn, C., Briscoe, K., Aghmesheh, M., Fox, P., Abdi, E., Clarke, S., Della-Fiorentina, S., Shannon, J., Gedye, C., Begbie, S., Simes, J., Stockler, M., 2020. Oral THC:CBD cannabis extract for refractory chemotherapy-induced nausea and vomiting: a randomised, placebo-controlled, phase II crossover trial. Ann. Oncol. 09237534. 31 (11), 1553–1560. doi:10.1016/j.annonc.2020.07.020.

Häuser, W., Finn, D.P., Kalso, E., Krcevski-Skvarc, N., Kress, H.G., Morlion, B., Perrot, S., Schäfer, M., Wells, C., Brill, S., 2018. European Pain Federation (EFIC) position paper on appropriate use of cannabis-based medicines and medical cannabis for chronic pain management. Eur. J. Pain. 15322149. 22 (9), 1547–1564. Available from: http://onlinelibrary.wiley.com/journal/10.1002/(ISSN)1532-2149. doi:10.1002/ejp.1297.

Holt, A., Nouhravesh, N., Strange, J.E., Nielsen, S.K., Schjerning, A.-M., Rasmussen, P.V., Torp-Pedersen, C., Gislason, G.H., Schou, M., McGettigan, P., Lamberts, M., 2024. Cannabis for chronic pain: cardiovascular safety in a nationwide Danish study. Eur. Heart J. 45 (6), 475–484. doi:10.1093/eurheartj/ehad834.

Jetly, R., Heber, A., Fraser, G., Boisvert, D., 2015. The efficacy of nabilone, a synthetic cannabinoid, in the treatment of PTSD-associated nightmares: a preliminary randomized, double-blind, placebo-controlled cross-over design study. Psychoneuroendocrinology. 03064530. 51, 585–588. doi:10.1016/j.psyneuen.2014.11.002.

Khurshid, H., Qureshi, I.A., Jahan, N., Went, T.R., Sultan, W., Sapkota, A., Alfonso, M., 2021. A systematic review of fibromyalgia and recent advancements in treatment: is medicinal cannabis a new hope? Cureus. 2168-8184. 13 (8). doi:10.7759/cureus.17332.

Laux, L.C., Bebin, E.M., Checketts, D., Chez, M., Flamini, R., Marsh, E.D., Miller, I., Nichol, K., Park, Y., Segal, E., Seltzer, L., Szaflarski, J.P., Thiele, E.A., Weinstock, A., 2019. Long-term safety and efficacy of cannabidiol in children and adults with treatmentresistant Lennox-Gastaut syndrome or Dravet syndrome: expanded access program results. Epilepsy Res. 18726844. 154, 13–20. Available from: www.elsevier.com/locate/epilepsyres. doi:10.1016/j.eplepsyres.2019.03.015.

Legare, C.A., Raup-Konsavage, W.M., Vrana, K.E., 2022. Therapeutic potential of cannabis, cannabidiol, and cannabinoid-based pharmaceuticals. Pharmacology. 14230313. 107 (3), 131–149. Available from: www.karger.com/journals/pha/pha_jh.htm. doi:10.1159/000521683.

MacCallum, C.A., Russo, E.B., 2018. Practical considerations in medical cannabis administration and dosing. Eur. J. Intern. Med. 18790828. 49, 12–19. Available from: www.elsevier.com/locate/ejim. doi:10.1016/j.ejim.2018.01.004.

Mlost, J., Bryk, M., Starowicz, K., 2020. Cannabidiol for pain treatment: focus on pharmacology and mechanism of action. Int. J. Mol. Sci. 14220067. 21 (22), 1–22. Available from: https://www.mdpi.com/1422-0067/21/22/8870/pdf. doi:10.3390/ijms21228870.

Pagano, C., Navarra, G., Coppola, L., Avilia, G., Bifulco, M., Laezza, C., 2022. Cannabinoids: therapeutic use in clinical practice. Int. J. Mol. Sci. 1422-0067. 23 (6), 3344. doi:10.3390/ijms23063344.

Poli, P., Crestani, F., Salvadori, C., Valenti, I., Sannino, C., 2018. Medical cannabis in patients with chronic pain: effect on pain relief, pain disability, and psychological aspects. A prospective non randomized single arm clinical trial. Clin. Ter. 19726007. 169 (3), E102. Available from: http://www.clinicaterapeutica.it/ojs/index.php/ClinicaTerapeutica. doi:10.7417/T.2018.2062.

Richards, J.R., Bing, M.L., Moulin, A.K., Elder, J.W., Rominski, R.T., Summers, P.J., Laurin, E.G., 2019. Cannabis use and acute coronary syndrome. Clin. Toxicol. 15569519. 57 (10), 831–841. Available from: http://www.tandfonline.com/loi/ictx18. doi:10.1080/15563650.2019.1601735.

Russo, E.B., 2011. Taming THC: potential cannabis synergy and phytocannabinoid-terpenoid entourage effects. Br. J. Pharmacol. 14765381. 163 (7), 1344–1364. doi:10.1111/j.1476-5381.2011.01238.x.

Russo, E.B., 2019. Cannabis and pain. Pain Med. 15264637. 20 (11), 2083–2085. Available from: http://oxfordjournals.org/en/press/announcements-from-oup/pain-medicine.html. doi:10.1093/pm/pnz227.

Schubart, C.D., Sommer, I.E.C., van Gastel, W.A., Goetgebuer, R.L., Kahn, R.S., Boks, M.P.M., 2011. Cannabis with high cannabidiol content is associated with fewer psychotic experiences. Schizophr. Res. 15732509. 130 (1–3), 216–221. doi:10.1016/j.schres.2011.04.017.

Sinclair, J., Collett, L., Abbott, J., Pate, D.W., Sarris, J., Armour, M., 2021. Effects of cannabis ingestion on endometriosis-associated pelvic pain and related symptoms. PLoS ONE. 19326203. 16 (10). Available from: https://journals.plos.org/plosone/article/file?id=10.1371/journal.pone.0258940&type=printable. doi:10.1371/journal.pone.0258940.

Solmi, M., De Toffol, M., Kim, J.Y., Choi, M.J., Stubbs, B., Thompson, T., Firth, J., Miola, A., Croatto, G., Baggio, F., Michelon, S., Ballan, L., Gerdle, B., Monaco, F., Simonato, P., Scocco, P., Ricca, V., Castellini, G., Fornaro, M., Murru, A., Vieta, E., Fusar-Poli, P., Barbui, C., Ioannidis, J.P.A., Carvalho, A.F., Radua, J., Correll, C.U., Cortese, S., Murray, R.M., Castle, D., Shin, J.I., Dragioti, E., 2023. Balancing risks and benefits of cannabis use: umbrella review of meta-analyses of randomised controlled trials and observational studies. BMJ. 1756-1833. 382, e072348. doi:10.1136/bmj-2022-072348.

Sukpiriyagul, A., Chartchaiyarerk, R., Tabtipwon, P., Smanchat, B., Prommas, S., Bhamarapravatana, K., Suwannarurk, K., 2023. Oral tetrahydrocannabinol (THC): Cannabinoid (CBD) cannabis extract adjuvant for reducing chemotherapy-induced nausea and vomiting (CINV): a randomized, double-blinded, placebo-controlled, crossover trial. Int. J. Womens Health. 11791411. 15, 1345–1352. Available from: https://www.dovepress.com/getfile.php?fileID=92011. doi:10.2147/IJWH.S401938.

Szaflarski, J.P., Bebin, E.M., Cutter, G., DeWolfe, J., Dure, L.S., Gaston, T.E., Kankirawatana, P., Liu, Y., Singh, R., Standaert, D.G., Thomas, A.E., Hoef, L.W.V., 2018. Cannabidiol improves frequency and severity of seizures and reduces adverse events in an open-label add-on prospective study. Epilepsy Behav. 15255069. 87, 131–136. Available from: http://www.elsevier.com/inca/publications/store/6/2/2/8/2/2/index.htt. doi:10.1016/j.yebeh.2018.07.020.

Ware, M.A., Wang, T., Shapiro, S., Robinson, A., Ducruet, T., Huynh, T., Gamsa, A., Bennett, G.J., Collet, J.P., 2010. Smoked cannabis for chronic neuropathic pain: a randomized controlled trial. CMAJ. 14882329. 182 (14), E694. Available from: http://www.cmaj.ca/cgi/reprint/182/14/E694. doi:10.1503/cmaj.091414.

Warr, D., Hesketh, P., 2020. Cannabinoids as antiemetics: everything that's old is new again. Ann. Oncol. 09237534. 31 (11), 1425–1426. doi:10.1016/j.annonc.2020.08.2104.

CHAPTER

Side effects of using cannabinoids

14

Orlando Florete Jr.
Orles Pain Management and Regenerative Medicine Group, Jacksonville, FL, United States

Introduction

The use of cannabinoids in modern medicine has surged over the past two decades, driven by shifting legal frameworks, increasing patient demand, and a growing body of evidence supporting therapeutic applications. Cannabinoids, primarily derived from the *Cannabis sativa* plant, include a wide range of compounds with diverse pharmacologic properties. While Δ9-tetrahydrocannabinol (THC) and cannabidiol (CBD) are the most well-known and studied constituents, the cannabinoid class also includes numerous synthetic and endogenous analogs. As clinical interest grows, it becomes imperative for healthcare providers to not only understand the therapeutic effects but also the potential risks associated with cannabinoid use (Pertwee, 2006; Zou and Kumar, 2018; ElSohly and Slade, 2005).

The primary pharmacological effects of cannabinoids are mediated through the endocannabinoid system, a complex signaling network consisting of cannabinoid receptors (CB1 and CB2), endogenous ligands, and enzymes responsible for ligand synthesis and degradation. CB1 receptors are abundant in the central nervous system and are responsible for the psychoactive effects of cannabinoids, while CB2 receptors are predominantly found in peripheral tissues, especially within the immune system. This widespread receptor distribution underpins the wide-ranging physiological and psychological effects of cannabinoids but also contributes to the breadth of potential adverse reactions (Pertwee, 2006).

Despite their therapeutic potential, cannabinoids are not devoid of side effects. Clinical and observational studies have identified a spectrum of adverse events, ranging from mild and transient to serious and potentially life-threatening. Common side effects include dizziness, somnolence, dry mouth, and cognitive impairment, while more severe reactions may involve psychosis, cardiovascular events, and respiratory complications (Zou and Kumar, 2018; ElSohly and Slade, 2005; Crean et al., 2011; Ramaekers et al., 2004; Volkow et al., 2014). These side effects can vary significantly depending on the type of cannabinoid, the route of administration, the dosage, and the individual patient's genetic, psychological, and medical profile (Fig. 14.1).

Compounding the clinical challenge is the wide variability in cannabinoid formulations and dosing regimens. Patients may be exposed to standardized pharmaceutical

Cannabinoids for Chronic Pain. https://doi.org/10.1016/B978-0-443-36773-1.00015-5

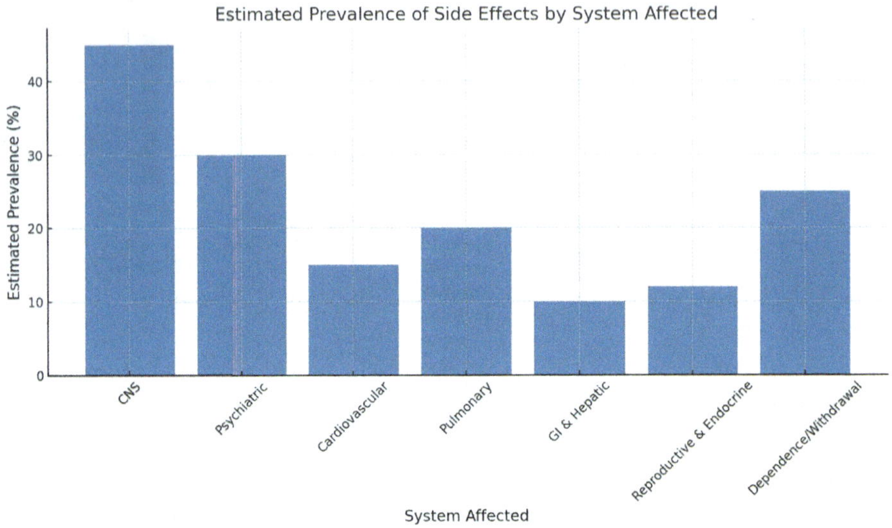

FIG. 14.1 Estimated prevalence of side effects by system affected.

preparations, artisanal CBD oils, or inhaled cannabis products of varying potency and composition. This lack of consistency in product quality and dosing introduces unpredictability in both therapeutic response and adverse effects. In addition, the presence of other psychoactive or toxic compounds, especially in unregulated markets, poses an added risk to users (ElSohly and Slade, 2005).

Given the expanding role of cannabinoids in treating conditions such as chronic pain, multiple sclerosis-related spasticity, chemotherapy-induced nausea, and epilepsy, understanding their side effect profile is no longer optional—it is essential. This chapter aims to provide a comprehensive overview of the known side effects of cannabinoids, categorized by organ system and clinical presentation. We strive to promote informed decision-making and enhance patient safety in cannabinoid-based therapy by equipping healthcare professionals with the available evidence-based knowledge (ElSohly and Slade, 2005).

Overview of cannabinoids

Cannabinoids are a diverse class of chemical compounds that exert their effects primarily through interactions with the endocannabinoid system. They are broadly categorized into endogenous cannabinoids (endocannabinoids), phytocannabinoids, and synthetic cannabinoids. Each group varies in origin, chemical structure, receptor affinity, and physiological effect, but all influence the cannabinoid receptors CB1 and CB2 to varying degrees. Understanding the distinctions among these types is essential for contextualizing therapeutic applications and side-effect profiles (Pertwee, 2006).

Endocannabinoids, such as anandamide (AEA) and 2-arachidonoylglycerol (2-AG), are produced naturally within the human body and play key roles in maintaining homeostasis. These molecules act as retrograde neurotransmitters, modulating synaptic transmission by binding to presynaptic CB1 and CB2 receptors. Endocannabinoids are rapidly synthesized on demand and metabolized by enzymes such as fatty acid amide hydrolase (FAAH) and monoacylglycerol lipase. Dysregulation of this system has been implicated in a variety of pathological conditions, including chronic pain, mood disorders, and neurodegenerative diseases (Zou and Kumar, 2018).

Phytocannabinoids are naturally occurring compounds found in the *Cannabis sativa* and *Cannabis indica* plants. The most prominent of these is THC, which is responsible for the psychoactive properties of cannabis. THC is a partial agonist at CB1 receptors, particularly in the brain, producing euphoria, altered perception, and increased appetite. CBD, another major phytocannabinoid, does not have significant affinity for CB1 or CB2 receptors but modulates endocannabinoid signaling through indirect mechanisms, including inhibition of FAAH and interaction with serotonin and TRPV1 receptors. Unlike THC, CBD is nonintoxicating and has demonstrated anticonvulsant, anxiolytic, and antiinflammatory properties (Pertwee, 2006; ElSohly and Slade, 2005).

Synthetic cannabinoids are artificially manufactured compounds designed to mimic the effects of natural cannabinoids. Pharmaceutical agents, such as dantrolene, are used to treat chemotherapy-induced nausea and AIDS-related anorexia, with standardized dosing and safety profiles. However, a more concerning class of synthetic cannabinoids—often found in products marketed as "Spice" or "K2"—has emerged in illicit markets. These compounds bind with high affinity to CB1 receptors, often resulting in unpredictable and severe toxicities, including seizures, psychosis, and cardiovascular collapse (ElSohly and Slade, 2005). Their use represents a significant public health concern due to the lack of regulation and quality control.

Cannabinoid pharmacokinetics and pharmacodynamics vary significantly based on the route of administration. Inhalation leads to rapid absorption and onset of effects, peaking within minutes and lasting for a few hours. Oral administration, while slower in onset, produces longer-lasting effects due to first-pass metabolism in the liver, where THC is converted into the more potent metabolite 11-hydroxy-THC. Sublingual and transdermal preparations are being developed to provide more consistent therapeutic levels without the peaks and troughs associated with inhalation. These differences in bioavailability and metabolism affect not only the efficacy but also the side-effect burden of cannabinoid therapies (Pertwee, 2006).

The therapeutic potential of cannabinoids continues to expand, with evidence supporting their use in a variety of clinical conditions, including epilepsy, chronic pain, multiple sclerosis, and inflammatory disorders. However, the pharmacologic complexity of cannabinoids, combined with the lack of standardized dosing guidelines, necessitates a careful and informed approach to their use (Crean et al., 2011). As this field evolves, clinicians must remain cognizant of both the promise and the limitations of cannabinoid-based treatments, particularly regarding adverse effects, drug interactions, and patient-specific considerations.

Central nervous system side effects

Cannabinoids, particularly those that interact with CB1 receptors in the brain, profoundly impact central nervous system (CNS) function. These effects span a range from mild cognitive changes to severe psychiatric manifestations, depending on the specific cannabinoid compound, dose, route of administration, and individual patient susceptibility. While some CNS effects may be transient and benign, others can significantly impair quality of life or complicate the management of comorbid conditions (Crean et al., 2011).

Cognitive impairment is one of the most frequently reported CNS side effects associated with THC-rich cannabis products. Acute use can impair attention, working memory, and executive functioning. These effects are dose-dependent and more pronounced in inexperienced users. Chronic use, particularly beginning during adolescence, may result in more persistent cognitive deficits, with evidence suggesting potential alterations in hippocampal structure and function. CBD, by contrast, has not been consistently associated with cognitive impairment and may even attenuate some of THC's deleterious cognitive effects (Crean et al., 2011).

Short-term memory dysfunction is a hallmark feature of THC-induced cognitive disruption. The mechanism involves inhibiting glutamate release in the hippocampus, a brain region critical for encoding and consolidating new memories. Users may find it difficult to retain new information or follow complex conversations. These effects typically resolve within several hours but can be distressing or impair academic and occupational performance, especially in high-stakes settings (Crean et al., 2011; Ramaekers et al., 2004).

Psychomotor slowing and decreased coordination are additional CNS-related concerns. These impairments can negatively impact activities requiring motor precision and timing, such as driving or operating machinery. Numerous studies have documented an increased risk of motor vehicle accidents among individuals who have recently consumed THC. These effects are exacerbated when cannabinoids are used in conjunction with alcohol or other sedatives, highlighting the importance of patient education regarding concurrent substance use (Ramaekers et al., 2004).

Sedation and somnolence are common effects of both THC and CBD, though they manifest through different mechanisms. THC-induced sedation is thought to result from modulation of GABAergic and dopaminergic signaling in the basal ganglia and limbic system. CBD, while typically considered non-intoxicating, can also cause drowsiness, particularly at higher doses or in combination with other central depressants. These effects are especially relevant in older adults or patients with neurologic comorbidities (Crean et al., 2011).

Headache and dizziness are frequently reported CNS-related side effects, particularly during initial exposure or dose escalation. These symptoms are thought to be related to cerebral vasodilation and altered autonomic regulation (Volkow et al., 2014). While often self-limiting, they may deter continued use in sensitive individuals. Careful dose titration and patient counseling can help mitigate these effects, particularly in cannabinoid-naïve individuals.

Disorientation and confusion are also noted, especially with high-THC formulations. These symptoms are typically transient but can cause considerable distress, particularly in older adults or individuals with preexisting cognitive impairment. The potential for cannabinoids to unmask or exacerbate delirium in hospitalized or critically ill patients is an area of increasing clinical concern. For this reason, the use of THC-rich products in elderly populations should be approached with caution (Crean et al., 2011; Volkow et al., 2014).

Seizure threshold modulation is a complex and paradoxical aspect of cannabinoid pharmacology. While CBD is FDA-approved for the treatment of rare forms of epilepsy (e.g., Dravet syndrome, Lennox–Gastaut syndrome), some synthetic cannabinoids and high doses of THC have been associated with seizure provocation. The underlying mechanisms likely involve dysregulation of neuronal excitability and CB1 receptor overactivation. This duality necessitates careful monitoring in patients with seizure disorders who are using cannabinoid therapies (Crean et al., 2011).

Depersonalization and derealization are subjective symptoms reported by some individuals after cannabinoid use, particularly with high-potency THC. These alterations in self-perception and environment can be unsettling and may mimic dissociative disorders. Although usually transient, persistent symptoms may signal an underlying vulnerability to psychiatric illness. Clinicians should evaluate such symptoms carefully, especially in adolescent or young adult users (Crean et al., 2011).

Tolerance and CNS adaptation occur with repeated cannabinoid exposure. Neuroadaptations in CB1 receptor density and function can lead to diminished responsiveness over time, prompting escalation in dose and increasing the risk of side effects. This process is particularly problematic in recreational or nonmedically supervised use, where dosage control is minimal. Abrupt cessation after chronic use may also lead to withdrawal symptoms, including irritability, restlessness, and insomnia, all of which are centrally mediated.

The central nervous system side effects of cannabinoids are diverse and clinically significant. While some effects, such as sedation or mild euphoria, may be therapeutically desirable in specific contexts, others—like cognitive impairment, psychomotor slowing, or neurologic dysregulation—pose risks that must be carefully managed (Di Forti et al., 2019). A thorough understanding of these effects allows clinicians to tailor cannabinoid-based therapies to minimize harm while optimizing therapeutic benefit. Further research is needed to elucidate long-term CNS consequences, particularly with emerging synthetic compounds and high-potency cannabis formulations.

Psychiatric side effects

The psychiatric consequences of cannabinoid use represent one of the most complex and controversial areas of cannabinoid research. The endocannabinoid system is crucial in emotional regulation, stress response, and affective balance. Accordingly, exogenous cannabinoids—particularly those affecting CB1 receptors—can profoundly influence psychiatric states. These effects may be transient or persistent and can range from mild anxiety to full-blown psychotic episodes, depending on the compound used, its potency, route of administration, and an individual's underlying vulnerability (Di Forti et al., 2019; Crippa et al., 2009).

Acute anxiety and panic reactions are among the most frequently reported psychiatric effects, especially following the use of high-THC cannabis. These symptoms often appear rapidly following inhalation or ingestion and include increased heart rate, chest tightness, feelings of doom, and hypervigilance. Naïve users or those with preexisting anxiety disorders are particularly susceptible. While symptoms typically resolve within a few hours, they may be severe enough to result in emergency medical visits or discourage future use (Crippa et al., 2009).

Paranoia and altered perception are also typical during acute intoxication, particularly with THC-dominant strains or products. Users may report feeling watched, judged, or persecuted. In some cases, this can progress to delusional thinking, blurring the line between drug-induced symptoms and a primary psychotic disorder. The risk of such reactions appears to increase with dose, rapid-onset formulations (e.g., vaping), and higher THC-to-CBD ratios (Crippa et al., 2009).

Cannabis-induced psychosis is a clinically recognized phenomenon characterized by hallucinations, delusions, disorganized thinking, and perceptual disturbances occurring in the context of recent cannabis use. These episodes can be indistinguishable from primary psychotic disorders like schizophrenia, though they typically resolve within days to weeks of abstinence. However, repeated episodes or continued use may lead to a more chronic course. The pathophysiology is not fully understood, but it likely involves excessive CB1 receptor stimulation, dopaminergic dysregulation, and interactions with genetic predispositions, particularly variants in the *COMT* and *AKT1* genes (Di Forti et al., 2019; Crippa et al., 2009).

Cannabis use and schizophrenia share a bidirectional relationship. Epidemiological studies suggest that early and frequent cannabis use—especially during adolescence—is associated with a two- to fivefold increase in the risk of developing schizophrenia (Di Forti et al., 2019). Conversely, individuals with schizophrenia are more likely to use cannabis, possibly to alleviate negative symptoms, medication side effects, or dysphoria. However, cannabis use in this population is associated with worsened clinical outcomes, including increased hospitalizations, reduced medication adherence, and poorer long-term prognosis.

Mood disorders, including depression and bipolar disorder, may also be affected by cannabinoid use. While some individuals report mood elevation or stress relief with low-dose cannabis, chronic or heavy use has been linked to increased rates of depressive symptoms and suicidal ideation, particularly in adolescents and young adults. In bipolar disorder, cannabis use has been associated with increased manic episodes and reduced treatment adherence. The mood-altering properties of THC may temporarily improve mood but appear to disrupt long-term emotional regulation and circadian rhythm stability (Lev-Ran et al., 2014).

Amotivational syndrome refers to a hypothesized pattern of apathy, reduced goal-directed behavior, and emotional blunting associated with chronic cannabis use (Lev-Ran et al., 2014). Though controversial and inconsistently supported by empirical data, some long-term users exhibit decreased productivity, academic disengagement, and occupational underperformance. These patterns may be due to dopaminergic suppression in mesolimbic pathways, as well as a psychological dependence on the anxiolytic and hedonic effects of cannabis.

Depersonalization and derealization, described as feeling detached from one's body or the external environment, can occur with cannabinoid intoxication. While often transient, these symptoms can be distressing and may trigger panic attacks or reinforce preexisting psychiatric vulnerabilities. In rare cases, chronic depersonalization may persist beyond the period of acute intoxication, necessitating psychiatric evaluation and intervention.

Suicidality and self-harm behaviors have been reported in association with heavy cannabis use, particularly among adolescents and individuals with concurrent mood or personality disorders. While causality remains difficult to establish, multiple longitudinal studies suggest a dose-response relationship between the frequency of cannabis use and the risk of suicidal thoughts, attempts, and deaths. These findings underscore the need for thorough psychiatric screening in individuals using cannabinoids, especially those with a personal or family history of affective illness (Lev-Ran et al., 2014).

Personality changes have been described anecdotally and in some longitudinal studies of chronic cannabis users. Traits such as increased impulsivity, emotional lability, and reduced conscientiousness have been noted. While some of these changes may reflect underlying personality predispositions, the neuroadaptive effects of long-term cannabinoid exposure on prefrontal cortical function cannot be ruled out. These effects are particularly concerning for adolescents and young adults, whose brains are still developing significantly.

Substance use disorders frequently cooccur with cannabinoid use. Individuals who use cannabis are at increased risk of using and misusing other substances, including alcohol, nicotine, and opioids. This may be due to overlapping neural reward pathways or the use of cannabinoids to mitigate withdrawal symptoms from other substances. Synthetic cannabinoids, in particular, have been associated with the rapid onset of dependence and severe psychiatric symptoms, including suicidal behavior, aggression, and hallucinations.

Withdrawal-related psychiatric symptoms often emerge after the cessation of regular cannabinoid use. Symptoms include irritability, anxiety, insomnia, mood swings, and reduced appetite, typically beginning within 24–72 h and lasting up to two weeks. These symptoms can mimic primary psychiatric disorders and may lead to relapse in individuals attempting to discontinue use. Effective management includes psychoeducation, behavioral support, and, in some cases, short-term pharmacologic intervention (Thomas et al., 2014).

Cannabinoids exert diverse and significant psychiatric effects that must be carefully considered in both clinical and recreational contexts. While some individuals may benefit from their anxiolytic or mood-stabilizing properties, others—particularly those with psychiatric comorbidities or genetic predispositions—may experience significant harm. A nuanced understanding of these effects is essential for guiding appropriate use, counseling patients, and mitigating psychiatric risks associated with cannabinoid therapies.

Cardiovascular side effects

Cannabinoids exert multiple effects on the cardiovascular system, primarily through interactions with the autonomic nervous system and direct modulation of vascular and myocardial tissue via cannabinoid receptors. CB1 receptors are expressed in endothelial cells, vascular smooth muscle, and cardiac myocytes, while CB2 receptors are more prevalent in immune cells and may influence vascular inflammation. Acute and chronic exposure to cannabinoids—particularly THC—can result in a broad spectrum of cardiovascular effects, some benign and others potentially life-threatening, depending on individual susceptibility and coexisting conditions (Thomas et al., 2014).

Tachycardia, or increased heart rate, is one of the most consistent cardiovascular effects observed after THC consumption. It is primarily mediated by increased sympathetic nervous system activity and the inhibition of parasympathetic tone. This effect is dose-dependent and typically begins within minutes of inhalation, lasting up to three hours. While often benign in healthy individuals, THC-induced tachycardia can be concerning in patients with underlying cardiac conditions, particularly those with arrhythmias or ischemic heart disease, where increased myocardial oxygen demand may provoke adverse events.

Orthostatic hypotension is another frequently reported effect, characterized by a drop in blood pressure upon standing, often leading to dizziness, lightheadedness, or syncope. This is thought to be due to cannabinoid-induced vasodilation and reduced baroreceptor sensitivity. Older adults are particularly vulnerable to this side effect due to age-related changes in autonomic regulation and vascular compliance. In clinical settings, orthostatic hypotension may result in falls or reduced adherence to cannabinoid-based therapies (Thomas et al., 2014).

Hypertension, paradoxically, has also been documented in some users, particularly with synthetic cannabinoids, which can exert more potent and prolonged effects on CB1 receptors. These agents have been associated with severe hypertensive episodes, including hypertensive crises, which can result in acute end-organ damage. The mechanisms likely involve excessive sympathetic activation and direct vascular smooth muscle constriction, highlighting the dangers of unregulated or high-potency cannabinoid compounds (Rezkalla et al., 2003).

Myocardial infarction (MI) has been linked to acute cannabis use, especially within the first hour following consumption. Several large observational studies and case reports suggest that cannabis can act as a trigger for MI, particularly in young individuals without traditional cardiovascular risk factors. Proposed mechanisms include increased heart rate, reduced oxygen-carrying capacity of blood (due to carboxyhemoglobin formation from inhaled smoke), vasospasm, and platelet activation (Thomas et al., 2014; Singh et al., 2018). The risk is significantly elevated in individuals with preexisting coronary artery disease or those who combine cannabis use with other vasoactive substances such as cocaine or amphetamines.

Arrhythmias, including atrial fibrillation, premature ventricular contractions, and ventricular tachycardia, have been observed following cannabinoid use. These effects may be mediated through changes in cardiac conduction, autonomic tone, or ischemia-induced excitability. Synthetic cannabinoids, in particular, have been implicated in cases of sudden cardiac death, particularly among young adults with otherwise normal cardiovascular anatomy. Clinicians should be alert to the potential for arrhythmic complications in patients presenting with palpitations or syncope after cannabinoid exposure.

Peripheral vasodilation is a common hemodynamic effect of cannabinoids and contributes to the sensation of warmth or flushing experienced by many users. This vasodilation is mediated by nitric oxide release and CB1 receptor activation on vascular smooth muscle cells. While generally benign, it may exacerbate hypotension in vulnerable populations or interfere with the autoregulatory mechanisms needed to maintain cerebral perfusion, particularly during periods of orthostatic stress.

Stroke, including both ischemic and hemorrhagic subtypes, has been reported in association with cannabinoid use, although the absolute risk remains low. Proposed mechanisms include reversible cerebral vasoconstriction syndrome, cardioembolic events secondary to arrhythmias, and vascular inflammation. Case series have documented strokes in young adults with recent cannabis use and no other identifiable risk factors, suggesting that cannabinoids may act as acute cerebrovascular triggers in susceptible individuals.

Endothelial dysfunction has been observed in both human and animal studies following cannabinoid exposure. THC and synthetic cannabinoids can impair endothelial-dependent vasodilation, promote oxidative stress, and increase inflammatory markers within the vascular endothelium. These changes may contribute to the long-term atherogenic potential of chronic cannabinoid use, particularly in individuals with existing cardiovascular risk factors such as diabetes, hypertension, or dyslipidemia.

Cannabinoids can produce a range of cardiovascular side effects, some of which may be serious or life-threatening. While many effects, such as mild tachycardia or hypotension, are transient and manageable in healthy individuals, others—including myocardial infarction, arrhythmias, and stroke—require careful consideration in clinical practice. Healthcare providers should screen for cardiovascular risk factors prior to initiating cannabinoid-based therapy, educate patients on potential symptoms, and consider alternative treatments for those with known cardiovascular disease or high-risk profiles.

Pulmonary side effects

Cannabinoid use, particularly through inhalation, has notable implications for pulmonary health. While much attention has focused on the systemic effects of cannabinoids, the respiratory consequences of smoking or vaping cannabis warrant equal consideration, especially given their widespread prevalence. Cannabis smoke, like tobacco smoke, contains a complex mixture of particulates, volatile compounds, and carcinogens that interact with respiratory tissue in potentially harmful ways.

Furthermore, the increasing popularity of vaping cannabinoid oils introduces new respiratory risks related to aerosolized additives and delivery systems.

Chronic bronchitis is the most consistently documented pulmonary consequence of habitual cannabis smoking. Symptoms include chronic cough, wheezing, sputum production, and shortness of breath (Tashkin, 2013). These outcomes are associated with airway inflammation, goblet cell hyperplasia, and increased mucus secretion. Although the prevalence and severity of chronic bronchitis symptoms may be less than in tobacco smokers, the overlap in histopathological changes—such as airway epithelial damage and mucosal hyperplasia—is significant. Importantly, these effects appear dose-dependent and may regress with cessation of cannabis use.

Airway inflammation is another hallmark of cannabis inhalation. Cannabis smoke contains many of the same toxicants as tobacco smoke, including ammonia, hydrogen cyanide, and polycyclic aromatic hydrocarbons. Bronchoscopy studies have demonstrated increased erythema, edema, and vascular congestion in habitual cannabis users. Inflammatory cell infiltration, especially neutrophils and macrophages, has also been observed in bronchoalveolar lavage specimens. This chronic inflammatory environment may predispose users to respiratory infections and impaired mucociliary clearance.

Impaired pulmonary function has been observed in long-term cannabis smokers, though the extent and clinical significance remain debated. Some studies suggest reduced forced expiratory volume (FEV1) and mid-expiratory flow rates in one second. In contrast, others find no significant long-term decline in lung function among exclusive cannabis users. Nonetheless, when combined with tobacco or other inhaled substances, cannabis smoking likely contributes to additive or synergistic pulmonary damage. Regular monitoring of lung function may be prudent in patients using inhaled cannabinoids chronically.

Bullous lung disease and spontaneous pneumothorax have been reported disproportionately among young, otherwise healthy cannabis users (Tashkin, 2013). These cases often involve large peripheral bullae predominantly in the upper lobes, suggesting barotrauma or localized parenchymal injury from deep inhalation and breath-holding practices. The risk may be heightened by techniques such as "hotboxing" or using high-resistance smoking devices like water pipes and vaporizers. Though rare, these complications can lead to recurrent pneumothoraces requiring surgical intervention.

Infectious complications, including fungal and bacterial pneumonias, have also been reported in association with smoked or vaporized cannabis. Contaminated cannabis products—especially those grown or stored in nonsterile conditions—can harbor pathogens such as *Aspergillus* or *Pseudomonas*. Immunocompromised individuals, including transplant recipients and patients with HIV/AIDS or cancer, are at particular risk. Inhalation of contaminated aerosols may result in fulminant infections that are difficult to treat and life-threatening.

EVALI (E-cigarette or vaping-associated lung injury) has emerged as a serious and potentially fatal condition linked to vaping cannabinoid products, particularly those obtained from illicit sources. First described in 2019, EVALI is characterized by acute respiratory distress, bilateral pulmonary infiltrates, and hypoxemia (Mammen et al., 2016). The exact etiology is still under investigation, but vitamin E acetate and other cutting agents used in unregulated THC cartridges have been strongly implicated. Hundreds of hospitalizations and several deaths have been attributed to EVALI, underscoring the risks of unregulated cannabinoid vaping.

Hypersensitivity pneumonitis, an immune-mediated lung disorder, has been associated with inhaled cannabis in rare cases. This condition results from an exaggerated immune response to inhaled antigens, leading to alveolar inflammation, diffuse infiltrates, and impaired gas exchange. Symptoms include fever, cough, and dyspnea, and diagnosis typically requires imaging, pulmonary function testing, and sometimes biopsy. While uncommon, it serves as a reminder that inhaled botanical products are not without the risk of allergic or immunologic pulmonary responses.

Asthma exacerbation may occur in susceptible individuals using inhaled cannabinoids. While some preclinical studies suggest bronchodilatory properties of THC, these effects are transient and offset by airway irritation and increased airway reactivity. Case reports have documented cannabis-induced bronchospasm in asthmatic patients, occasionally requiring emergency intervention. Vaporized cannabinoids may reduce some irritant effects compared to smoke, but their safety profile in asthmatic populations remains inadequately studied.

The pulmonary effects of cannabinoids are diverse and influenced mainly by the route of administration. Inhalation of combusted or aerosolized products introduces direct toxicants into the respiratory tract, leading to acute and chronic pulmonary pathology. While the long-term risks of cannabis smoke may not entirely mirror those of tobacco, they are nonetheless significant and clinically relevant. Healthcare providers should counsel patients on safer administration methods—such as oral or sublingual delivery—when pulmonary health is a concern and remain vigilant for signs of respiratory complications, especially in at-risk populations.

Gastrointestinal and hepatic side effects

Cannabinoids exert notable effects on the gastrointestinal (GI) tract and liver, mediated through central and peripheral mechanisms. Cannabinoid receptors—particularly CB1 and CB2—are expressed throughout the enteric nervous system, intestinal epithelium, gastric mucosa, and hepatocytes. These receptors modulate motility, secretion, appetite, emesis, and inflammatory responses (Layden et al., 2020) While some of these effects have therapeutic value, such as appetite stimulation or antiemetic properties, cannabinoids can also produce a range of adverse GI and hepatic outcomes that warrant clinical attention.

Cannabinoid hyperemesis syndrome (CHS) is a paradoxical condition associated with chronic, high-dose cannabis use. It is characterized by cyclic episodes of severe nausea, vomiting, and abdominal pain, often resistant to conventional antiemetics. A distinguishing feature of CHS is compulsive hot bathing or showering, which patients report provides temporary symptomatic relief, possibly due to heat-mediated desensitization of TRPV1 receptors. CHS typically resolves with sustained cannabis abstinence, though recurrence is common with resumption. The condition is often misdiagnosed as cyclic vomiting syndrome or gastroenteritis, delaying appropriate management.

Nausea and vomiting, while commonly alleviated by cannabinoid-based therapies in controlled settings (e.g., chemotherapy-induced nausea), may paradoxically worsen in specific populations or with inappropriate dosing. Acute THC exposure can delay gastric emptying and increase visceral hypersensitivity, potentially exacerbating dyspeptic symptoms. This biphasic response underscores the importance of dosing precision and individualized treatment plans, particularly when cannabinoids are used for GI symptom control.

GI motility disturbances are also observed with cannabinoid use. THC and other CB1 agonists reduce gastric and intestinal smooth muscle contractility, leading to delayed gastric emptying and prolonged intestinal transit time. These effects may contribute to symptoms such as bloating, constipation, and early satiety. In patients with preexisting motility disorders—such as gastroparesis or irritable bowel syndrome—cannabinoids may worsen symptoms or confound diagnostic evaluations.

Appetite modulation is among the most recognized GI effects of cannabinoids. THC stimulates appetite by interacting with hypothalamic pathways and enhancing olfactory sensitivity (Layden et al., 2020). This "munchies" phenomenon is utilized therapeutically in patients with cachexia or anorexia, such as those with cancer, HIV/AIDS, or advanced chronic illnesses. Conversely, CBD appears to have a more variable or even suppressive effect on appetite, depending on the dose and context. In nontarget populations, particularly recreational users, excessive appetite stimulation may contribute to unhealthy weight gain or disrupted eating patterns.

Diarrhea and abdominal discomfort have been reported in association with CBD, particularly in high doses or in purified pharmaceutical formulations such as Epidiolex. Clinical trials in epilepsy patients have documented a relatively high incidence of GI adverse events, which appear to be dose-dependent and reversible upon dose reduction or discontinuation (Sorensen et al., 2017). The mechanisms are not fully elucidated but may involve altered intestinal permeability or dysbiosis of the gut microbiome.

Cannabinoids may exacerbate gastroesophageal reflux symptoms because they affect lower esophageal sphincter (LES) tone. THC can reduce LES pressure, predisposing susceptible individuals to acid reflux, heartburn, and regurgitation. These symptoms are particularly relevant in individuals who consume cannabinoids in conjunction with alcohol, fatty meals, or sedatives—all of which can further impair LES function. Recognition of this interaction is essential when evaluating new or worsening GERD symptoms in cannabis users.

Liver enzyme elevations have been reported with CBD use, especially at higher doses or when combined with other hepatically metabolized medications. Clinical trials of CBD for epilepsy have documented increases in alanine transaminase and aspartate transaminase, sometimes exceeding five times the upper limit of normal (Devinsky et al., 2018). Most cases are asymptomatic and reversible, but periodic liver function monitoring is advised in patients using high-dose CBD, particularly those also taking valproate, which may potentiate hepatotoxicity.

Drug interactions via hepatic enzyme modulation are of increasing clinical relevance. Cannabinoids, particularly CBD, inhibit several cytochrome P450 (CYP) isoenzymes, including CYP2C19, CYP3A4, and CYP2D6. This inhibition can alter serum levels of coadministered medications, potentially leading to toxicity or therapeutic failure. Examples include antiepileptics, anticoagulants (e.g., warfarin), antidepressants, and immunosuppressants. Clinicians must be aware of these interactions when initiating or adjusting cannabinoid therapies, especially in polypharmacy contexts.

Cannabinoids can significantly affect both the GI tract and hepatic function through a variety of mechanisms. While these effects can be leveraged for therapeutic purposes, they also pose risks that must be carefully considered, particularly in patients with preexisting GI or liver conditions. Understanding the dose-dependent nature of these side effects and potential drug interactions is essential for optimizing treatment outcomes and minimizing harm. Clinicians should provide anticipatory guidance, conduct appropriate monitoring, and tailor cannabinoid therapies to each patient's unique physiological and pharmacological profile.

Reproductive and endocrine side effects

Cannabinoids interact with the hypothalamic-pituitary-gonadal axis and other hormonal systems, producing various reproductive and endocrine effects. CB1 and CB2 receptors are expressed in reproductive tissues, including the testes, ovaries, endometrium, and placenta, as well as within key regulatory centers of the brain (Gaston et al., 2017). These interactions influence hormone production, sexual function, fertility, and developmental processes. While some effects are reversible and dose-dependent, others may have longer-term or transgenerational implications.

In males, chronic cannabis use has been associated with alterations in reproductive hormones. Several studies have documented reduced serum testosterone levels, decreased luteinizing hormone, and altered gonadotropin-releasing hormone pulsatility following regular THC exposure. These hormonal disruptions are associated with impaired spermatogenesis, reduced sperm concentration, motility, and altered morphology. Animal models and human studies suggest a dose-response relationship, with heavier and longer-term use linked to greater reproductive impairment.

Sperm function and fertility may be directly affected by cannabinoid exposure. THC and other cannabinoids can bind to CB1 receptors located on sperm cells, impacting their ability to undergo capacitation, hyperactivation, and acrosome reaction—all critical processes for fertilization. DNA fragmentation and oxidative stress have also been observed in the sperm of habitual cannabis users, raising concerns about potential mutagenic effects. These findings are especially relevant for couples undergoing fertility evaluations or assisted reproductive technologies.

In females, cannabis use may disrupt menstrual cyclicity and ovulatory function. THC influences the hypothalamic-pituitary-ovarian axis, potentially reducing the secretion of follicle-stimulating hormone and luteinizing hormone. This hormonal dysregulation can result in anovulatory cycles, oligomenorrhea, or amenorrhea. Some studies have also suggested an earlier onset of menopause in chronic cannabis users, although the long-term clinical significance remains under investigation. These disruptions can complicate family planning and may mask underlying gynecological disorders.

Pregnancy outcomes may be adversely affected by cannabinoid use. THC crosses the placental barrier and accumulates in fetal tissues, which may interfere with neurodevelopment, cell differentiation, and placental function. Observational studies have linked prenatal cannabis exposure to lower birth weight, increased risk of preterm birth, and impaired neonatal neurobehavioral outcomes (Gundersen et al., 2015). Although confounding factors such as polysubstance use and socioeconomic status complicate these findings, the American College of Obstetricians and Gynecologists advises against cannabis use during pregnancy due to potential harms.

Lactation and breastfeeding are additional concerns, as THC is excreted into breast milk and may persist for extended periods due to its lipophilic nature. The effects of postnatal exposure on infants are not fully understood but may include sedation, poor feeding, and delayed motor development. There is insufficient evidence to establish a safe threshold for cannabinoid exposure during lactation, and most professional guidelines recommend abstaining from cannabis use while breastfeeding.

Pubertal development may be affected by adolescent cannabinoid exposure. Studies suggest that early and chronic use of cannabis during adolescence—a critical period for hormonal maturation—may delay the onset of puberty in males and alter the timing of menarche in females. These effects may be mediated by disruptions in sex steroid production or signaling through the endocannabinoid system, which plays a regulatory role in the timing of pubertal milestones. Given the importance of hormonal regulation during adolescence, these findings are of significant public health concern.

Thyroid function may also be influenced by cannabinoid use, although evidence is limited and mixed. Some animal models suggest that THC can suppress thyroid hormone levels (T3 and T4) and thyroid-stimulating hormone, potentially by inhibiting the hypothalamic-pituitary-thyroid axis. Human data are sparse, but isolated case reports have noted reversible hypothyroid-like symptoms in heavy users. Further research is needed to determine the clinical relevance of these findings and whether cannabinoids interact with thyroid medications.

Prolactin levels may be modulated by cannabinoid exposure, with some studies showing reduced secretion. Prolactin plays essential roles in lactation, reproductive function, and immune modulation. CB1 receptor activation in the hypothalamus is believed to inhibit prolactin-releasing hormone secretion. This may affect milk production in females, while it may alter testosterone feedback regulation in males. Clinical implications are generally mild but may become more pronounced in patients with pituitary disorders or endocrine-sensitive conditions.

Adrenal function may also be indirectly affected. THC can influence corticotropin-releasing hormone and adrenocorticotropic hormone secretion, altering cortisol levels. While acute stress responses may be blunted, chronic exposure may dysregulate the hypothalamic-pituitary-adrenal axis, affecting circadian cortisol rhythm, immune response, and stress tolerance. These endocrine shifts could have downstream effects on metabolic and psychiatric health, particularly in vulnerable individuals.

The reproductive and endocrine side effects of cannabinoids are complex and multifactorial. They can affect fertility, hormone regulation, sexual development, pregnancy outcomes, and endocrine homeostasis across both sexes and all life stages. While some effects are reversible with cessation, others—particularly those affecting development or involving epigenetic mechanisms—may have longer-lasting consequences. Clinicians should assess reproductive and endocrine histories when considering cannabinoid therapies and counsel patients accordingly, especially those who are pregnant, planning to conceive, or experiencing endocrine dysfunction.

Dependence and withdrawal

The increasing medical and recreational use of cannabinoids has brought heightened awareness of their potential for dependence and withdrawal (Calvigioni et al., 2014). While often perceived as less addictive than substances such as opioids or stimulants, cannabinoids—particularly those with high concentrations of Δ9-tetrahydrocannabinol (THC)—can induce physiological and psychological dependence in a subset of users. The development of cannabis use disorder (CUD) is now well recognized and defined in the Diagnostic and Statistical Manual of Mental Disorders, Fifth Edition (DSM-5).

CUD is characterized by a pattern of problematic cannabis use leading to clinically significant impairment or distress. The DSM-5 criteria include tolerance, withdrawal, unsuccessful attempts to cut down, continued use despite harm, and social or occupational impairment. The severity of CUD is graded as mild, moderate, or severe based on the number of criteria met. Prevalence rates vary widely, with estimates suggesting that approximately 9% of users will develop CUD, increasing to 17% among those who begin use in adolescence and over 25%–30% among daily users.

Tolerance to cannabinoids develops with repeated exposure, particularly to THC. Neurobiologically, chronic stimulation of CB1 receptors leads to receptor downregulation and desensitization in brain regions such as the prefrontal cortex, hippocampus, and basal ganglia. This tolerance manifests clinically as a diminished subjective high and reduced therapeutic efficacy, prompting dose escalation. Tolerance can reverse gradually with cessation, but the normalization rate depends on the duration and intensity of use.

Dependence is both psychological and physiological. Psychological dependence refers to the compulsion to use cannabinoids for their mood-altering, anxiolytic, or sedative effects. Physiological dependence occurs when abrupt discontinuation leads to withdrawal symptoms. The reinforcing properties of THC are believed to be mediated through increased dopamine release in the mesolimbic reward pathway, particularly the nucleus accumbens. This mechanism, while less potent than that of opioids or stimulants, is sufficient to sustain habitual use in vulnerable individuals.

Withdrawal symptoms occur in approximately 50%–90% of daily or near-daily users upon cessation. The syndrome typically begins within 24–72 h, peaks at 2–6 days, and resolves within 1–2 weeks. Common symptoms include irritability, anxiety, insomnia, depressed mood, decreased appetite, restlessness, and physical discomfort such as abdominal pain or chills (Calvigioni et al., 2014; Metz and Stickrath, 2015). While rarely medically dangerous, these symptoms can be distressing enough to interfere with daily functioning and trigger relapse.

Insomnia is one of the most persistent and troubling withdrawal symptoms. Users report difficulty falling asleep, frequent awakenings, and vivid or disturbing dreams. These changes may reflect a rebound in rapid eye movement sleep, which is suppressed during chronic cannabis use. In some individuals, sleep disturbances may persist for weeks, underscoring the role of cannabinoids in modulating circadian rhythms and sleep architecture.

Mood disturbances are also prominent during withdrawal. Anxiety, irritability, and depressive symptoms can mirror primary mood disorders, making diagnosis challenging in patients with cooccurring psychiatric conditions. These symptoms are thought to result from endocannabinoid system dysregulation and altered neurotransmitter activity, including serotonin and norepinephrine. In some individuals, particularly adolescents or those with mood disorders, the mood destabilization during withdrawal can be profound.

Appetite changes often accompany cessation of cannabis use. THC is a known orexigenic agent, and its absence can lead to reduced appetite, nausea, and weight loss. These symptoms are usually transient but may be distressing for patients who use cannabinoids therapeutically to stimulate appetite in conditions such as cancer or HIV/AIDS. Clinicians should anticipate this effect and provide dietary or pharmacologic support during withdrawal.

Craving is a central feature of cannabis dependence. Craving intensity correlates with withdrawal severity and predicts relapse risk. Environmental cues, stress, mood, and availability of cannabis influence it. Neuroimaging studies suggest that craving involves the activation of brain regions associated with reward, memory, and impulse control, including the anterior cingulate and orbitofrontal cortex. Behavioral interventions, such as cognitive-behavioral therapy and mindfulness-based strategies, have been shown to reduce craving and improve abstinence rates.

Relapse is common, particularly within the first month of cessation. Factors contributing to relapse include high baseline use, lack of social support, concurrent psychiatric disorders, and poor coping skills. Unlike opioid or alcohol use disorders, there are no FDA-approved pharmacotherapies for cannabis dependence, though several agents (e.g., gabapentin, N-acetylcysteine, and dronabinol) are under investigation. Behavioral therapies remain the mainstay of treatment.

Adolescents and young adults are at particularly high risk of developing cannabis dependence. Early exposure during critical periods of brain development may sensitize the endocannabinoid and dopaminergic systems to its reinforcing effects.

Furthermore, adolescents may lack the insight, coping strategies, and executive control necessary to regulate use, thereby increasing their vulnerability. Prevention and early intervention programs targeted at youth are critical for reducing the long-term burden of CUD.

Synthetic cannabinoids, such as those found in products like "Spice" or "K2," are associated with a higher risk of severe dependence and withdrawal symptoms. These compounds are full agonists at CB1 receptors and often bind with greater affinity than THC. Withdrawal from synthetic cannabinoids can include intense agitation, aggression, hallucinations, and even seizures (Budney and Hughes, 2006). The clinical management of such cases is more complex and may require inpatient stabilization.

Cooccurring substance use disorders are common in individuals with cannabis dependence. Tobacco and alcohol are frequently coused, and cannabis may be used to modulate the effects or withdrawal symptoms of other drugs. This polydrug use complicates the clinical picture and increases the difficulty of achieving sustained abstinence. Integrated treatment approaches that address all substances simultaneously are more effective than single-substance interventions.

Social and functional impairment is a hallmark consequence of chronic cannabis dependence. Users may experience decreased motivation, poor academic or occupational performance, strained relationships, and reduced participation in meaningful activities. These consequences can be subtle but accumulate over time, especially in individuals who rely on cannabis as a coping mechanism. Addressing underlying psychosocial factors is essential in the long-term management of CUD.

Recovery and remission from cannabis dependence are achievable with appropriate support and intervention. Many individuals achieve abstinence or reduced use over time, especially with structured behavioral therapy. Motivational enhancement therapy, contingency management, and family-based interventions have all demonstrated efficacy. Educating patients on the course of withdrawal, offering nonjudgmental support, and providing relapse prevention strategies can significantly improve outcomes.

Dependence and withdrawal are clinically significant aspects of cannabinoid use, particularly in the context of chronic, high-dose THC exposure. While often underestimated in severity, these effects can impair quality of life and complicate medical treatment. Early recognition, patient education, and comprehensive treatment strategies are critical for addressing CUD and supporting long-term recovery.

Summary and clinical implications

Cannabinoids, encompassing phytocannabinoids, synthetic analogs, and endocannabinoid modulators, have demonstrated therapeutic potential across various conditions. However, as clinical and recreational use expands globally, so too does the responsibility of healthcare providers to understand and manage the full spectrum of cannabinoid-related side effects. These adverse effects, while often underappreciated, can impact nearly every organ system and vary widely based on individual vulnerability, product composition, administration route, and duration of use.

The central nervous system is a significant site of cannabinoid action and also of adverse effects. Short-term use can cause cognitive impairments, psychomotor slowing, and disorientation, while chronic exposure may result in more persistent deficits. These effects are especially concerning in adolescents, older adults, and those in safety-sensitive occupations. Clinicians should screen for signs of neurologic dysfunction and educate patients on the risks, particularly when cannabinoids are used alongside other CNS depressants.

Psychiatric side effects represent one of the most clinically significant risks of cannabinoid use. Anxiety, panic attacks, and psychosis can occur even in healthy individuals, with a greater risk in those with psychiatric vulnerabilities or a family history of mental illness. The strong association between early cannabis use and the later development of schizophrenia, as well as the worsening of mood and anxiety disorders with chronic use, reinforces the need for pretreatment psychiatric screening and cautious patient selection.

Cardiovascular complications, though less commonly recognized, can be serious and sometimes life-threatening. Tachycardia, orthostatic hypotension, and an increased risk of myocardial infarction and arrhythmias have been documented, especially with high-potency THC or synthetic cannabinoids. Patients with existing cardiovascular disease or multiple risk factors should be evaluated carefully before initiating cannabinoid therapy and monitored closely thereafter.

Pulmonary effects are primarily linked to the inhalation route, with cannabis smoke containing many of the same harmful constituents as tobacco. Chronic bronchitis, reduced lung function, and rare but severe complications such as pneumothorax and EVALI have all been reported. Where possible, noninhaled routes of administration should be prioritized, especially in patients with asthma, chronic obstructive pulmonary disease, or compromised respiratory function.

The GI and hepatic systems are also affected. CHS, appetite dysregulation, and altered motility are key concerns, particularly with chronic use. Although typically well tolerated, CBD has been linked to liver enzyme elevations and potential drug–drug interactions through cytochrome P450 inhibition. Monitoring liver function and assessing medication interactions are critical, especially in patients on polypharmacy regimens.

Reproductive and endocrine effects of cannabinoids are increasingly recognized. These include disrupted menstrual cycles, reduced sperm quality, hormonal alterations, and potential impacts on fetal development during pregnancy. THC crosses the placenta and is excreted in breast milk, making cannabis use during pregnancy and lactation particularly concerning. Reproductive counseling should be part of cannabinoid risk-benefit discussions in patients of childbearing potential.

Dependence and withdrawal, long underestimated, are now well-documented phenomena. CUD affects a significant minority of users, particularly those who begin in adolescence or use daily. Withdrawal symptoms can be distressing and contribute to relapse. Management strategies should include behavioral therapies, patient education, and relapse prevention planning. Recognizing the signs of CUD early allows for timely intervention and better outcomes.

Clinicians must also be aware of the challenges surrounding dosing, product variability, and regulation. The potency and composition of cannabinoid products—especially in unregulated or recreational markets—can vary widely, complicating efforts to predict or control side effects. Standardized pharmaceutical preparations, where available, offer better safety profiles but are not accessible to all patients. Until regulation improves, clinicians must emphasize harm reduction and informed use (Hasin et al., 2015).

In conclusion, while cannabinoids offer meaningful therapeutic options in selected populations, they also pose a range of side effects that can impact nearly every system in the body. A nuanced, individualized approach is essential, incorporating a careful assessment of benefits, risks, patient history, and monitoring needs. Education, open dialog, and evidence-based practice will be necessary as the medical use of cannabinoids continues to evolve.

References

Budney, A.J., Hughes, J.R., 2006. The cannabis withdrawal syndrome. Curr. Opin. Psychiatry. 09517367. 19 (3), 233–238. doi:10.1097/01.yco.0000218592.00689.e5.

Calvigioni, D., Hurd, Y.L., Harkany, T., Keimpema, E., 2014. Neuronal substrates and functional consequences of prenatal cannabis exposure. Eur. Child Adolesc. Psychiatry. 1435165X. 23 (10), 931–941. Available from: http://www.springerlink.com/content/1018-8827. doi:10.1007/s00787-014-0550-y.

Crean, R.D., Crane, N.A., Mason, B.J., 2011. An evidence-based review of acute and long-term effects of cannabis use on executive cognitive functions. J. Addict. Med. 19353227. 5 (1), 1–8. doi:10.1097/ADM.0b013e31820c23fa.

Crippa, J.A., Zuardi, A.W., Martín-Santos, R., Bhattacharyya, S., Atakan, Z., McGuire, P., Fusar-Poli, P., 2009. Cannabis and anxiety: a critical review of the evidence. Hum. Psychopharmacol. 10991077. 24 (7), 515–523. Available from: http://www3.interscience.wiley.com/cgi-bin/fulltext/122541929/PDFSTART. doi:10.1002/hup.1048Brazil.

Devinsky, O., Patel, A.D., Cross, J.H., Villanueva, V., Wirrell, E.C., Privitera, M., Greenwood, S.M., Roberts, C., Checketts, D., VanLandingham, K.E., Zuberi, S.M., 2018. Effect of cannabidiol on drop seizures in the lennox–gastaut syndrome. N. Engl. J. Med. 15334406. 378 (20), 1888–1897. Available from: http://www.nejm.org/medical-index. doi:10.1056/NEJMoa1714631.

Di Forti, M., Quattrone, D., Freeman, T.P., Tripoli, G., Gayer-Anderson, C., Quigley, H., Rodriguez, V., Jongsma, H.E., Ferraro, L., Cascia, C.L., Barbera, D.L., Tarricone, I., Berardi, D., Szöke, A., Arango, C., Tortelli, A., Velthorst, E., Bernardo, M., Del-Ben, C.M., Menezes, P.R., Selten, J.P., Jones, P.B., Kirkbride, J.B., Rutten, B.P., de Haan, L., Sham, P.C., van Os, J., Lewis, C.M., Lynskey, M., Morgan, C., Murray, R.M., Amoretti, S., Arrojo, M., Baudin, G., Beards, S., Bernardo, M., Bobes, J., Bonetto, C., Cabrera, B., Carracedo, A., Charpeaud, T., Costas, J., Cristofalo, D., Cuadrado, P., Díaz-Caneja, C.M., Ferchiou, A., Franke, N., Frijda, F., Garcínu Eernardo, E., Garcia-Portilla, P., González, E., Hubbard, K., Jamain, S., Jiménez-López, E., Leboyer, M., López Montoya, G., Lorente-Rovira, E., Loureiro, C.M.,

Marrazzo, G., Martínez, C., Matteis, M., Messchaart, E., Moltó, M.D., Nacher, J., Olmeda, M.S., Parellada, M., González Peñas, J., Pignon, B., Rapado, M., Richard, J.R., Rodríguez Solano, J.J., Roldán Díaz, L., Ruggeri, M., Sáiz, P.A., Sánchez, E., Sanjuán, J., Sartorio, C., Schürhoff, F., Seminerio, F., Shuhama, R., Sideli, L., Stilo, S.A., Termorshuizen, F., Tosato, S., Tronche, A.M., van Dam, D., van der Ven, E., 2019. The contribution of cannabis use to variation in the incidence of psychotic disorder across Europe (EU-GEI): a multicentre case-control study. Lancet Psychiatry. 22150374. 6 (5), 427–436. Available from: https://www.journals.elsevier.com/the-lancet-psychiatry. doi:10.1016/S2215-0366(19)30048-3.

ElSohly, M.A., Slade, D., 2005. Chemical constituents of marijuana: the complex mixture of natural cannabinoids. Life Sci. 00243205. 78 (5), 539–548. doi:10.1016/j.lfs.2005.09.011.

Gaston, T.E., Bebin, E.M., Cutter, G.R., Liu, Y., Szaflarski, J.P., 2017. Interactions between cannabidiol and commonly used antiepileptic drugs. Epilepsia. 15281167. 58 (9), 1586–1592. Available from: http://www3.interscience.wiley.com/journal/117957420/issueyear?year=2009. doi:10.1111/epi.13852.

Gundersen, T.D., Jørgensen, N., Andersson, A.M., Bang, A.K., Nordkap, L., Skakkebæk, N.E., Priskorn, L., Juul, A., Jensen, T.K., 2015. Association between use of marijuana and male reproductive hormones and semen quality: a study among 1,215 healthy young men. Am. J. Epidemiol. 14766256. 182 (6), 473–481. Available from: http://aje.oxfordjournals.org/. doi:10.1093/aje/kwv135.

Hasin, D.S., Saha, T.D., Kerridge, B.T., Goldstein, R.B., Chou, S.P., Zhang, H., Jung, J., Pickering, R.P., Ruan, J., Smith, S.M., Huang, B., Grant, B.F., 2015. Prevalence of marijuana use disorders in the United States between 2001-2002 and 2012-2013. JAMA Psychiatry. 2168622X. 72 (12), 1235–1242. Available from: http://archpsyc.jamanetwork.com/data/Journals/PSYCH/934694/yoi150057.pdf. doi:10.1001/jamapsychiatry.2015.1858.

Layden, J.E., Ghinai, I., Pray, I., Kimball, A., Layer, M., Tenforde, M.W., Navon, L., Hoots, B., Salvatore, P.P., Elderbrook, M., Haupt, T., Kanne, J., Patel, M.T., Saathoff-Huber, L., King, B.A., Schier, J.G., Mikosz, C.A., Meiman, J., 2020. Pulmonary illness related to e-cigarette use in Illinois and Wisconsin – final report. N. Engl. J. Med. 15334406. 382 (10), 903–916. Available from: http://www.nejm.org/medical-index. doi:10.1056/NEJMoa1911614.

Lev-Ran, S., Roerecke, M., Foll, B.L., George, T.P., McKenzie, K., Rehm, J., 2014. The association between cannabis use and depression: a systematic review and meta-analysis of longitudinal studies. Psychol. Med. 0033-2917. 44 (4), 797–810. doi:10.1017/s0033291713001438.

Mammen, G., Rehm, J., Rueda, S., 2016. Vaporizing cannabis through e-cigarettes: prevalence and socio-demographic correlates among Ontario high school students. Can. J. Public Health. 19207476. 107 (3), e337. Available from: http://journal.cpha.ca/index.php/cjph/article/download/5747/3455. doi:10.17269/CJPH.107.5747.

Metz, T.D., Stickrath, E.H., 2015. Marijuana use in pregnancy and lactation: a review of the evidence. Am. J. Obstet. Gynecol. 10976868. 213 (6), 761–778. Available from: http://www.elsevier.com/inca/publications/store/6/2/3/2/7/7/index.htt. doi:10.1016/j.ajog.2015.05.025.

Pertwee, R.G., 2006. The pharmacology of cannabinoid receptors and their ligands: an overview. Int. J. Obes. 0307-0565. 30 (S1), S13. doi:10.1038/sj.ijo.0803272.

Ramaekers, J.G., Berghaus, G., Van Laar, M., Drummer, O.H., 2004. Dose related risk of motor vehicle crashes after cannabis use. Drug Alcohol Depend. 03768716. 73 (2), 109–119. Available from: www.elsevier.com/locate/drugalcdep. doi:10.1016/j.drugalcdep.2003.10.008.

Rezkalla, S.H., Sharma, P., Kloner, R.A., 2003. Coronary no-flow and ventricular tachycardia associated with habitual marijuana use. Ann. Emerg. Med. 01960644. 42 (3), 365–369.

Available from: http://www.elsevier.com/inca/publications/store/6/2/3/2/7/8/index.htt. doi:10.1016/S0196-0644(03)00426-8.

Singh, A., Saluja, S., Kumar, A., Agrawal, S., Thind, M., Nanda, S., Shirani, J., 2018. Cardiovascular Complications of Marijuana and Related Substances: a Review. Cardiol. Ther. 2193-8261. 7 (1), 45–59. doi:10.1007/s40119-017-0102-x.

Sorensen, C.J., DeSanto, K., Borgelt, L., Phillips, K.T., Monte, A.A., 2017. Cannabinoid hyperemesis syndrome: diagnosis, pathophysiology, and treatment—a systematic review. J. Med. Toxicol. 19376995. 13 (1), 71–87. Available from: http://www.springer.com/biomed/pharmaceutical+science/journal/13181. doi:10.1007/s13181-016-0595-z.

Tashkin, D.P., 2013. Effects of marijuana smoking on the lung. Ann. Am. Thorac. Soc. 23256621. 10 (3), 239–247. Available from: http://www.atsjournals.org/doi/pdf/10.1513/AnnalsATS.201212-127FR. doi:10.1513/AnnalsATS.201212-127FR.

Thomas, G., Kloner, R.A., Rezkalla, S., 2014. Adverse cardiovascular, cerebrovascular, and peripheral vascular effects of marijuana inhalation: what cardiologists need to know. Am. J. Cardiol. 18791913. 113 (1), 187–190. doi:10.1016/j.amjcard.2013.09.042.

Volkow, N.D., Wang, G., Telang, F., et al., 2014. Decreased dopamine brain reactivityin marijuna abusers is associated with negative emotionality and addictionseverity. Proc. Natl. Acad. Sci. USA 111 (30), E3149–E3156.

Zou, S., Kumar, U., 2018. Cannabinoid receptors and the endocannabinoid system: signaling and function in the central nervous system. Int. J. Mol. Sci. 1422-0067. 19 (3), 833. doi:10.3390/ijms19030833.

CHAPTER

Perioperative management of patients using cannabinoids

15

Jon B. Lee, Sajan H. Shah, and Jennifer M. Hah

Department of Anesthesiology, Perioperative and Pain Medicine, Division of Pain Medicine, Stanford University, Stanford, CA, United States

Introduction

Cannabinoid use, both medicinal and recreational, has significantly increased, likely driven by evolving societal attitudes, expanding legalization, and growing evidence of therapeutic benefit. According to the United Nations Office on Drugs and Crime (2024), cannabinoid use has grown by 28% in the past decade. In the general population aged 15–64, cannabis use has grown from an estimated 3.5% of the population in 2000 to 4.4% of the population in 2022. Cannabinoids interact with the endocannabinoid system and can influence numerous physiological processes relevant to anesthesia and surgery, including cardiovascular stability, respiratory function, coagulation, and pain perception (Echeverria-Villalobos et al., 2019). This chapter aims to provide a comprehensive overview of the existing evidence and practical recommendations to guide anesthesiologists, surgeons, and pain management specialists in effectively managing patients who use cannabinoids in the perioperative period.

Cannabis use disorder in the perioperative period has been associated with potential increases in morbidity, perioperative complications, in-house mortality, and total hospitalization costs (Potnuru et al., 2023). Another study found that active cannabis use was associated with an increased perioperative risk of myocardial infarction (Goel et al., 2020). Furthermore, cannabis users undergoing surgery have demonstrated an increased incidence of airway hyperreactivity and bronchospasm (Alexander and Joshi, 2019). Given the growing prevalence of cannabis use worldwide and the developing literature on its health-related impacts, clinicians involved in perioperative care must be aware of the potential implications of cannabinoids on anesthetic planning, postoperative pain control, and patient outcomes.

Preoperative assessment

The increasing prevalence underscores the importance of systematic screening and assessment of cannabinoid use during preoperative evaluation, highlighting the necessity for informed perioperative management strategies. In 2020, the American

Cannabinoids for Chronic Pain. https://doi.org/10.1016/B978-0-443-36773-1.00001-5

Society of Regional Anesthesia and Pain Medicine formed a set of consensus guidelines on the perioperative management of patients on cannabis and cannabinoids. The consensus group recommends universal screening for cannabinoids before surgery, which should include the type of cannabis, time of last consumption, route of administration, amount, and frequency of use. Unless clinically indicated due to concern for acute intoxication, universal toxicology screening is generally not necessary. Clinicians should specifically ask about cannabinoid use, as many patients may not volunteer this information without prompting. Validated screening tools such as the Cannabis Use Disorders Identification Test-Revised (CUDIT-R) and the Tobacco, Alcohol, Prescription medication, and other Substance use (TAPS-1) are available to assist clinicians in systematically evaluating cannabis use and identifying patients who may have cannabis use disorder (CUD) (Adamson et al., 2010). The CUDIT-R is an 8-item, self-report screening tool designed to identify individuals at risk for CUD. Each item is scored 0–4, with a total score range of 0–32. A score of 8 or more on the CUDIT-R suggests hazardous cannabis use, while a score of 12 or more indicates a probable cannabis use disorder (Kritikos et al., 2025). The TAPS-1 is a four-item screener that asks about past 12-month use of tobacco, alcohol, prescription medications (for nonmedical reasons), and illicit drugs. TAPS-1 demonstrates high sensitivity for detecting unhealthy substance use. For marijuana specifically, sensitivity is 0.82 and specificity is 0.93, and for other substances, specificity is generally 0.85 or higher (McNeely et al., 2016). Incorporating these validated screening tools into routine preoperative assessments can help clinicians identify at-risk patients, facilitate early intervention, and optimize perioperative outcomes.

Patient education and counseling play a crucial role in the preoperative optimization of patients who use cannabinoids. Clinicians should communicate the potential risks associated with perioperative cannabis use, including cardiovascular and respiratory complications, as well as challenges in postoperative pain management (Echeverria-Villalobos et al., 2019). In addition, patients should be informed of other long-term health effects related to cannabis use, such as cognitive impairment, anxiety, dependence, sleep disturbance, and psychosis (Gorelick, 2023). Comprehensive education about these broader health considerations can empower patients to make informed decisions, enhance adherence to perioperative recommendations, and improve surgical outcomes.

Preoperative management

Counseling should emphasize the importance of preoperative optimization through cannabis cessation or reduction to enhance patient safety and improve surgical outcomes. Recommendations regarding the optimal timing for cannabis discontinuation vary considerably in the literature. Some authors recommend cessation of cannabis at least 72 h before elective surgery (Echeverria-Villalobos et al., 2019; Huson et al., 2018). In contrast, others suggest a more conservative approach, withholding cannabis for up to 10 days before surgery to maximize patient safety and minimize perioperative risk (Stewart and Fong, 2021).

Major professional societies have also guided perioperative cannabis management. The Society for Perioperative Assessment and Quality Improvement (SPAQI) recommends stopping cannabis use at least 72 h before elective surgery, noting that a more extended period of abstinence (up to 2 weeks) is ideal for patients who smoke cannabis regularly, to reduce airway irritability (Cummings et al., 2021). The Perioperative Pain and Addiction Interdisciplinary Network (PAIN) suggests tapering or cessation at least 7 days before surgery for patients consuming high daily doses of cannabis (>1.5 g/day smoked cannabis, > 300 mg/day CBD oil, or > 20 mg/day THC oil), while cautioning against abrupt cessation within 24 h of surgery (Ladha et al., 2021). The American Society of Regional Anesthesia and Pain Medicine (ASRA) advises postponing elective surgery for patients who are acutely intoxicated from cannabis due to compromised decision-making capacity, or who have smoked cannabis within 2 h of surgery, to minimize the risk of acute myocardial infarction. However, ASRA does not provide a definitive recommendation regarding longer-term cannabis cessation, instead emphasizing individualized counseling and a thorough risk-benefit assessment (Shah et al., 2023). There are select FDA-approved formulations of synthetic cannabinoids, such as dronabinol, nabilone, and Epidiolex, for which the above guidelines may not apply. Dronabinol and nabilone are FDA-approved for chemotherapy-induced nausea and vomiting, while Epidiolex is approved for specific forms of epilepsy (Lennox–Gastaut syndrome, Tuberous Sclerosis, and Dravet syndrome). Abrupt cessation of these FDA-approved formulations is not recommended, as this may adversely affect the patient's perioperative course (Cummings et al., 2021; Shah et al., 2023).

Summary and practical recommendations

In summary, the existing recommendations for perioperative cannabis cessation vary considerably, ranging from at least 72 h to as long as two weeks before elective surgery, depending on the patient's pattern of cannabis use and the surgical context. Given the limited evidence, further research is needed to establish clear, evidence-based guidelines regarding the ideal timing and duration for cessation of THC-containing products. Until more definitive data become available, clinicians should provide tailored recommendations based on patient-specific factors, including frequency, dose, route of cannabis use, underlying medical conditions, and surgical risk. Clear communication, shared decision-making, and patient education are essential to optimize perioperative outcomes and patient safety.

For patients who choose not to discontinue cannabis use pre-operatively, they should be counseled regarding the risk of cannabis withdrawal syndrome. Patients on high doses of cannabinoids, specifically THC-containing products, are at a higher risk than patients consuming CBD-dominant strains. Although the level of available evidence is low, dronabinol, a synthetic THC analog, could be used perioperatively to treat severe cannabis withdrawal symptoms (Shah et al., 2023; Levin et al., 2011). This formulation is FDA-approved for refractory chemotherapy-induced nausea and vomiting and would be considered off-label for use in mitigating cannabis withdrawal syndrome.

For patients motivated to discontinue cannabis use preoperatively, clinicians should provide targeted counseling and support. Effective psychosocial interventions include cognitive-behavioral therapy (CBT) and motivational enhancement therapy, which can help patients identify triggers, develop coping strategies, and maintain abstinence (Gates et al., 2016). Pharmacotherapy may also be considered in selected cases, particularly for patients experiencing significant withdrawal symptoms or difficulty maintaining abstinence. Although no medications are explicitly FDA-approved for cannabis cessation, off-label use of pharmacologic agents such as gabapentin, N-acetylcysteine, or short-term anxiolytics may be considered to manage withdrawal symptoms and support abstinence efforts (Nielsen et al., 2019). Multidisciplinary collaboration involving anesthesiologists, surgeons, primary care providers, psychologists, and addiction specialists can enhance the effectiveness of these interventions, improve patient adherence, and ultimately optimize perioperative outcomes.

Intraoperative cannabis management

The rising prevalence of cannabis use among surgical patients has made it imperative for anesthesiologists and perioperative physicians to understand its intraoperative implications. The pharmacologic profile of cannabinoids is complex. Delta-9-tetrahydrocannabinol (THC) and cannabidiol (CBD), the two most well-studied compounds, exert their effects via CB1 and CB2 receptors, producing a wide range of physiologic responses (Lee et al., 2024). Both agents are highly lipophilic, leading to prolonged half-lives and accumulation in fatty tissue, particularly in chronic users. Their hepatic metabolism—primarily through cytochrome P450 isoenzymes CYP2C9 and CYP3A4—also introduces potential drug interactions with common anesthetic agents (Lee et al., 2024).

Anesthetic requirements

Chronic cannabis users may demonstrate increased tolerance to sedatives and general anesthetics, including propofol and volatile agents. This is presumed to stem from receptor-level adaptations, although clinical data are mostly limited to retrospective studies. Several reports suggest that regular cannabis users require higher doses of induction agents to achieve adequate anesthetic depth, while acute intoxication may paradoxically lower anesthetic needs (Shah et al., 2023; Lee et al., 2024). Complicating this picture is the fact that cannabis alters EEG patterns and increases sympathetic tone—factors that can obscure traditional markers of anesthetic depth and increase the risk of intraoperative awareness. Until higher-quality prospective data become available, intraoperative anesthetic titration in cannabis users should rely heavily on multimodal monitoring and clinical judgment.

Airway and respiratory considerations

Inhaled cannabis has been associated with airway hyperreactivity, increased secretions, and impaired mucociliary clearance, contributing to elevated risks of bronchospasm, laryngospasm, and perioperative hypoxia (Echeverria-Villalobos et al., 2019; Shah et al., 2023). Chronic smoking may produce obstructive pulmonary changes, with reduced FEV1/FVC ratios on spirometry. These patients may benefit from preinduction bronchodilators, humidified oxygen, and consideration of regional techniques when feasible. Given overlapping risk factors with tobacco smoking, it is reasonable to approach chronic cannabis smokers with similar respiratory precautions.

Hemodynamic effects

Cannabis exerts dose-dependent cardiovascular effects that vary by acute versus chronic use. Acute THC exposure typically activates the sympathetic nervous system, leading to tachycardia and episodic hypertension. Conversely, chronic use may be associated with parasympathetic dominance, manifesting as bradycardia, hypotension, and orthostatic intolerance (Lee et al., 2024). THC also promotes vasodilation via CB1-mediated nitric oxide release, raising concern for intraoperative hypotension. Although rare, severe events such as arrhythmias and coronary vasospasm have been documented (Kritikos et al., 2025). Intraoperative management should include vigilant hemodynamic monitoring, volume responsiveness assessment, and preparedness for vasopressor support, particularly in chronic users.

Temperature regulation

Impairments in thermoregulation have been observed in cannabis users, particularly with high-dose or chronic THC exposure. This may manifest as intraoperative hypothermia or paradoxical shivering. Through modulation of TRPV1 receptors, CBD may also influence central thermoregulatory centers (Shah et al., 2023). These temperature fluctuations can be exacerbated by general or neuraxial anesthesia. Routine use of forced-air warming systems, warmed IV fluids, and active temperature monitoring is advisable for cannabis users undergoing surgery.

Gastrointestinal and antiemetic considerations

Cannabinoids have long been recognized for their antiemetic properties and are FDA-approved for chemotherapy-induced nausea and vomiting. However, chronic cannabis use has been linked to delayed gastric emptying, an increased risk of postoperative nausea and vomiting (PONV), and cannabis hyperemesis syndrome (Mims et al., 2024). These paradoxical effects may be underrecognized in the perioperative setting. Daily users may require more aggressive PONV prophylaxis, and anesthetic agents that delay gastric emptying should be used judiciously. In high-risk cases, combining NK1 antagonists with serotonin antagonists and dexamethasone may offer benefits.

Summary and practical recommendations

Cannabis use is no longer a perioperative outlier. It is necessary to understand pharmacologic variability, the potential for tolerance or sensitization, and organ system effects associated with cannabinoids. Management should be individualized: titrate anesthetics carefully, monitor for airway reactivity and cardiovascular instability, and anticipate altered responses to temperature and antiemetics. SPQAI and ASRA recommend delaying elective surgery when acute cannabis use is suspected, optimizing nonopioid analgesics, and using enhanced recovery protocols (Cummings et al., 2021; Shah et al., 2023). Equally important is preoperative screening for cannabis use disorder, coupled with patient education to mitigate perioperative risks.

Postoperative management

Multimodal analgesia

Multimodal analgesia remains the standard of care for postoperative pain management and should be similarly optimized for patients who are using cannabis before surgery. To summarize, nonopioid analgesics, including acetaminophen and NSAIDs or COX-2 inhibitors, should be scheduled unless contraindicated.

The risks and benefits of additional nonopioid adjuvant medications should be considered carefully. Gabapentinoids have demonstrated equivocal analgesia (Canadian Perioperative Anesthesia Clinical Trials (PACT) Group, 2020) and modest opioid-sparing effects after surgery (Viderman et al., 2024). Gabapentinoids increase the risk of severe respiratory depression when combined with CNS depressants and opioids (Tambon et al., 2023). In addition, comorbidities such as respiratory disease, advanced age, and chronic kidney disease further increase the risk of respiratory events when prescribing gabapentinoids (Park et al., 2022). The risks and benefits of gabapentin and pregabalin should be weighed carefully in the context of the multimodal analgesia plan. Cannabinoid receptor activation results in the inhibition of voltage-gated calcium channels, enhancing the synergistic effects between gabapentinoids and THC in preclinical studies (Atwal et al., 2019). Therefore patients should be counseled regarding the risks of concomitant gabapentinoid and cannabinoid use, which can enhance the risks of respiratory depression and sedation (Shah et al., 2023).

Alpha-2 agonists, such as dexmedetomidine and clonidine, may provide opioid-sparing analgesic effects and antiemetic benefits in the immediate 12 to 24 h after surgery. Still, the benefits should be carefully weighed against the risks of hypotension, bradycardia, and sedation (Sanchez Munoz et al., 2017). Tizanidine is a centrally acting alpha-2 agonist and muscle relaxant with emerging evidence supporting its use in postoperative pain management. Among patients undergoing inguinal hernia repair, tizanidine 4 mg, administered 1 h preoperatively and twice daily for 1 week as part of a multimodal analgesic plan, significantly reduced postoperative pain and additional analgesic consumption while improving the return to daily activities and quality of life (Yazicioğlu et al., 2016). Further research is needed to determine the role of tizanidine in multimodal analgesic protocols. Although there is no direct evidence addressing pharmacological interactions between cannabis and alpha-2 agonists in the perioperative period, both agents have the potential for additive sedative effects, especially in combination with other CNS depressants.

For patients not receiving regional or neuraxial analgesia, intravenous lidocaine is another modality to reduce pain and opioid use in the immediate postoperative phase, 24 to 72 h after surgery. Care should be taken to monitor for neurotoxic adverse effects, including dizziness, blurred vision, lightheadedness, myoclonus, circumoral numbness, tinnitus, metallic taste in the mouth, confusion, and slurred speech. To date, there is limited evidence addressing any interactions between cannabis and regional anesthesia, neuraxial anesthesia, and systemic lidocaine. In addition, the effect of these multimodal analgesia techniques, specifically among cannabis users, has not been reported.

Subanesthetic ketamine represents another multimodal analgesia component with the potential for decreasing acute pain intensity and postoperative opioid consumption in the acute phase. Preclinical evidence suggests ketamine's central antinociceptive effects are partially mediated by endogenous cannabinoid signaling, and augmenting endocannabinoid tone may potentiate the analgesic effect of ketamine (Pacheco et al., 2019). Clinically, there is potential for additive CNS effects, such as sedation and psychomotor impairment, when combining NMDA antagonists with cannabis. Similarly, both ketamine and cannabis may promote tachycardia and hypertension, which could lead to hemodynamic instability, particularly among patients with cardiovascular disease. However, no direct evidence addresses the impact of subanesthetic ketamine doses on cannabis users after surgery.

Given conflicting research on the impact of cannabis use on postoperative pain and opioid use requirements, future research is needed to determine whether cannabis has a clear role as a component in multimodal analgesia and whether it can be considered a safe, opioid-sparing, pain-alleviating therapy. Postoperative pain management among cannabis users should focus on optimizing the administration of multimodal analgesia.

Postoperative opioid prescribing considerations

Chronic cannabis use may be a risk factor for elevated postoperative pain. Similarly, regular cannabis use can blunt the analgesic effect of opioids, and patients may require higher doses for pain control, thereby increasing the risk of opioid-related adverse events. In a retrospective study of 34,521 patients undergoing elective surgery at a single center, 1683 reported cannabis use in the 30 days before their operation. Cannabis use was associated with increased postoperative opioid consumption and increased pain in the first 24 h, but it was not associated with an increased risk of pulmonary complications during hospitalization (Ekrami et al., 2024). Of note, patients receiving regional or neuraxial anesthesia and those with chronic pain diagnoses were excluded from the study. Future research is needed to examine the impact of these critical factors on the surgical outcomes of patients with varying degrees of preoperative cannabis use.

Contrary findings have been reported in disparate surgical populations, demonstrating the lack of association between preoperative cannabis use and increased postoperative pain and opioid use (Puri et al., 2024). Beyond the immediate postoperative period, preoperative cannabis use has not demonstrated an association with increased opioid use in the first 3 to 7 days after liver transplant (Runge et al., 2025). Among patients undergoing reduction mammoplasty, preoperative cannabis use was not associated with postoperative pain, opioid use, or surgical complications (Shamamian et al., 2024). In a retrospective cohort study of 1216 patients undergoing one to three-level lumbar fusions, cannabis users reported lower rates of opioid use two to six months after surgery and lower rates of high-dose opioid use (≥100 mg morphine equivalents per day) in the first month after surgery (Mirpuri et al., 2024). In this context, more conservative opioid prescribing to cannabis users may be warranted in specific surgical populations.

Aside from the equivocal impact of cannabis on postoperative pain and opioid use, cannabis use following surgery is associated with many increased risks. For example, cannabis use may be associated with an increased risk of deep venous thrombosis, revision surgery, pneumonia, cerebrovascular accidents, and cardiac complications after total joint arthroplasty (Turan et al., 2024). Chronic cannabis users undergoing open aortic aneurysm repair have an increased risk of postoperative stroke and respiratory complications (Bernal et al., 2024). Cannabis use is associated with a constellation of adverse surgical outcomes, including tachycardia, myocardial infarction, arrhythmias, stroke, coronary and cerebral vasospasm, coagulopathy, cardiac arrest, cardiomyopathy, bronchial hyperreactivity, airway inflammation, delayed gastrointestinal motility, and postoperative nausea and vomiting (Shah et al., 2023). Cannabis use disorder specifically has been associated with a modestly increased risk of perioperative morbidity and mortality after major, elective, noncardiac surgery (Potnuru et al., 2023).

Given the equivocal effect of cannabis use on postoperative pain and opioid use, opioid prescribing should be reserved for breakthrough pain, with a focus on prioritizing nonopioid multimodal analgesia for perioperative pain management. Further, patients should be educated regarding the potential for adverse surgical outcomes in the context of continued cannabis use after surgery.

Monitoring for cannabis withdrawal syndrome

After surgery, patients may be at risk for cannabis withdrawal, as symptoms can manifest one to two days after drastic reductions or cessation of heavy or chronic cannabis use. Symptoms often peak within 2 to 6 days and persist for one to two weeks, although some patients may experience a slow decrease in symptoms over many weeks. Patients may experience sleep disturbances, irritability, anger, anxiety, depression, headache, restlessness, decreased appetite, abdominal pain, muscle aches, tremors, sweating, fevers, or chills. Transient psychosis is rare and usually resolves within two weeks of abstinence. The DSM-5 diagnosis requires the presence of three or more of the following clinical criteria: irritability, anger or aggression, nervousness or anxiety, sleep difficulty (e.g., insomnia or vivid dreaming), decreased appetite or weight loss, restlessness, or depressed mood; combined with at least one physical symptom causing discomfort: abdominal pain, shakiness/tremors, sweating, fevers, chills, and/or headache (Echeverria-Villalobos et al., 2019). Cannabis users with comorbid opioid dependence rarely present with cannabis withdrawal syndrome.

Although most episodes of cannabis withdrawal are unlikely to require intervention and are self-limited, patients with moderate to severe withdrawal may benefit from treatment. There are no FDA-approved medications for the treatment of cannabis withdrawal, but dronabinol, nabiximols, and gabapentin have been reported as efficacious in small clinical trials. Nonmedication interventions may include exercise, meditation, relaxation techniques, herbal preparations, and cognitive behavioral therapy. Cognitive behavioral therapy and counseling are the mainstays of treatment for cannabis withdrawal. Medications may also target withdrawal symptoms, using zolpidem for insomnia and benzodiazepines for anxiety symptoms (Gorelick, 2023).

Summary and practical recommendations

The general management of postoperative pain, regardless of the presence of cannabis use, should focus on the administration of nonopioid multimodal analgesia. The standard emphasis on scheduled acetaminophen and NSAIDs or COX-2 inhibitors is appropriate. Although direct interactions between components of multimodal analgesia and the effects of perioperative cannabis use have been minimally examined, the potential for additive CNS effects should be considered with appropriate continued monitoring. Additional multimodal analgesia components may include gabapentin, pregabalin, alpha-2 agonists, neuraxial anesthesia, regional anesthesia/analgesia, intravenous lidocaine, and subanesthetic ketamine infusions.

There is conflicting evidence regarding the association between cannabis use and increased postoperative pain and persistent postoperative opioid use. A constellation of adverse surgical outcomes warrants prioritization of opioid-sparing strategies for postoperative pain management. Cannabis users may experience increased postoperative morbidity and mortality, ranging from reoperations, pneumonia, stroke, myocardial infarction, deep venous thrombosis, to postoperative nausea and vomiting. Opioid prescribing should be limited to the treatment of breakthrough pain in the context of optimized nonopioid management. Monitoring for cannabis withdrawal syndrome is warranted after surgery, and symptoms are typically self-limited, requiring minimal intervention.

Conclusion

There are many considerations spanning the perioperative management of patients using cannabinoids. There is wide variation in recommendations for preoperative cannabis cessation, ranging from 72 h to two weeks before elective surgery. Preoperative education regarding the risks of continued cannabis use and cannabis withdrawal syndrome should be provided. For patients motivated to discontinue cannabis use before surgery, effective interventions include cognitive behavioral therapy and motivational enhancement therapy. The various effects of cannabis use should be considered in the intraoperative phase, with careful titration of anesthetics, monitoring for airway reactivity, treatment of cardiovascular instability, anticipation of altered thermoregulation, and consideration of postoperative nausea and vomiting. After surgery, pain management should focus on optimizing nonopioid multimodal analgesia with conservative opioid prescribing. Continued monitoring for cannabis withdrawal syndrome during hospitalization and after discharge is essential and may serve as a key intervention point to reduce chronic cannabis use.

References

Adamson, S.J., Kay-Lambkin, F.J., Baker, A.L., Lewin, T.J., Thornton, L., Kelly, B.J., Sellman, J.D., 2010. An improved brief measure of cannabis misuse: the cannabis use disorders identification test-revised (CUDIT-R). Drug Alcohol Depend. 03768716. 110 (1-2), 137–143. doi:10.1016/j.drugalcdep.2010.02.017.

Alexander, J.C., Joshi, G.P., 2019. A review of the anesthetic implications of marijuana use. Baylor Univ. Med. Cent. Proc. 15253252. 32 (3), 364–371. Available from: http://www.tandfonline.com/loi/ubmc20. doi:10.1080/08998280.2019.1603034.

Atwal, N., Casey, S.L., Mitchell, V.A., Vaughan, C.W., 2019. THC and gabapentin interactions in a mouse neuropathic pain model. Neuropharmacology. 00283908. 144, 115–121. doi:10.1016/j.neuropharm.2018.10.006.

Bernal, L.R., Tanaka, A., Ikeno, Y., Afifi, R.O., Sandhu, H.K., Miller, C.C., Estrera, A.L., 2024. Outcomes after open repair of aortic aneurysms and dissections in cannabis consumers. JTCVS Open. 26662736. 22, 107–113. Available from https://www.sciencedirect.com/science/journal/26662736. doi:10.1016/j.xjon.2024.08.002.

Canadian Perioperative Anesthesia Clinical Trials (PACT) Group, 2020. Perioperative use of gabapentinoids for the management of postoperative acute pain: erratum. Anesthesiology 133 (5), 1159.

Cummings, K.C., Keshock, M., Ganesh, R., Sigmund, A., Kashiwagi, D., Devarajan, J., Grant, P.J., Urman, R.D., Mauck, K.F., 2021. Preoperative Management of Surgical Patients Using Dietary Supplements: society for Perioperative Assessment and Quality Improvement (SPAQI) Consensus Statement. Mayo Clin. Proc. 19425546. 96 (5), 1342–1355. Available from: http://www.journals.elsevier.com/mayo-clinic-proceedings. doi:10.1016/j.mayocp.2020.08.016.

Echeverria-Villalobos, M., Todeschini, A.B., Stoicea, N., Fiorda-Diaz, J., Weaver, T., Bergese, S.D., 2019. Perioperative care of cannabis users: a comprehensive review of pharmacological and anesthetic considerations. J. Clin. Anesth. 09528180. 57, 41–49. doi:10.1016/j.jclinane.2019.03.011.

Ekrami, E., Sari, S., Kopac, O., Wang, D., Mascha, E.J., Stamper, S., Esa, W.A.S., Nair, H., Ruetzler, K., Turan, A., 2024. Association between cannabis use and opioid consumption, pain, and respiratory complications after surgery: a retrospective cohort analysis. Anesth. Analg. 0003-2999. 139 (4), 724–733. doi:10.1213/ane.0000000000006785.

Gates, P.J., Sabioni, P., Copeland, J., Foll, B.L., Gowing, L., 2016. Psychosocial interventions for cannabis use disorder. Cochrane Database Syst. Rev. 1469493X. 2016 (5). Available from: http://as.wiley.com/WileyCDA/Brand/id-6.html. doi:10.1002/14651858.CD005336.pub4.

Goel, A., McGuinness, B., Jivraj, N.K., Wijeysundera, D.N., Mittleman, M.A., Bateman, B.T., Clarke, H., Kotra, L.P., Ladha, K.S., 2020. Cannabis use disorder and perioperative outcomes in major elective surgeries: a retrospective cohort analysis. Anesthesiology. 15281175. 132 (4), 625–635. Available from: http://journals.lww.com/anesthesiology/pages/default.aspx. doi:10.1097/ALN.0000000000003067.

Gorelick, D.A., 2023. Cannabis-related disorders and toxic effects. N. Engl. J. Med. 15334406. 389 (24), 2267–2275. Available from: http://www.nejm.org/medical-index. doi:10.1056/NEJMra2212152.

Huson, H.B., Granados, T.M., Rasko, Y., 2018. Surgical considerations of marijuana use in elective procedures. Heliyon. 24058440. 4 (9). Available from: http://www.journals.elsevier.com/heliyon/. doi:10.1016/j.heliyon.2018.e00779.

Kritikos, A.F., Taylor, B., Lamuda, P., Pollack, H., Schneider, J.A., 2025. Patterns of past month cannabis consumption and cannabis use disorder – insights from a nationally representative survey. Drug Alcohol Depend. 03768716. 272, 112680. doi:10.1016/j.drugalcdep.2025.112680.

Ladha, K.S., McLaren-Blades, A., Goel, A., Buys, M.J., Farquhar-Smith, P., Haroutounian, S., Kotteeswaran, Y., Kwofie, K., Foll, B.L., Lightfoot, N.J., Loiselle, J., Mace, H., Nicholls, J., Regev, A., Rosseland, L.A., Shanthanna, H., Sinha, A., Sutherland, A., Tanguay, R., Yafai, S., Glenny, M., Choi, P., Ladak, S.S.J., Leroux, T.S., Kawpeng, I., Samman, B., Singh, R., Clarke, H., 2021. Perioperative Pain and Addiction Interdisciplinary Network (PAIN): consensus recommendations for perioperative management of cannabis and cannabinoid-based medicine users by a modified Delphi process. Br. J. Anaesth. 14716771. 126 (1), 304–318. Available from: https://www.journals.elsevier.com/british-journal-of-anaesthesia. doi:10.1016/j.bja.2020.09.026.

Lee, B.H., Sideris, A., Ladha, K.S., Johnson, R.L., Wu, C.L., 2024. Cannabis and cannabinoids in the perioperative period. Anesth. Analg. 15267598. 138 (1), 16–30. Available from: http://journals.lww.com/anesthesia-analgesia/toc/publishahead. doi:10.1213/ANE.0000000000006070.

Levin, F.R., Mariani, J.J., Brooks, D.J., Pavlicova, M., Cheng, W., Nunes, E.V., 2011. Dronabinol for the treatment of cannabis dependence: a randomized, double-blind, placebo-controlled trial. Drug Alcohol Depend. 18790046. 116 (1-3), 142–150. doi:10.1016/j.drugalcdep.2010.12.010.

McNeely, J., Li-Tzy, W., Subramaniam, G., Sharma, G., Cathers, L.A., Svikis, D., Sleiter, L., Russell, L., Nordeck, C., Sharma, A., O'Grady, K.E., Bouk, L.B., Cushing, C., King, J., Wahle, A., Schwartz, R.P., 2016. Performance of the Tobacco, Alcohol, Prescription Medication, and Other Substance Use (TAPS) Tool for Substance Use Screening in Primary Care Patients. Ann. Intern. Med. 0003-4819. 165 (10), 690–699. doi:10.7326/m16-0317.

Mims, M.M., Parikh, A.C., Sandhu, Z., Demoss, N., Mhawej, R., Queimado, L., 2024. Surgery-related considerations in treating people who use cannabis: a review. JAMA

Otolaryngol. Head Neck Surg. 21686181. 150 (10), 918–924. Available from: http://archotol.jamanetwork.com/issues.aspx. doi:10.1001/jamaoto.2024.2545.

Mirpuri, P., Khalid, S.I., King, P., Roy, J.M., Elsamadicy, A., Mehta, A.I., Adogwa, O., 2024. Effect of cannabis use history on postoperative opioid utilization in lumbar fusion patients: an American retrospective study. Asian Spine J. 1976-1902. 18 (5), 639–646. doi:10.31616/asj.2024.0194.

Munoz, M.C.S., De Kock, M., Forget, P., 2017. What is the place of clonidine in anesthesia? Systematic review and meta-analyses of randomized controlled trials. J. Clin. Anesth. 18734529. 38, 140–153. Available from: www.elsevier.com/locate/jclinane. doi:10.1016/j.jclinane.2017.02.003.

Nielsen, S., Gowing, L., Sabioni, P., Foll, B.L., 2019. Pharmacotherapies for cannabis dependence. Cochrane Database Syst. Rev. 1469493X. 2019 (1). Available from: http://as.wiley.com/WileyCDA/Brand/id-6.html. doi:10.1002/14651858.CD008940.pub3.

Pacheco, D.D.F., Romero, T.R.L., Duarte, I.D.G., 2019. Ketamine induces central antinociception mediated by endogenous cannabinoids and activation of CB 1 receptors. Neurosci. Lett. 18727972. 699, 140–144. Available from: www.elsevier.com/locate/neulet. doi:10.1016/j.neulet.2019.01.059.

Park, C.M., Inouye, S.K., Marcantonio, E.R., Metzger, E., Bateman, B.T., Lie, J.J., Lee, S.B., Levin, R., Kim, D.H., 2022. Perioperative Gabapentin Use and In-Hospital Adverse Clinical Events among Older Adults after Major Surgery. JAMA Intern. Med. 21686114. 182 (11), 1117–1127. Available from: http://archinte.jamanetwork.com/issues.aspx. doi:10.1001/jamainternmed.2022.3680.

Potnuru, P.P., Jonna, S., Williams, G.W., 2023. Cannabis Use Disorder and Perioperative Complications. JAMA Surg. 21686262. 158 (9), 935–944. Available from: http://archsurg.jamanetwork.com/issues.aspx. doi:10.1001/jamasurg.2023.2403.

Puri, S., Ong, C., Chiu, Y.-F., Lebowitz, J., Sideris, A., Valle, A.G.D., Chalmers, B.P., 2024. Preoperative Cannabis Use Did Not Increase Opioid Utilization After Primary Total Knee Arthroplasty in a Propensity Score–Matched Model. HSS J. 1556-3316. 20 (2), 268–273. doi:10.1177/15563316221151158.

Runge, A., Loeb, B., Shui, A.M., Fenton, C., Lai, J., Rubin, J., 2025. Cannabis and liver transplant in the era of legalization: effects of pretransplant cannabis use on postoperative opioid use and transplant outcomes. Liver Transpl. 1527–6465. doi:10.1097/lvt.0000000000000585.

Shah, S., Schwenk, E.S., Sondekoppam, R.V., Clarke, H., Zakowski, M., Rzasa-Lynn, R.S., Yeung, B., Nicholson, K., Schwartz, G., Michael Hooten, W., Wallace, M., Viscusi, E.R., Narouze, S., 2023. ASRA Pain Medicine consensus guidelines on the management of the perioperative patient on cannabis and cannabinoids. Reg. Anesth. Pain Med. 1098-7339. 48 (3), 97–117. doi:10.1136/rapm-2022-104013.

Shamamian, P.E., Kwon, D.Y., Oleru, O., Seyidova, N., Kim, E., Genet, S., Gyasi, A., Wang, C.Y., Henderson, P.W., 2024. The Use of Recreational Cannabis Among Breast Reduction Patients: characteristics, Complications, and Immediate Postoperative Analgesic Needs. Plast. Surg. 2292–5503. doi:10.1177/22925503241276549.

Stewart, C., Fong, Y., 2021. Perioperative cannabis as a potential solution for reducing opioid and benzodiazepine dependence. JAMA Surg. 21686262. 156 (2), 181–190. Available from: http://archsurg.jamanetwork.com/issues.aspx. doi:10.1001/jamasurg.2020.5545.

Tambon, M., Montarnal, B., Lepetit, M., Lapeyre-Mestre, M., 2023. Non-opioid antinociceptive drugs : risk of respiratory depression and death related to concomitant use of gabapentinoids in addition to opioids. Expert Opin. Drug Saf. 1474-0338. 22 (3), 183–194. doi:10.1080/14740338.2023.2193397.

Turan, O.A., Rullan, P.J., Pasqualini, I., Pumo, T.J., Surace, P.A., Murray, T.G., Piuzzi, N.S., 2024. Cannabis use following total joint arthroplasty is associated with increased risks? A meta-analysis. J. Arthroplasty. 15328406. 39 (2), 533. Available from: https://www.sciencedirect.com/science/journal/08835403. doi:10.1016/j.arth.2023.07.008.

Viderman, D., Aubakirova, M., Salamat, A., Kaldybayev, D., Sadir, N., Tankacheyev, R., Abdildin, Y.G., 2024. The Impact of Gabapentinoids on Pain-Related Outcomes after Knee and Hip Surgery: a Systematic Review with Meta-Analysis of Randomized Controlled Trials. J. Clin. Med. 2077-0383. 13 (14), 4205. doi:10.3390/jcm13144205.

Yazicioğlu, D., Caparlar, C., Akkaya, T., Mercan, U., Kulaçoğlu, H., 2016. Tizanidine for the management of acute postoperative pain after inguinal hernia repair. Eur. J. Anaesthesiol. 0265-0215. 33 (3), 215–222. doi:10.1097/eja.0000000000000371.

CHAPTER

16 Cannabis and children's health: Navigating the intersection of medicine, ethics, and policy

Robert Moghim
Interventional Pain & Spine, Colorado Pain Care, Denver, CO, United States

Introduction

Cannabis, once relegated to the fringes of both medicine and society, now occupies a complex but increasingly legitimized role in healthcare. Its medical use has expanded in recent years, driven by patient advocacy, evolving laws, and growing evidence of therapeutic potential. However, no aspect of this discussion is more sensitive than its application in children. As clinicians, we are entrusted with the health and well-being of our most vulnerable population. Decisions about pediatric care demand an especially high burden of proof, particularly when the treatment involves a substance as historically controversial and politically charged as cannabis.

Having practiced pain medicine in Colorado since before recreational cannabis was legalized, I have observed firsthand both the benefits and risks associated with its integration into healthcare. As a physician, educator, and policy advisor, I approach this subject with a commitment to data, safety, and responsible innovation. The goal of this chapter is not to take a polarized stance but to examine the evolving landscape of cannabis in pediatric medicine with rigor and empathy.

Historical background: From pharmacopoeia to prohibition

Cannabis has a long history in medicine, dating back thousands of years. Ancient Chinese texts reference its use for treating ailments such as gout, malaria, and rheumatism. In India, it was used for epilepsy and as an analgesic. By the 19th century, cannabis extracts were commonly found in Western medicine and were listed in the United States Pharmacopoeia from 1850 until 1942. Physicians prescribed these preparations for conditions ranging from labor pain to migraines to insomnia.

This period of medical acceptance ended abruptly in the early 20th century. A combination of xenophobic fear, misinformation, and bureaucratic influence contributed to the Marihuana Tax Act of 1937, which effectively criminalized cannabis in the United States. The Controlled Substances Act of 1970 later classified cannabis as

Cannabinoids for Chronic Pain. https://doi.org/10.1016/B978-0-443-36773-1.00013-1

a Schedule I drug, defined as having no accepted medical use and a high potential for abuse. This designation created substantial barriers to research, clinical access, and medical legitimacy for decades to follow (Grinspoon and Bakalar, 1997).

Meaningful reconsideration of cannabis in medicine did not reemerge until the AIDS crisis of the 1980s and 1990s, when patients suffering from cachexia and severe nausea found cannabis to be one of the few effective remedies. Their experiences and advocacy paved the way for medical marijuana initiatives in California and other states, ultimately contributing to broader policy reform and renewed research interest.

Charlotte's web: The catalyst for pediatric medical cannabis

While the medical cannabis movement gained traction among adults with cancer and chronic illness, the turning point in pediatric acceptance emerged from a single case that captured national attention: that of Charlotte Figi. Charlotte was diagnosed with Dravet syndrome, a rare and severe form of epilepsy that begins in infancy. Despite multiple medications, ketogenic diets, and experimental therapies, her seizures persisted—sometimes occurring hundreds of times per week—resulting in significant cognitive decline and physical distress.

In a final effort to find relief, her parents sought a high-CBD, low-THC cannabis extract cultivated by the Stanley brothers in Colorado. Within weeks of initiating treatment, Charlotte's seizure frequency and intensity declined dramatically. The extract, later named "Charlotte's Web," became a symbol of hope for families across the United States (Maa and Figi, 2014).

Charlotte's case elevated pediatric cannabis into the public consciousness in a way no study or publication could. Media coverage, including Dr. Sanjay Gupta's widely viewed CNN documentary "Weed," influenced public opinion and prompted legislative changes across numerous states to allow medical cannabis use in children with epilepsy (Gupta, 2013).

Her story illustrates a central theme of this chapter: anecdotal evidence, while not sufficient on its own, can serve as a powerful catalyst for scientific investigation and policy reform.

The endocannabinoid system and pediatric implications

The human body contains a complex regulatory network known as the endocannabinoid system (ECS), which plays a crucial role in maintaining physiological balance. This system includes cannabinoid receptors (CB1 and CB2), endogenous ligands such as anandamide and 2-arachidonoylglycerol, and enzymes responsible for synthesizing and breaking down these compounds.

CB1 receptors are densely concentrated in the brain and central nervous system, whereas CB2 receptors are more prevalent in peripheral organs, particularly within the immune system. Cannabidiol (CBD) and tetrahydrocannabinol (THC), the two most studied phytocannabinoids, interact with the ECS in different ways. THC binds directly to CB1 receptors and produces psychoactive effects, while CBD modulates receptor activity indirectly and does not induce intoxication (Ibeas Bih et al., 2015).

Understanding the ECS is essential when considering pediatric use. The adolescent brain is still developing, and CB1 receptors are abundant in regions responsible for memory, executive function, and emotional regulation. Exposure to THC during these critical developmental periods may interfere with these processes and has been associated with cognitive deficits and an increased risk of psychiatric disorders (Volkow et al., 2014).

CBD, by contrast, does not appear to produce these adverse effects and has been the focus of most pediatric research. Nonetheless, because the ECS plays a significant role in neural development, any cannabinoid—even those generally considered safe—should be approached with caution in children.

Evidence-based use cases: Where cannabis shows promise

Epilepsy and seizure disorders

The strongest clinical evidence supporting cannabis in pediatric medicine pertains to treatment-resistant epilepsy. The drug Epidiolex, approved by the US Food and Drug Administration (FDA), is a purified CBD extract and has demonstrated significant efficacy in reducing seizure frequency in patients with Dravet syndrome and Lennox–Gastaut syndrome (Devinsky et al., 2017).

Randomized controlled trials have shown a 30%–50% reduction in seizure frequency compared with placebo, with an acceptable safety profile. Common adverse effects include somnolence, decreased appetite, and elevated liver enzymes, all typically reversible with dosage adjustments (Devinsky et al., 2018).

The approval of Epidiolex marked the first instance of an FDA-authorized cannabis-derived medication and established a framework for future cannabinoid-based therapeutics.

Autism spectrum disorder

Emerging evidence suggests that cannabis, particularly CBD, may reduce certain symptoms associated with autism spectrum disorder (ASD), including anxiety, aggression, and self-injurious behavior. A small number of observational studies have reported favorable outcomes, although randomized trials remain limited in both number and scope (Barchel et al., 2019).

An open-label study conducted in Israel found that CBD-enriched cannabis oil significantly improved behavior in children with ASD, though placebo-controlled data are needed to validate these findings. The mechanism may involve CBD's influence on serotonin and GABA pathways, in addition to the modulation of the ECS (Aran et al., 2019).

Pediatric oncology and palliative care

In children undergoing cancer treatment, cannabis has been used to manage chemotherapy-induced nausea, neuropathic pain, and anxiety. While the bulk of supporting evidence comes from adult populations, pediatric oncologists in several states have incorporated cannabis into multidisciplinary care plans, often with encouraging results.

In terminal or palliative contexts, cannabis may enhance quality of life by addressing complex symptom burdens that are inadequately managed by standard therapies. Nonetheless, ethical issues surrounding cognition, communication, and end-of-life decision-making require thoughtful consideration.

Spasticity and movement disorders

Some children with cerebral palsy or traumatic brain injury experience spasticity that does not respond to conventional treatments. Small case series and anecdotal reports indicate that cannabis may offer meaningful relief, though robust randomized data are lacking. Clinicians in Canada and Israel have begun using medical cannabis for these indications under strictly regulated protocols (Koppel et al., 2014).

Risks and considerations in pediatric use

Despite its promise, cannabis use in children raises serious concerns. Key risks are discussed as follows.

Cognitive and psychiatric effects

Studies have associated adolescent THC exposure with reduced IQ, poorer academic outcomes, and an increased risk of depression, anxiety, and psychosis. These associations are particularly strong in individuals with genetic vulnerabilities or preexisting mental health conditions (Bechtold et al., 2015).

Although CBD does not appear to share these risks, long-term pediatric safety data are limited, and the full implications of chronic exposure remain uncertain.

Product variability and contamination

Nonpharmaceutical cannabis products commonly exhibit wide variability in cannabinoid content. Some products labeled as "CBD" have been found to contain substantial amounts of THC or contaminants, such as pesticides and heavy metals. This variability poses an unacceptable risk for pediatric patients and underscores the need for pharmaceutical-grade formulations and strict regulatory oversight (Bonn-Miller et al., 2017).

Dosing challenges

Standardized dosing guidelines for most pediatric cannabis indications are lacking. This complicates safe titration and increases the risk of undertreatment or adverse effects. Close clinical monitoring, baseline laboratory testing, and collaboration with pharmacists are essential when initiating therapy.

Legal and ethical frameworks

Legally, pediatric cannabis use is permitted only in specific contexts and under strict regulation. For example, in Colorado, a child may qualify for medical cannabis only with recommendations from two licensed physicians, at least one of whom must be a pediatric specialist. Products must be administered by a caregiver, and schools retain the authority to restrict on-campus use.

Ethically, decisions about cannabis in children should be guided by the principles of beneficence, nonmaleficence, autonomy, and justice. Informed consent is critical and must include a frank discussion of known risks and areas of uncertainty. Clinicians should ensure that cannabis is chosen for a legitimate clinical rationale rather than as a last resort solely because conventional therapies have failed.

Clinical approach: When to consider cannabis

My clinical algorithm for considering cannabis in pediatric patients begins with a thorough diagnostic workup and the exhaustion of first-line therapies. If cannabis is being considered, follow these steps:

- Select CBD-only formulations unless there is a clear rationale for THC.
- Start at the lowest effective dose and titrate slowly.
- Use only lab-tested, pharmaceutical-grade products.
- Monitor liver function, developmental milestones, and cognitive function over time.
- Collaborate with relevant subspecialists and ensure multidisciplinary oversight.

This approach aims to balance clinical innovation with patient safety and reflects the seriousness of prescribing cannabis to children.

Future directions

The future of cannabis in pediatric medicine will depend on progress in several critical areas:

1. Expanded Research Funding: Federal descheduling of cannabis would enable more rigorous research, including multicenter pediatric trials.
2. Pharmaceutical Innovation: Development of targeted, receptor-specific cannabinoid medications could enhance therapeutic benefits while minimizing risks.
3. Standardized Education: Medical schools and continuing education programs must incorporate cannabis pharmacology, clinical applications, and ethics into their curricula.
4. International Collaboration: Countries such as Israel and Canada are at the forefront of pediatric cannabis research. Cross-border collaboration may accelerate scientific advancements and improve clinical guidance.

Conclusion

Cannabis is neither a miracle cure nor inherently dangerous. It is a complex botanical substance with both therapeutic potential and genuine risks, particularly for children. As physicians, our responsibility is to evaluate each case individually, guided by scientific evidence and compassion. For some families, cannabis may provide a meaningful improvement in quality of life; for others, it may represent unnecessary risk or offer little benefit.

My role is not to advocate for or against cannabis but to deliver thoughtful, evidence-based care. The central question is not whether cannabis should be used in pediatric medicine, but rather how, when, and for whom. The answers to these questions will emerge through continued research, careful clinical observation, and an unwavering commitment to the well-being of our youngest patients.

References

Aran, A., et al., 2019. Brief report: cannabidiol-rich cannabis in children with autism spectrum disorder and severe behavioral problems—a retrospective feasibility study. J. Autism Dev. Disord. 49 (3), 1284–1288.

Barchel, D., et al., 2019. Oral cannabidiol use in children with autism spectrum disorder. Front. Pharmacol. 9, 1521.

Bechtold, J., et al., 2015. Chronic adolescent marijuana use and later emotional and behavioral outcomes: a prospective study. Psychol. Addict. Behav. 29 (3), 552–563.

Bonn-Miller, M.O., et al., 2017. Labeling accuracy of cannabidiol extracts sold online. JAMA 318 (17), 1708–1709.

Devinsky, O., et al., 2017. Trial of cannabidiol for drug-resistant seizures in the Dravet syndrome. N. Engl. J. Med. 376 (21), 2011–2020.
Devinsky, O., et al., 2018. Effect of cannabidiol on drop seizures in the Lennox–Gastaut syndrome. N. Engl. J. Med. 378 (20), 1888–1897.
Grinspoon, L., Bakalar, J.B., 1997. Marihuana: The Forbidden Medicine, Yale University Press.
Gupta, S., 2013. Weed: Dr. Sanjay Gupta Reports, CNN.
Ibeas Bih, C., et al., 2015. Molecular targets of cannabidiol in neurological disorders. Neurotherapeutics 12 (4), 699–730.
Koppel, B.S., et al., 2014. Systematic review: efficacy and safety of medical marijuana in selected neurologic disorders. Neurology 82 (17), 1556–1563.
Maa, E., Figi, P., 2014. The case for medical marijuana in epilepsy. Epilepsia 55 (6), 783–786.
Volkow, N.D., et al., 2014. Adverse health effects of marijuana use. N. Engl. J. Med. 370 (23), 2219–2227.

CHAPTER

Cannabinoids in pregnancy

17

Orlando Florete Jr.

Orles Pain Management and Regenerative Medicine Group, Jacksonville, FL, United States

Introduction

Cannabinoids, including the psychoactive delta-9-tetrahydrocannabinol (THC) and the nonpsychoactive cannabidiol (CBD), are increasingly used for both medicinal and recreational purposes. These compounds act on the endocannabinoid system (ECS), which regulates appetite, mood, and neurodevelopment through CB1 and CB2 receptors. THC primarily targets CB1 receptors in the brain, while CBD modulates receptor activity and inhibits anandamide degradation (Lu and MacKie, 2016). Given the ECS's role in early embryonic development, the introduction of exogenous cannabinoids during pregnancy raises concerns about potential disruption of critical physiological pathways (Fride, 2004). Cannabis is the most widely used illicit substance among pregnant individuals, with prevalence estimates ranging from 3% to 10% in the United States (Young-Wolff et al., 2017). Usage is often underreported due to stigma and legal risks, but toxicology testing reveals higher rates (Metz et al., 2019). Motivations for use include managing nausea, anxiety, or pain, and the rising popularity of CBD products—often perceived as safer due to the absence of intoxication—further complicates exposure patterns during gestation (Westfall et al., 2006).

THC crosses the placenta and is detectable in fetal tissues, potentially accumulating due to its lipophilic properties (HURD et al., 2005). Maternal cannabis use has been variably linked to adverse outcomes such as anemia and reduced uterine blood flow (Gunn et al., 2016). However, evidence remains inconsistent, and distinguishing cannabis-specific effects from those of tobacco or other substances is challenging.

From a fetal standpoint, prenatal exposure to cannabinoids has been associated with increased risks of preterm birth, lower birth weight, and growth restriction (Conner et al., 2016). The ECS is crucial in neuronal development, and interference by THC may impair processes like synaptogenesis and axonal growth (Tortoriello et al., 2014). Animal and human studies suggest long-term consequences for attention, memory, and behavior, though socioeconomic and environmental confounders complicate causality (Goldschmidt et al., 2004).

Obstetric guidelines recommend universal screening for substance use in pregnancy, including cannabis. When identified, clinicians should use nonjudgmental, patient-centered counseling to encourage cessation (ACOG (The American College

Cannabinoids for Chronic Pain. https://doi.org/10.1016/B978-0-443-36773-1.00012-X

of Obstetricians and Gynecologists), 2017). Motivational interviewing and behavioral support are practical approaches. Patients should be informed of the lack of safety data, particularly regarding CBD, and advised against all cannabis use during pregnancy.

Legal implications vary widely. Some jurisdictions classify prenatal cannabis exposure as child neglect, potentially involving child protective services. These policies can deter patients from seeking prenatal care and may disproportionately affect marginalized populations. Ethical clinical practice requires balancing public health responsibilities with patient trust and access to care.

There is a growing need for robust research to understand cannabinoid effects in pregnancy. Studies should address modern high-potency cannabis strains, quantify maternal and fetal cannabinoid levels, and follow exposed offspring long-term. Research on isolated CBD use is especially lacking, despite its increasing prevalence and marketing as a wellness product.

While cannabis may offer symptomatic relief in nonpregnant populations, its use during pregnancy poses potential risks to fetal development and birth outcomes. Given the available evidence, clinical consensus supports advising against cannabinoid use in pregnancy. As legalization and cultural acceptance expand, continued education, screening, and research are essential to ensure safe maternal–fetal care.

Pharmacology of cannabinoids

Cannabinoids are a class of compounds that exert their effects primarily through the ECS, a complex cell-signaling network involved in homeostasis, neurodevelopment, pain modulation, immune function, and reproduction. The ECS comprises cannabinoid receptors (CB1 and CB2), endogenous ligands (anandamide [AEA] and 2-arachidonoylglycerol [2-AG]), and enzymes such as fatty acid amide hydrolase (FAAH) and monoacylglycerol lipase (MAGL) that regulate ligand degradation (Lu and MacKie, 2016).

The CB1 receptor is predominantly expressed in the central nervous system, especially in areas responsible for cognition, memory, motor control, and reward processing, including the hippocampus, basal ganglia, and cerebellum (Fride, 2004). CB2 receptors are primarily found in peripheral tissues, particularly immune cells, where they modulate inflammatory responses. Both receptors are G protein-coupled and initiate downstream signaling cascades that influence neurotransmitter release and cellular function (Young-Wolff et al., 2017).

The primary psychoactive component of cannabis, THC, acts as a partial agonist at both CB1 and CB2 receptors, though with a higher affinity for CB1. THC's binding results in the inhibition of adenylyl cyclase activity, reduced cAMP levels, and altered calcium and potassium ion channel activity, culminating in suppressed neurotransmitter release and altered synaptic function (Metz et al., 2019). These actions underlie THC's characteristic effects on mood, appetite, memory, and motor coordination.

CBD, by contrast, has low affinity for CB1 and CB2 receptors but influences the ECS through indirect mechanisms. It inhibits FAAH, thereby increasing anandamide levels, and modulates other receptor systems, including TRPV1 (transient receptor potential vanilloid 1), 5-HT1A serotonin receptors, and peroxisome proliferator-activated receptors gamma (PPARsγ) nuclear receptors (Westfall et al., 2006). CBD also acts as a negative allosteric modulator at the CB1 receptor, potentially mitigating some of THC's psychoactive effects (HURD et al., 2005).

In the context of pregnancy, the pharmacokinetics of cannabinoids are especially significant. THC is highly lipophilic, rapidly distributing to fatty tissues, including the brain and placenta. It readily crosses the placental barrier and can accumulate in fetal tissues. Its long half-life (ranging from 25 to 36 h with chronic use) is further prolonged in pregnancy due to increased adipose stores and altered metabolism (Gunn et al., 2016). THC is primarily metabolized in the liver by cytochrome P450 enzymes (CYP2C9, CYP2C19, and CYP3A4) into active and inactive metabolites, which are excreted via feces and urine.

Endocannabinoid signaling is active during early embryogenesis, suggesting critical roles in implantation, placentation, and fetal neurodevelopment. Anandamide levels are tightly regulated during early pregnancy, and dysregulation—either through exogenous THC exposure or inhibition of degrading enzymes—may impair trophoblast function and embryo implantation (Tortoriello et al., 2014). Elevated anandamide levels have been associated with miscarriage and pregnancy failure in animal and human studies.

Emerging research shows that exogenous cannabinoids can interfere with ECS signaling during fetal brain development. Animal models indicate that prenatal THC exposure alters the expression of neural adhesion molecules, axonal guidance proteins, and genes involved in synaptogenesis (Conner et al., 2016). These disruptions may result in altered cortical connectivity, behavioral abnormalities, and cognitive impairments observed in human epidemiologic studies of cannabis-exposed offspring.

The pharmacologic effects of cannabinoids are further complicated by evolving cannabis products with increasingly potent THC concentrations and varying CBD:THC ratios. Novel delivery methods alter absorption profiles and bioavailability, including vaping and edibles. Such factors pose additional challenges to assessing the pharmacologic impact of cannabinoids in pregnancy and underscore the need for updated pharmacokinetic and safety data in this population.

Maternal outcomes

Cannabis use in pregnancy is increasingly common, often perceived by patients as a safer alternative to prescription medications for common pregnancy-related symptoms. Despite changing social norms and expanding legalization, cannabis contains biologically active compounds that can influence maternal physiology. The most notable of these is THC, which is highly lipophilic and crosses the placenta, potentially affecting both maternal and fetal systems.

One of the most frequently cited reasons for cannabis use during pregnancy is the management of nausea and vomiting, particularly during the first trimester. Some individuals report self-medicating with cannabis when conventional antiemetics are ineffective or unavailable. Limited observational studies and anecdotal reports support these claims, but robust clinical trials are lacking (Fride, 2004). Importantly, most professional guidelines advise against cannabis use due to potential risks to the developing fetus, and the therapeutic benefit remains unproven.

Cannabis may also affect maternal nutrition and weight gain during pregnancy. Studies have identified an association between prenatal cannabis use and inadequate gestational weight gain, which is a known risk factor for fetal growth restriction and other complications (Young-Wolff et al., 2017). While cannabis can increase appetite through central CB1 receptor activation, its influence on weight gain in pregnancy appears to be inconsistent. It may be modified by other cooccurring factors, such as nausea severity and socioeconomic status.

Emerging evidence also suggests a link between prenatal cannabis use and maternal anemia. THC may interfere with nutrient absorption or appetite, leading to nutritional deficiencies contributing to anemia (Metz et al., 2019). Anemia in pregnancy is associated with an increased risk of preterm birth, low birth weight, and postpartum hemorrhage. While the mechanisms remain unclear, this potential association warrants further investigation, especially in populations with limited access to prenatal care.

Beyond hematologic effects, cannabis may also impact cardiovascular function in pregnant individuals. THC has been shown to increase heart rate and may alter vascular tone, potentially affecting blood pressure and uterine perfusion. Animal and limited human studies suggest that cannabis can lead to vasoconstriction of uterine arteries, which could impair fetal oxygenation (Westfall et al., 2006). However, data remain inconsistent and are often confounded by tobacco co-use.

Mental health comorbidities are frequently present in pregnant cannabis users. Higher rates of depression, anxiety, and posttraumatic stress disorder have been reported among this population (HURD et al., 2005). These psychiatric conditions may drive cannabis use as a form of self-medication, particularly in the absence of accessible mental health services. Unfortunately, cannabis use may also obscure or delay the diagnosis of underlying mental illness, and THC can exacerbate anxiety in some users.

Substance use patterns during pregnancy are often complex, with cannabis rarely being the only substance used. Many individuals who use cannabis also consume tobacco, alcohol, or other drugs. This pattern of poly-substance use introduces substantial confounding when studying maternal outcomes, as these substances have overlapping and sometimes synergistic effects on maternal and fetal health (Gunn et al., 2016). It also highlights the importance of comprehensive screening and intervention strategies in prenatal care.

Finally, social and structural determinants of health must be considered when evaluating cannabis use in pregnancy. Individuals who use cannabis may also face higher rates of poverty, unstable housing, food insecurity, and limited access to prenatal care. These factors contribute to adverse maternal outcomes independently and may compound the risks associated with cannabis use. Clinicians should adopt a nonjudgmental, supportive approach to care, focusing on harm reduction and linkage to social services as appropriate (Tortoriello et al., 2014).

Fetal and neonatal outcomes

Cannabis use during pregnancy has raised significant concerns regarding fetal growth and development. THC, the principal psychoactive component of cannabis, crosses the placental barrier and can be detected in fetal tissues. Due to its lipophilic properties, THC accumulates in fatty organs, including the developing brain, which may result in prolonged fetal exposure even after maternal cessation (Lu and MacKie, 2016). This raises the possibility of sustained disruption to crucial developmental processes.

One of the most consistently reported associations with prenatal cannabis use is reduced birth weight. Numerous observational studies and metaanalyses have demonstrated that infants born to mothers who used cannabis during pregnancy are at increased risk for low birth weight and small-for-gestational-age status (Fride, 2004). These outcomes may result from compromised placental function or restricted fetal growth, though the underlying mechanisms are not fully understood.

In addition to growth restriction, prenatal cannabis exposure has been associated with preterm birth. While findings vary across studies, some research suggests that cannabis use, especially during the first and second trimesters, may increase the risk of delivery before 37 weeks of gestation (Young-Wolff et al., 2017). This may be mediated through inflammation, hormonal disruption, or vascular effects, but more research is needed to clarify causality and isolate cannabis-specific effects from confounders.

Structural anomalies associated with cannabis use are less well established. The available evidence does not support a strong teratogenic effect of cannabis; however, subtle morphologic changes and increased rates of congenital anomalies have been reported in some studies, particularly among individuals who used cannabis in combination with other substances (Metz et al., 2019). Because most cannabis-using pregnancies are also exposed to tobacco, alcohol, or other drugs, attributing defects to cannabis alone is difficult.

Neurodevelopmental concerns are among the most critical outcomes associated with prenatal cannabis exposure. The fetal ECS plays a key role in neuronal proliferation, migration, axonal pathfinding, and synaptogenesis. Disruption of these processes by exogenous cannabinoids, particularly THC, may result in altered brain circuitry and long-term cognitive or behavioral impairments (Westfall et al., 2006). Animal studies consistently show neurodevelopmental changes following prenatal cannabinoid exposure.

Human longitudinal studies suggest associations between prenatal cannabis exposure and adverse neurobehavioral outcomes in infancy, childhood, and adolescence.

These include executive function, attention, memory, and impulse control impairments (HURD et al., 2005). Some cohorts have reported higher rates of externalizing behaviors, such as aggression and hyperactivity, in cannabis-exposed children. However, these findings are modest and may be confounded by environmental and genetic influences.

In the neonatal period, infants exposed to cannabis in utero may exhibit altered autonomic regulation, such as changes in arousal, startle response, and sleep patterns (Gunn et al., 2016). These signs may reflect early withdrawal or central nervous system dysregulation. While not life-threatening, such symptoms may interfere with feeding, bonding, and early development. Breastfeeding considerations are also relevant, as THC is excreted in human milk and may affect the infant if exposure continues postpartum.

Overall, while definitive causal relationships remain difficult to establish due to confounding factors, the accumulating body of evidence supports caution. Cannabis use during pregnancy has been associated with growth restriction, preterm birth, and neurodevelopmental concerns. Given the evolving potency of cannabis products and the increasing frequency of use, further longitudinal research is essential to understand long-term fetal and neonatal effects and to guide clinical recommendations (Tortoriello et al., 2014).

Screening and counseling

Universal screening for substance use in pregnancy, including cannabis, is recommended by major medical organizations such as the American College of Obstetricians and Gynecologists. Early identification enables timely intervention, yet screening practices remain inconsistent across clinical settings. Many patients do not voluntarily disclose cannabis use due to stigma, legal repercussions, or mistrust of the medical system. Combining patient-centered interviews with validated screening tools and, when appropriate, toxicology testing, can improve detection.

A nonjudgmental and supportive approach is essential when discussing substance use with pregnant individuals. Clinicians should use open-ended questions, normalize the screening process, and emphasize the shared goal of optimizing maternal and fetal health. Motivational interviewing is an evidence-based technique that enhances patient engagement and encourages behavior change by exploring ambivalence and reinforcing autonomy (Fride, 2004). Establishing trust is crucial, particularly when patients fear punitive consequences.

Counseling should focus on education regarding the potential risks of cannabis use during pregnancy and lactation. Many patients perceive cannabis, especially CBD products, as natural and safe and may be unaware of its association with fetal growth restriction, neurodevelopmental impacts, and long-term behavioral outcomes. Clinicians should provide accurate, balanced information and clarify that the absence of definitive evidence does not imply safety. Personalized risk communication can help patients make informed decisions.

Harm reduction strategies may be appropriate for individuals unable or unwilling to discontinue use immediately. These include reducing the frequency or dose of use, switching from smoked to noninhaled forms, or avoiding use during critical windows of development, such as the first trimester. While abstinence remains the ideal goal, harm reduction approaches acknowledge real-world challenges and prioritize gradual risk reduction over disengagement from care.

Referral to behavioral health support services, addiction specialists, or substance use treatment programs may be indicated, particularly for patients with cooccurring mental health conditions or poly-substance use. Integration of behavioral health into prenatal care can improve engagement and outcomes. Peer support groups, parenting programs, and social services can also play a role in supporting behavior change and addressing the social determinants of health that often underlie substance use.

Providers should be aware of the legal context surrounding cannabis use in pregnancy, which varies by jurisdiction. In some regions, prenatal substance use may trigger mandatory reporting to child welfare agencies. These policies can create fear and deter women from seeking care. Clinicians must balance legal obligations with ethical principles, advocating for policies that support, rather than punish, pregnant individuals seeking help.

Legal and ethical issues

The intersection of cannabis use in pregnancy with legal and ethical frameworks is increasingly complex, particularly as cannabis legalization expands. While medical and recreational cannabis use is permitted in many US states and countries, the implications of its use during pregnancy are governed by a patchwork of local, state, and institutional policies. These legal variations significantly influence clinical care, patient decision-making, and provider–patient relationships.

In several jurisdictions, prenatal cannabis use is considered child abuse or neglect, prompting mandatory reporting to child protective services. Positive toxicology screens—whether in the mother or the newborn—may trigger social service investigations or even custody loss. These policies are not applied uniformly across substances or populations and can disproportionately impact low-income individuals and communities of color, raising concerns about systemic bias and reproductive justice.

Mandatory reporting laws may have unintended consequences, including discouraging pregnant individuals from disclosing substance use or from seeking prenatal care altogether. Fear of legal repercussions or loss of custody can erode trust in healthcare providers and drive patients away from essential services. Providers must be aware of the legal requirements in their state or country and strive to protect patient confidentiality and promote engagement with care.

Ethical dilemmas arise when providers must balance their duty to advocate for fetal health with respect for maternal autonomy. While clinicians may be concerned about the potential harms of cannabis exposure to the fetus, punitive approaches may do more harm than good. Ethical care emphasizes noncoercive, respectful communication that empowers patients to make informed decisions without fear of judgment or legal retaliation (Metz et al., 2019).

Informed consent is another important ethical consideration. Patients must be provided with accurate, the available information about the risks of cannabis use during pregnancy, including areas of uncertainty. Clinicians should avoid overstating or understating potential harms without definitive clinical trial data. Ethical counseling should be rooted in transparency, cultural sensitivity, and acknowledgment of the limitations of existing research (Westfall et al., 2006).

Institutional policies also vary widely. Some hospitals routinely conduct toxicology screening in labor and delivery units, while others do so only when clinically indicated. Lack of standardization can lead to inconsistent care, mistrust, and legal vulnerability for providers and patients. Ethically, screening should be applied equitably and accompanied by clear explanations to patients, with consent obtained whenever possible.

The legalization of cannabis for medical purposes introduces another layer of complexity. Some pregnant individuals use cannabis with a medical prescription, believing it to be a safer or more natural alternative to conventional pharmaceuticals. However, most medical cannabis programs do not offer pregnancy-specific safety guidance, and prescribing clinicians may be unaware of the potential reproductive risks. Ethical practice requires that providers counsel patients carefully, even when cannabis is legally prescribed.

Ethical frameworks such as beneficence, nonmaleficence, autonomy, and justice should guide clinical responses to cannabis use in pregnancy. Policies and practices that emphasize punishment over support may violate these principles. A more just and practical approach would prioritize harm reduction, access to prenatal and mental health care, and systemic reforms that reduce stigma and improve outcomes for pregnant individuals and their children.

Research gaps and future directions

Despite increasing cannabis use among pregnant individuals, critical gaps remain in our understanding of its safety and long-term consequences. Much of the literature is observational, with studies often limited by small sample sizes, inconsistent measurement of exposure, and inadequate control for confounding variables such as tobacco use, socioeconomic status, and mental health conditions. These limitations hinder the ability to draw clear conclusions about causality and effect magnitude.

One of the most urgent needs in cannabinoid research is the development of standardized and accurate methods for assessing prenatal exposure. Many studies rely on self-reported cannabis use, which is subject to underreporting due to stigma or fear of legal consequences. While toxicology testing of maternal urine, cord blood, or meconium can provide objective measures, these tools vary in sensitivity, timing of detection, and interpretation. More refined biomarkers and longitudinal tracking systems are needed.

Dose-response relationships also remain poorly defined. There is limited data on how frequency, dose, route of administration, and cannabinoid composition affect maternal or fetal outcomes. The variability in modern cannabis products—ranging from low-THC strains to highly concentrated extracts and edibles—further complicates this issue. Research must adapt to reflect contemporary patterns of cannabis use, including the impact of high-potency products on fetal neurodevelopment.

Another significant research gap pertains to the pharmacokinetics of cannabinoids in pregnancy. Physiological changes during pregnancy—including increased body fat, altered hepatic metabolism, and changes in gastrointestinal absorption—likely influence how cannabinoids are processed in maternal and fetal systems. Yet, there are few studies specifically examining the metabolism, distribution, and clearance of THC, CBD, or their metabolites in pregnant individuals.

The role of CBD in pregnancy is especially understudied. CBD is widely marketed as a nonintoxicating compound with therapeutic potential for anxiety, pain, and nausea. Pregnant individuals may perceive CBD products as safer than THC-containing cannabis, yet there is a lack of robust safety data supporting this assumption. Given the rising popularity of CBD, especially in over-the-counter formulations, targeted research into its reproductive and developmental effects is essential.

Most available data on neurodevelopmental outcomes come from cohorts exposed to lower-potency cannabis during earlier decades. These studies have documented subtle but persistent effects on attention, executive function, and behavior into adolescence and young adulthood (HURD et al., 2005). However, it is unknown whether more potent modern cannabis products will result in more pronounced or different neurodevelopmental trajectories. Updated longitudinal studies are critically needed.

Moreover, there is a lack of research on the potential epigenetic and transgenerational effects of prenatal cannabis exposure. Emerging animal studies suggest that THC can alter DNA methylation and histone modification patterns in the developing brain, potentially affecting gene expression and behavior beyond the immediate generation (Gunn et al., 2016). Investigating whether these changes occur in humans and their long-term implications represents an important frontier.

Social and structural determinants of cannabis use in pregnancy are also understudied. Little research has focused on how race, poverty, trauma, and access to healthcare intersect with cannabis use patterns and outcomes (Tortoriello et al., 2014). Understanding these dynamics is vital to designing equitable and effective interventions, especially given the disproportionate legal and health consequences marginalized populations face.

There is also a need to explore the therapeutic potential of cannabinoids in carefully controlled settings. While the available guidance advises against cannabis use in pregnancy, future research may identify circumstances under which specific cannabinoids could be safely used to treat conditions such as hyperemesis gravidarum or chronic pain. However, such applications must be grounded in rigorous safety and efficacy data (Conner et al., 2016).

Preclinical animal models remain a valuable tool for understanding mechanisms of cannabinoid action in the developing fetus. However, translating findings from rodents or nonhuman primates to humans requires caution due to species differences in ECS development, metabolism, and behavior (Goldschmidt et al., 2004). Harmonizing preclinical and human studies with consistent endpoints and exposure parameters could enhance the translational value of animal research.

Ethical research practices are essential in this domain. Pregnant individuals have historically been excluded from clinical trials, leading to data gaps and reliance on extrapolated evidence. Ethical frameworks must balance the need for safety with the inclusion of pregnant populations in research that directly affects them (ACOG (The American College of Obstetricians and Gynecologists), 2017). Community engagement, culturally sensitive study design, and protections against legal repercussions can help facilitate ethical and inclusive research participation.

Public health messaging and policy must be informed by strong scientific evidence. As cannabis legalization and commercialization evolve, so too must our research infrastructure. Cross-disciplinary collaboration among obstetricians, neuroscientists, toxicologists, epidemiologists, and public health experts will be essential in closing knowledge gaps, improving maternal–fetal outcomes, and guiding evidence-based practice and policy (Howlett et al., 2002).

References

ACOG (The American College of Obstetricians and Gynecologists), 2017. Committee opinion no. 722: marijuana use during pregnancy and lactation. Obstet. Gynecol. 0029-7844. 130 (4), e205. doi:10.1097/aog.0000000000002354.

Conner, S.N., Bedell, V., Lipsey, K., Macones, G.A., Cahill, A.G., Tuuli, M.G., 2016. Maternal marijuana use and adverse neonatal outcomes: a systematic review and meta-analysis. Obstet. Gynecol. 1873233X. 128 (4), 713–723. Available from: http://journals.lww.com/greenjournal. doi:10.1097/AOG.0000000000001649.

Fride, E., 2004. The endocannabinoid-CB1 receptor system in pre- and postnatal life. Eur. J. Pharmacol. 00142999. 500 (1-3), 289–297. doi:10.1016/j.ejphar.2004.07.033.

Goldschmidt, L., Richardson, G.A., Cornelius, M.D., Day, N.L., 2004. Prenatal marijuana and alcohol exposure and academic achievement at age 10. Neurotoxicol. Teratol. 26 (4), 521–532. doi:10.1016/j.ntt.2004.04.003.

Gunn, J.K.L., Rosales, C.B., Center, K.E., Nuñez, A., Gibson, S.J., Christ, C., Ehiri, J.E., 2016. Prenatal exposure to cannabis and maternal and child health outcomes: a systematic review and meta-analysis. BMJ Open. 2044-6055. 6 (4), e009986. doi:10.1136/bmjopen-2015-009986.

Howlett, A.C., Barth, F., Bonner, T.I., Cabral, G., Casellas, P., Devane, W.A., Felder, C.C., Herkenham, M., Mackie, K., Martin, B.R., Mechoulam, R., Pertwee, R.G., 2002. International union of pharmacology. XXVII. Classification of cannabinoid receptors. Pharmacol. Rev. 00316997. 54 (2), 161–202. doi:10.1124/pr.54.2.161.

HURD, Y., WANG, X., ANDERSON, V., BECK, O., MINKOFF, H., DOWEDWARDS, D., 2005. Marijuana impairs growth in mid-gestation fetuses. Neurotoxicol. Teratol. 08920362. 27 (2), 221–229. doi:10.1016/j.ntt.2004.11.002.

Lu, H.C., MacKie, K., 2016. An introduction to the endogenous cannabinoid system. Biol. Psychiatry. 18732402. 79 (7), 516–525. Available from: www.elsevier.com/locate/biopsychiat. doi:10.1016/j.biopsych.2015.07.028.

Metz, T.D., Silver, R.M., McMillin, G.A., Allshouse, A.A., Jensen, T.L., Mansfield, C., Heard, K., Kinney, G.L., Wymore, E., Binswanger, I.A., 2019. Prenatal marijuana use by self-report and umbilical cord sampling in a state with marijuana legalization. Obstet. Gynecol. 1873233X. 133 (1), 98–104. Available from: http://journals.lww.com/greenjournal. doi:10.1097/AOG.0000000000003028.

Tortoriello, G., Morris, C.V., Alpar, A., Fuzik, J., Shirran, S.L., Calvigioni, D., Keimpema, E., Botting, C.H., Reinecke, K., Herdegen, T., Courtney, M., Hurd, Y.L., Harkany, T., 2014. Miswiring the brain: Δ9-tetrahydrocannabinol disrupts cortical development by inducing an SCG10/stathmin-2 degradation pathway. EMBO J. 14602075. 33 (7), 668–685. Available from: http://www.nature.com/emboj/index.html. doi:10.1002/embj.201386035.

Westfall, R.E., Janssen, P.A., Lucas, P., Capler, R., 2006. Survey of medicinal cannabis use among childbearing women: patterns of its use in pregnancy and retroactive self-assessment of its efficacy against 'morning sickness'. Complement. Ther. Clin. Pract. 17443881. 12 (1), 27–33. doi:10.1016/j.ctcp.2005.09.006.

Young-Wolff, K.C., Tucker, L.Y., Alexeeff, S., Armstrong, M.A., Conway, A., Weisner, C., Goler, N., 2017. Trends in self-reported and biochemically tested Marijuana use among pregnant females in California from 2009-2016. JAMA. 15383598. 318 (24), 2490–2491. Available from: https://jamanetwork.com/journals/jama/articlepdf/2667052/jama_youngwolff_2017_ld_170046.pdf. doi:10.1001/jama.2017.17225.

Cannabis in the older population

18

Roshni Patel
RNP Regenerative Medicine, Center of Excellence in Pain and Regenerative Medicine, Quantum Brain Health, Farmington, CT, United States

Introduction

The use of cannabinoid compounds has increased significantly among older adults. This population is more physiologically vulnerable and susceptible to both the potential benefits and risks associated with its use. Within the United States, past-year cannabis use among those aged 65 and older has increased more than tenfold, from 0.4% in 2006 to over 7% in 2023, with concurrent cannabis and binge drinking prevalence rising by 450% from 2015 to 2019 in this age group (Han and Palamar, 2020; Kepner et al., 2023; NSDUH, 2023). This sharp rise has been accompanied by an 1,808% increase in cannabis-related emergency department visits among California adults aged 65 and older from 2005 to 2019, underscoring the clinical urgency of understanding cannabinoid use in this population (Han et al., 2023). The increase is primarily due to the desire to find alternatives to prescription medication, including opioid medications, because of concerns about side effects, dependence, and abuse, while also seeking relief from conditions such as chronic pain and sleep disturbances (Ware et al., 2015; Sznitman et al., 2020).

In the age group above 65, there is limited long-term efficacy data on tetrahydrocannabinol (THC)-containing compounds (Whiting et al., 2015; Abuhasira et al., 2019; Velayudhan et al., 2024). Most of the older population who use cannabis are doing so without medical supervision, which increases the potential risks and drug interactions (Abuhasira et al., 2019).

Randomized controlled studies have shown that THC-containing products are dose-dependent and associated with a significantly higher rate of neuropsychiatric adverse events, including dizziness and delirium, when compared to placebo or younger cohorts (Velayudhan et al., 2024).

The increasing prevalence of cannabis use among older adults has been accompanied by rising rates of cannabis use disorder, particularly among those with comorbid psychiatric conditions. This is especially concerning given the vulnerability of this population to adverse effects (Han et al., 2023; National Survey on Drug Use and Health [NSDUH], 2023).

Cannabinoids for Chronic Pain. https://doi.org/10.1016/B978-0-443-36773-1.00020-9

Even with the sharp increase in cannabis use among the older adults, there is only a small fraction of cannabinoid clinical trials that include participants aged 65 and older, leaving substantial gaps in the evidence base for this demographic (Abuhasira et al., 2019).

Pharmacokinetic changes, reduced hepatic metabolism, and increased THC bioavailability in the aging population lead to an increased risk of adverse outcomes in older adults (Han et al., 2023).

Long-term cannabis use in older adults has been associated with memory impairment and may interact with age-related cognitive decline (Whiting et al., 2015; Scott et al., 2019).

This leads to a tough challenge: finding a balance between retaining access to cannabinoids while mitigating the risks for this susceptible population—an ongoing struggle between policymakers and healthcare.

Pharmacokinetics of cannabinoids in older adults

Absorption and bioavailability

Oral administration of THC in older adults results in significantly lower bioavailability (4%–12%) compared to inhaled routes (10%–35% in younger adults), which leads to a delayed onset of action and variable therapeutic effects, with T_{max} values of 120–180 min versus 60 min in younger adults (Abuhasira et al., 2019; Grotenhermen, 2003). Age-related reductions in gastrointestinal motility and gastric acid secretion further slow absorption. This leads to increased systemic exposure; the area under the curve in older adults can be more than double that seen in younger cohorts (e.g., 3.51 vs. 1.67 ng · h/mL) (Ahmed et al., 2014; Grotenhermen, 2003).

Due to these pharmacokinetic changes in the aging population, the onset and duration of effects are unpredictable, which complicates titration and leads to increased adverse events (Grotenhermen, 2003).

Distribution

In the older population, the high lipophilic nature of THC leads to an extensive distribution into adipose tissue, which is typically increased in older adults. This results in a two-times higher volume of distribution and a prolonged elimination half-life of 5–13 days compared to 1–3 days in younger individuals (Grotenhermen, 2003). This increased tissue reservoir can cause drug accumulation, especially with repeated dosing, resulting in detectable plasma THC levels (>1 μg/L) persisting for >24 h after a single dose in seniors (Ahmed et al., 2014). There is an age-related decrease in total body water and an increase in body fat, which contributes to the alterations in the pharmacokinetics of cannabinoids, possibly increasing the exposure and toxicity within the central nervous system (Abuhasira et al., 2019).

Metabolism and excretion

There is a 40%–60% reduction in hepatic cytochrome P450 (CYP450) enzyme activity, especially with CYP2C9 and CYP3A4, the primary enzymes responsible for THC metabolism, with aging (Abuhasira et al., 2019).

This reduction in metabolic activity slows THC clearance and increases the risk of drug-drug interactions, especially with commonly prescribed medications such as warfarin, benzodiazepines, and antipsychotics (Grotenhermen, 2003).

Renal excretion of hydroxylated THC metabolites (e.g., 11-OH-THC) is also impaired in older adults due to age-related declines in glomerular filtration rate, increasing the risk of toxicity and prolonged psychoactive effects (Ahmed et al., 2014).

Polypharmacy and drug interaction risks

Among older adults, polypharmacy is increasingly common, with over 60% taking five or more medications, which increases the likelihood of clinically significant drug interactions with cannabinoids (Abuhasira et al., 2019).

Concomitant use of central nervous system depressants, such as benzodiazepines or opioids, can amplify the risk of dizziness (incidence rate ratio: 0.05/mg/day THC) and delirium (odds ratio: 1.8–2.4) (Velayudhan et al., 2024).

THC's anticholinergic properties may exacerbate cognitive impairment in patients on cholinesterase inhibitors like donepezil or rivastigmine (Abuhasira et al., 2019).

Mechanisms of action of cannabinoids in aging

Aging is associated with a 30%–50% reduction in hippocampal CB1 receptor density, which impairs synaptic plasticity and increases susceptibility to THC-induced memory deficits in older adults (Ginsburg and Hensler, 2022).

Transient compensatory upregulation of CB1 binding sites may occur in middle age, but this is followed by a progressive decline in later years, further diminishing endocannabinoid signaling capacity (Nidadavolu et al., 2022). These changes are significant for cognitive health because the hippocampus is critical for memory formation and is highly sensitive to cannabinoid modulation (Scott et al., 2019).

Interactions with chronic THC exposure: In older adults with long-term cannabis use, neuroimaging has shown an association with increased hippocampal atrophy and episodic memory decline in individuals over 50 years of age (Scott et al., 2019). Preclinical model studies potentially show that cannabidiol (CBD) may mitigate amyloid-β toxicity and reduce neuroinflammation, but robust evidence for neuroprotection in human trials remains lacking (Nidadavolu et al., 2022). With these complex interactions between cannabinoids and neurodegenerative pathways, there is a potential for both harm and benefit depending on the specific cannabinoid profile, dose, and patient vulnerability (Scott et al., 2019).

Differential Effects of THC and CBD: THC acts as a partial agonist at CB1 receptors, especially in the prefrontal cortex, which can worsen executive dysfunction in seniors with baseline cognitive impairment (mini-mental state examination, MMSE < 26) (Scott et al., 2019). CBD inhibits the fatty acid amide hydrolase enzyme, leading to increased endocannabinoid anandamide levels and potentially reducing neuroinflammation and oxidative stress (Nidadavolu et al., 2022). These different mechanisms of action highlight the importance of cannabinoid selection and ratio in clinical practice, especially for older adults at risk for cognitive decline or neuropsychiatric complications (Abuhasira et al., 2019).

Key pharmacokinetic and mechanistic comparisons

Clinical implications

- Dosing: Starting low and titrating slowly is recommended for older adults due to delayed absorption, prolonged half-life, and increased risk of accumulation (Grotenhermen, 2003; Abuhasira et al., 2019).
- Monitoring: In geriatric patients, close monitoring is recommended for adverse effects due to an increased risk of neuropsychiatric symptoms and drug interactions (Velayudhan et al., 2024).
- Selection of Formulation: It is recommended that preference should be given to a CBD-predominant or balanced THC:CBD formulation to decrease the cognitive and neuropsychiatric risks (Nidadavolu et al., 2022; Scott et al., 2019).
- Multiple comorbidities: Special caution is needed in patients with cognitive impairment, cardiovascular disease, or polypharmacy, as these factors increase their vulnerability to adverse events (Abuhasira et al., 2019).

Parameter	Younger adults	Older adults	Source
THC half-life	1–3 days	5–13 days	(Grotenhermen, 2003)
T_{max} (Oral)	60 min	120–180 min	(Ahmed et al., 2014)
CB1 receptor density	100%	50%–70%	(Ginsburg and Hensler, 2022)

Indications for cannabinoid use in older adults

The therapeutic indications for cannabinoids in older adults are an area of active research, with the available evidence drawn from randomized controlled trials, systematic reviews, and expert consensus statements. Though cannabinoids are used increasingly by geriatric populations, robust data specific to this demographic remain limited, and most recommendations are extrapolated from studies in younger adults or mixed-age cohorts (Abuhasira et al., 2019; Whiting et al., 2015).

Due to the unique safety considerations, comorbidities, and polypharmacy, careful clinical decision-making must be balanced with the modest benefits of clinical trials (Abuhasira et al., 2019).

Chronic pain, particularly of neuropathic or cancer-related origin, is the most frequently cited indication for medical cannabis in older adults (Ware et al., 2015). Systematic reviews and metaanalyses have shown that cannabinoids can achieve a statistically significant, modest reduction in pain intensity compared to placebo (Whiting et al., 2015).

The COMPASS study found that older adults using cannabis for chronic pain reported improvements in pain levels and quality of life but also experienced much higher rates of adverse events, requiring careful patient selection and monitoring (Ware et al., 2015).

Spasticity: Cannabinoids have shown efficacy in reducing spasticity, especially in conditions such as multiple sclerosis (Koppel et al., 2014). The National Academies of Sciences, Engineering, and Medicine (2017) concluded that substantial evidence supports the use of oral cannabinoids for spasticity in adults with multiple sclerosis. Due to a sparsity of studies specific to the geriatric population, clinical guidance is often inferred from younger cohorts.

Sleep: Sleep disturbances are very common in the older population, and cannabinoids are frequently used off-label to address insomnia and related symptoms. Older adults with chronic pain who used medical cannabis experienced improvements in sleep quality, though these effects were often secondary to pain relief (Sznitman et al., 2020).

The quality of evidence for cannabinoids in primary sleep disorders remains low, and further research is needed to clarify their role in geriatric sleep medicine (Walsh et al., 2017).

Behavioral and Psychological Symptoms of Dementia: A significant challenge in dementia care is the management of agitation and behavioral disturbances. The clinical trials and observational studies suggest that synthetic THC (dronabinol) and THC/CBD combinations may reduce agitation and aggression in patients with Alzheimer's disease and other dementias (Shelef et al., 2016; Van Den Elsen et al., 2015). These interventions have generally been well-tolerated in short-term studies. Still, cannabinoids are not considered first-line therapy and should be reserved for refractory cases or when standard treatments are contraindicated (Whiting et al., 2015; National Academies of Sciences, Engineering, and Medicine, 2017).

Appetite Stimulation and Weight Loss: Cannabinoids, specifically dronabinol, have been approved for appetite stimulation in patients with cancer or HIV/AIDS-related cachexia (Beal et al., 1995). The evidence for benefit in older adults is inconsistent, with some studies failing to show any significant improvements in appetite or weight gain compared to placebo (Abuhasira et al., 2019).

Nausea and Vomiting: Cannabinoids are established antiemetics in managing chemotherapy-induced nausea and vomiting, as supported by multiple systematic reviews and metaanalyses (Tramèr et al., 2001). The risk of adverse effects may be higher in this population due to altered pharmacokinetics and polypharmacy, and further studies are needed, especially for this vulnerable demographic (Abuhasira et al., 2019).

Palliative and End-of-Life Care: In the palliative care setting, cannabinoids are used to address a myriad of symptoms, including pain, nausea, appetite loss, and sleep disturbances, to improve the quality of life in patients with advanced illness (Bar-Sela et al., 2022). The evidence for these indications in the older population is minimal, and individualized risk-benefit assessments are essential due to the high prevalence of comorbidities and polypharmacy (Abuhasira et al., 2019).

Clinical use of cannabinoids in older adults should be highly individualized, with careful attention to comorbid conditions, concomitant medications, and frailty (Abuhasira et al., 2019). There remains a critical need for large, well-powered randomized controlled trials focusing specifically on older adults to better define the efficacy and safety of cannabinoids for all primary indications (Whiting et al., 2015).

Contraindications for cannabinoid use in older adults

Absolute contraindications

- Acute Psychosis and Unstable Psychiatric Disorders: Cannabinoids, particularly those containing THC, are contraindicated in individuals with acute psychosis or unstable psychiatric conditions due to the possibility of exacerbating symptoms or precipitating psychotic episodes (Abuhasira et al., 2019; Velayudhan et al., 2024).
- Personal or Family History of Schizophrenia or Psychosis: THC-containing products should not be prescribed to individuals with a personal or family history of schizophrenia or psychotic disorders due to an increased risk of psychiatric complications (Whiting et al., 2015; Koppel et al., 2014).
- Known hypersensitivity to cannabis or cannabinoid products: Any history of allergic or hypersensitivity reactions to cannabis or its components is an absolute contraindication (Abuhasira et al., 2019).

Relative contraindications

- Severe Cardiovascular Disease: Caution is advised in patients with severe cardiovascular conditions, including unstable angina, myocardial infarction, arrhythmias, or poorly controlled hypertension, as THC can increase heart rate, blood pressure, and myocardial oxygen demand, potentially triggering ischemic events or arrhythmias (Grotenhermen, 2003; Abuhasira et al., 2019).
- Severe Liver or Kidney Disease: Impaired hepatic or renal function can alter cannabinoid metabolism and excretion, increasing the risk of toxicity and adverse effects (Ahmed et al., 2014; Abuhasira et al., 2019).
- History of Substance Use Disorder: Individuals with a history of substance use disorder, including cannabis use disorder, are at a higher risk for dependence and should generally avoid cannabinoid therapy (Abuhasira et al., 2019; National Academies of Sciences, Engineering, and Medicine, 2017).
- Gait Instability, Falls, or Neurological Impairment: Older adults with a history of falls, gait instability, or significant neurological impairment are at an increased risk of injury due to the psychomotor and cognitive effects of cannabinoids (Velayudhan et al., 2024; Scott et al., 2019).
- Polypharmacy and Drug Interactions: Caution is warranted in patients taking multiple medications due to cannabinoids potentially interacting with a wide range of drugs, including anticoagulants, CNS depressants, and drugs metabolized by cytochrome P450 enzymes, increasing the risk of adverse effects or toxicity (Abuhasira et al., 2019; Grotenhermen, 2003).

Additional considerations

- Respiratory Disease: Smoking or vaping cannabis is not recommended for patients with chronic respiratory diseases (e.g., chronic obstructive pulmonary disease—COPD, asthma) due to the risk of lung injury and exacerbation of respiratory symptoms (Grotenhermen, 2003).

- Cognitive Impairment or Dementia: Cannabinoids may worsen cognitive function and are generally not recommended in patients with moderate to severe cognitive impairment or dementia unless the benefits outweigh the risks (Scott et al., 2019).
- Pregnancy and Breastfeeding: Although rare in older adults, cannabinoids are contraindicated in pregnancy and breastfeeding due to potential harm to the fetus or infant (National Academies of Sciences, Engineering, and Medicine, 2017).

Summary table: Contraindications

Clinical note

All cannabinoid prescribing in older adults should be individualized, with a thorough risk-benefit assessment and close monitoring for adverse effects, especially in the presence of comorbidities or polypharmacy (Abuhasira et al., 2019; Grotenhermen, 2003).

Contraindication	Absolute/ relative	Rationale/notes	Key author(s) and Year
Acute psychosis/ unstable psychiatric dx	Absolute	Risk of exacerbation of symptoms	(Abuhasira et al., 2019; Velayudhan et al., 2024)
Personal/family history of psychosis	Absolute	Increased psychiatric risk	(Whiting et al., 2015; Koppel et al., 2014)
Hypersensitivity to cannabis	Absolute	Allergy risk	(Abuhasira et al., 2019)
Severe cardiovascular disease	Relative	Arrhythmia, MI, stroke risk	(Grotenhermen, 2003; Abuhasira et al., 2019)
Severe liver/kidney disease	Relative	Impaired metabolism/ excretion	(Ahmed et al., 2014; Abuhasira et al., 2019)
Substance use disorder history	Relative	Risk of dependence/CUD	(National Academies of Sciences, Engineering, and Medicine, 2017; Abuhasira et al., 2019)
Gait instability/falls/ neurological dx	Relative	Increased fall/ injury risk	(Velayudhan et al., 2024; Scott et al., 2019)
Polypharmacy/drug interactions	Relative	CYP450, CNS depressants, anticoagulants	(Abuhasira et al., 2019; Grotenhermen, 2003)
Respiratory disease (smoking/vaping)	Relative	Lung injury risk	(Grotenhermen, 2003)
Cognitive impairment/ dementia	Relative	Worsening cognition	(Scott et al., 2019)
Pregnancy/ breastfeeding	Absolute	Fetal/infant harm	(National Academies of Sciences, Engineering, and Medicine, 2017)

Side effects of cannabinoids in older adults

Older adults are more vulnerable to side effects from cannabinoids, especially those containing THC, because of physiological changes that come with aging, the use of multiple medications, and the presence of other health conditions (Abuhasira et al., 2019; Velayudhan et al., 2024).

The likelihood of side effects can vary based on the type of cannabinoid used, the dosage, how it's taken, and individual patient factors. Products high in THC can carry a greater risk of adverse effects compared to CBD-dominant or more balanced formulations (Grotenhermen, 2003; Scott et al., 2019).

A comprehensive metaanalysis demonstrated a clear dose-dependent relationship between cumulative THC exposure and the risk of several adverse effects in adults aged 50 years and older (Velayudhan et al., 2024).

As weekly doses increased, the incidence of the following side effects also rose:

1. Dizziness and Lightheadedness: The side effect of dizziness increased significantly with higher THC doses, making this one of the most common and dose-sensitive adverse events (Velayudhan et al., 2024).
2. Somnolence and Drowsiness: Sedation and excessive sleepiness are frequently reported, especially at moderate to high THC exposures (Velayudhan et al., 2024).
3. Impaired Mobility, Balance, and Coordination: The incidence of motor impairment, balance problems, and unsteady gait increases with higher THC intake, elevating the risk of falls (Velayudhan et al., 2024).
4. Cognitive and Perceptual Disturbances: Higher THC doses were associated with increased rates of confusion, disorganized thinking, and perception disorders, including dissociation or hallucinations (Velayudhan et al., 2024).
5. Dry Mouth: This was a common side effect with a dose-dependent relationship; as the dose increased, so did the incidence of dry mouth (Velayudhan et al., 2024).
6. Other Neurological and Psychiatric Effects: Dose-related increases were noted in anxiety, paranoia, and, less commonly, psychosis, particularly in those with predisposing factors (Velayudhan et al., 2024).

The risk of these adverse events was substantially higher with THC-dominant formulations as compared to CBD-dominant or balanced THC:CBD products, and CBD alone did not significantly increase the incidence of all-cause adverse events in older adults (Velayudhan et al., 2024).

Additional common side effects:

1. Fatigue: Older patients reported increased generalized tiredness and reduced energy commonly, which increased with higher doses or in combination with other sedating medications (Abuhasira et al., 2019).
2. Gastrointestinal Effects: Nausea, vomiting, diarrhea, and appetite changes are more common with oral and inhaled formulations, especially at higher doses or if the compound is primarily CBD (Beal et al., 1995; Tramèr et al., 2001).

3. Orthostatic Hypotension: Cannabinoids can lower blood pressure upon standing, which leads to dizziness, fainting, and an increased risk of falls (Ahmed et al., 2014).
4. Tachycardia and Hypertension: Acutely, THC can increase heart rate and blood pressure, which may be dangerous for those with underlying cardiovascular disease (Grotenhermen, 2003; Abuhasira et al., 2019).
5. Liver Effects: High-dose, long-term CBD use may be associated with liver enzyme elevations and, rarely, liver injury, particularly in those with preexisting liver disease or polypharmacy (Chesney et al., 2020; Lo et al., 2023).
6. Allergic Reactions: Rare, but hypersensitivity to cannabis or its components is a contraindication (Abuhasira et al., 2019).

Serious and less common side effects

1. Serious Adverse Events: While all-cause and treatment-related adverse events are increased with THC and THC:CBD combinations, the rates of serious adverse events, withdrawals, and deaths are not significantly higher than controls in most studies (Whiting et al., 2015).
2. Psychosis and Delirium: Products that contain THC may precipitate or increase psychotic symptoms and/or delirium in susceptible individuals, particularly those with a personal or family history of psychosis or cognitive impairment (Koppel et al., 2014; Marconi et al., 2016).
3. Falls and Fractures: Cannabis use is associated with an increased risk of falls, fractures, and related injuries in older adults due to impaired balance, coordination, and attention (Velayudhan et al., 2024; Scott et al., 2019).
4. Hospitalizations and Emergency Visits: There has been a significant increase in cannabis-related emergency room visits in older adults, with an 1800% rise in California from 2005 to 2019, most often related to neuropsychiatric symptoms, cardiovascular events, or accidental overdose (Han et al., 2023).

Comparative side-effect profile: THC versus CBD

Clinical implications

1. Risk Mitigation: Start with the lowest effective dose, titrate slowly, and monitor closely for side effects, especially in frail or cognitively impaired patients (Abuhasira et al., 2019; Grotenhermen, 2003).
2. Formulation Selection: Prefer CBD-predominant or balanced THC:CBD products to minimize neuropsychiatric and cardiovascular risks (Scott et al., 2019).
3. Monitoring: Regularly assess for falls, cognitive changes, cardiovascular symptoms, and drug interactions, particularly in those with polypharmacy or comorbidities (Velayudhan et al., 2024).
4. Patient Education: Older adults should be counseled on the risks of delayed onset with oral formulations, the potential for cumulative effects, and the importance of avoiding dose escalation without medical supervision (Abuhasira et al., 2019).

Side effect	THC-Dominant	THC:CBD Combo	CBD-Dominant	Notes	Referenced authors
Dizziness/Lightheadedness	High	Moderate	Low	Dose-dependent, higher with THC	(Velayudhan et al., 2024)
Somnolence/Drowsiness	High	Moderate	Moderate	All can cause, but THC > CBD	(Velayudhan et al., 2024)
Cognitive impairment	High	Moderate	Low	THC is primary driver	(Scott et al., 2019)
Anxiety/Paranoia	High	Moderate	Low	More with THC	(Whiting et al., 2015)
Cardiovascular events	High	Moderate	Low	Especially with preexisting disease	(Grotenhermen, 2003)
GI Upset	Moderate	Moderate	Moderate	Nausea, diarrhea, more with oral use	(Tramèr et al., 2001)
Liver enzyme elevation	Low	Low	Moderate	High-dose CBD, especially with polypharmacy	(Chesney et al., 2020)
Falls/impaired balance	High	Moderate	Low	Increased fall risk with THC	(Velayudhan et al., 2024)
Psychosis/delirium	High	Moderate	Low	Especially in those with risk factors	(Koppel et al., 2014)
Fatigue	High	Moderate	Moderate	More common with higher doses	(Abuhasira et al., 2019)
Allergic reactions	Rare	Rare	Rare	Hypersensitivity to cannabis	(Abuhasira et al., 2019)

Drug interactions of cannabinoids in older adults

Older adults are at an increased risk for clinically significant drug interactions with cannabinoids due to age-related changes in drug metabolism, polypharmacy, and the prevalence of comorbidities (Beedham et al., 2020; Velayudhan et al., 2024).

Both THC and CBD interact with the liver's cytochrome P450 (CYP) enzymes, which are responsible for metabolizing many commonly prescribed medications. This interaction can lead to increased drug toxicity, reduced effectiveness, and unpredictable side effects (Beedham et al., 2020; Velayudhan et al., 2024; Chesney et al., 2020).

THC and CBD inhibit several CYP450 isoenzymes: CYP2C9, CYP2C19, CYP3A4, and CYP1A2, which are responsible for the metabolism of many drugs used in older adults (Beedham et al., 2020; Chesney et al., 2020; Velayudhan et al., 2024). Cannabis that is smoked can induce CYP1A2, potentially reducing the efficacy of drugs metabolized by this pathway, while CBD is a particularly potent inhibitor of CYP2C19 and CYP3A4, increasing serum concentrations of coadministered drugs metabolized by these enzymes (Chesney et al., 2020; Velayudhan et al., 2024). Both cannabinoids can also inhibit UGT enzymes, affecting drugs such as acetaminophen and certain statins (Beedham et al., 2020).

Clinically relevant drug interactions

Anticoagulants

THC and CBD can increase warfarin levels by inhibiting CYP2C9 and CYP3A4, raising the risk of bleeding (Beedham et al., 2020; Grayson et al., 2018). Case reports document significant international normalized ratio (INR) elevation and bleeding events in patients using warfarin and cannabinoids concurrently, though some instances show minimal INR fluctuation, suggesting variability based on dose and route (Beedham et al., 2020; Grayson et al., 2018). Direct oral anticoagulants, such as rivaroxaban and apixaban, metabolized by CYP3A4, may also have increased serum levels when used with cannabinoids, though clinical evidence is limited (Beedham et al., 2020).

Central nervous system depressants

Additive sedative effects can occur with benzodiazepines, opioids, and other CNS depressants, increasing the risk of falls, confusion, and respiratory depression (Beedham et al., 2020; Velayudhan et al., 2024). CBD can increase serum levels of benzodiazepines (e.g., clobazam, diazepam, lorazepam) by inhibiting CYP2C19, necessitating dose adjustments (Chesney et al., 2020).

Antiepileptics

CBD increases serum levels of clobazam and its active metabolite, N-desmethylclobazam, via CYP2C19 inhibition, raising the risk of sedation and toxicity; monitoring of antiepileptic drug levels and liver function tests is recommended when coadministered with CBD (Chesney et al., 2020).

Antidepressants and antipsychotics

CBD inhibits CYP2D6, CYP2C19, and CYP3A4, increasing serum concentrations of many antidepressants (e.g., sertraline, citalopram, amitriptyline, paroxetine) and antipsychotics (e.g., risperidone, olanzapine, haloperidol), which can potentiate adverse effects such as QT prolongation, anticholinergic syndrome, and sedation (Chesney et al., 2020; Velayudhan et al., 2024). THC may also alter the metabolism and efficacy of these medications, sometimes reducing their effectiveness or increasing side effects (Beedham et al., 2020).

Statins, immunosuppressants, and other cardiovascular drugs

Cannabinoids may increase serum levels of statins (e.g., simvastatin, atorvastatin) via CYP3A4 inhibition, increasing the risk of toxicity (Chesney et al., 2020). CBD can increase the blood levels of immunosuppressants (e.g., tacrolimus, cyclosporine), leading to toxic levels and increased side effects, including the risk of severe infections (Chesney et al., 2020). Cannabis can interact with medications such as nifedipine, verapamil, diltiazem, and carvedilol, potentially increasing their effects and risk of adverse events (Beedham et al., 2020).

Over-the-counter and other medications

THC may increase the risk of gastrointestinal (GI) bleeds or ulcers when combined with nonsteroidal anti-inflammatory drugs (NSAIDs), and additive sedative effects may occur with antihistamines (Beedham et al., 2020). CBD can interfere with the metabolism of proton pump inhibitors (PPIs), potentially increasing their side effects (Chesney et al., 2020). Supplements such as St. John's Wort, melatonin, valerian root, and echinacea may interact with THC and CBD, altering their metabolism and effects (Beedham et al., 2020).

Polypharmacy and clinical implications

Polypharmacy is prevalent among older adults using cannabinoids, compounding the risk of drug-drug interactions and adverse events (Beedham et al., 2020; Velayudhan et al., 2024). A comprehensive medication review is essential before initiating cannabinoids in older adults, and clinicians should start with the lowest effective dose, titrate slowly, and monitor for signs of toxicity or reduced efficacy of concomitant medications (Beedham et al., 2020; Velayudhan et al., 2024). Regular reassessment of the medication regimen is recommended, especially when adding or discontinuing cannabinoids (Chesney et al., 2020).

Clinical recommendations

- Conduct a comprehensive medication review before initiating cannabinoids in older adults (Beedham et al., 2020).
- Start with the lowest effective dose, titrate slowly, and monitor for signs of toxicity or reduced efficacy of concomitant medications (Velayudhan et al., 2024).

- Regularly reassess the medication regimen, especially when adding or discontinuing cannabinoids (Chesney et al., 2020).
- Educate patients about the potential for drug interactions and the importance of reporting all medications and supplements they take (Beedham et al., 2020).

Recommended doses

The recommended daily dose of cannabis for older adults varies depending on the cannabinoid composition, patient tolerance, and clinical indication. Generally, a cautious approach is advised, starting with low doses and titrating slowly. For THC-predominant products, initial doses often range from 1 to 2.5 mg/day, typically administered at bedtime, with gradual increases of 1–2.5 mg every 2–3 days. Most older adults achieve therapeutic benefit at doses between 5 and 10 mg THC per day, while doses above 20–30 mg/day are associated with increased adverse events without additional efficacy. For CBD-predominant products, starting doses of 5 mg once or twice daily are standard, with titration by 5–10 mg every 2–3 days up to 40 mg/day as tolerated. If symptom control is inadequate, low-dose THC may be added cautiously (MacCallum and Russo, 2018; Bhaskar et al., 2021; Minerbi et al., 2019).

Different cannabis formulations influence dosing strategies due to variations in pharmacokinetics and onset of action:

- Oral and Sublingual Products: These have a slower onset (1–3 h) and a longer duration of action, requiring lower starting doses and slower titration to avoid accumulation and delayed adverse effects. CBD-predominant oils and tinctures are preferred for initial therapy due to better tolerability (MacCallum and Russo, 2018; Minerbi et al., 2019).
- Inhaled Cannabis (Smoking or Vaporization): Provides rapid onset (minutes) and shorter duration, allowing for more immediate dose adjustments. However, inhaled forms have higher variability in dosing and an increased risk of acute adverse effects, making them less suitable for frail older adults (MacCallum and Russo, 2018).
- Oromucosal Sprays (e.g., Nabiximols): Deliver a balanced THC:CBD ratio with moderate onset and duration, allowing for flexible dosing. Typical doses range from 6 to 8 sprays per day, with a maximum of 12 sprays daily (MacCallum and Russo, 2018).

Specific dosage guidelines for chronic conditions in older adults are limited due to a paucity of large, geriatric-focused clinical trials. However, consensus recommendations suggest:

- Chronic Pain: Start with low-dose CBD or balanced THC:CBD formulations, titrating slowly to effect. THC doses above 10 mg/day should be used cautiously due to increased neuropsychiatric risks (Bhaskar et al., 2021; Minerbi et al., 2019).

- Behavioral and Psychological Symptoms of Dementia: Use of synthetic THC or low-dose THC:CBD combinations may be considered, starting at very low doses (1 mg THC) with slow titration, while monitoring for sedation and cognitive effects (Shelef et al., 2016; Van Den Elsen et al., 2015).
- Sleep Disorders: Low-dose THC or CBD may improve sleep quality, but dosing should be individualized, starting low to minimize next-day sedation (Walsh et al., 2017).
- Appetite Stimulation: Dronabinol dosing for anorexia typically starts at 2.5 mg twice daily, titrated as needed, but evidence in older adults is limited (Beal et al., 1995).

Age-related physiological changes significantly affect cannabinoid metabolism and pharmacokinetics, necessitating dose adjustments in older adults:

- Reduced Hepatic Metabolism: Declines in cytochrome P450 enzyme activity (notably CYP2C9, CYP3A4) slow THC clearance, prolonging the half-life and increasing systemic exposure, which raises the risk of accumulation and toxicity (Abuhasira et al., 2019; Ahmed et al., 2014).
- Increased Volume of Distribution: Higher body fat content in seniors leads to greater THC sequestration and prolonged elimination half-life (5–13 days versus 1–3 days in younger adults), necessitating lower and less frequent dosing (Grotenhermen, 2003).
- Polypharmacy and Drug Interactions: Common in older adults, polypharmacy can alter cannabinoid metabolism and increase adverse effects, requiring careful medication review and dose titration (Velayudhan et al., 2024).

Safety considerations in dosing cannabis for older adults include:

- Neuropsychiatric Risks: Higher THC doses increase the risks of dizziness, confusion, delirium, and psychosis, especially in those with cognitive impairment or a psychiatric history. Starting low and titrating slowly mitigates these risks (Velayudhan et al., 2024; Scott et al., 2019).
- Fall Risk: THC-induced dizziness and impaired balance increase the risk of falls and fractures. Monitoring and dose adjustments are critical to minimize harm (Velayudhan et al., 2024).
- Cardiovascular Effects: THC can cause tachycardia, hypertension, and orthostatic hypotension, which may exacerbate underlying cardiovascular disease. Lower doses and cautious titration are advised (Grotenhermen, 2003).
- Drug Interactions: Cannabinoids interact with many medications metabolized by CYP450 enzymes, necessitating dose adjustments and close monitoring in polypharmacy (Chesney et al., 2020).
- Delayed Onset and Accumulation: Oral cannabinoids have a delayed onset and prolonged effects, increasing the risk of overdose if doses are escalated too quickly (MacCallum and Russo, 2018).

Summary table: Recommended dosing and safety considerations in older adults

Aspect	Recommendation/Consideration	References
Starting dose (THC)	1–2.5 mg/day, titrate slowly	(MacCallum and Russo, 2018; Minerbi et al., 2019)
Starting dose (CBD)	5 mg once or twice daily; titrate up to 40 mg/day.	(MacCallum and Russo, 2018; Bhaskar et al., 2021)
Formulation preference	CBD-predominant or balanced THC:CBD for initial therapy	(Minerbi et al., 2019)
Age-related metabolism	Reduced clearance, prolonged half-life, increased sensitivity	(Abuhasira et al., 2019; Ahmed et al., 2014)
Safety monitoring	Monitor cognition, falls, cardiovascular status, and drug interactions	(Velayudhan et al., 2024; Scott et al., 2019)
Regulatory considerations	Follow FDA-approved indications and state-specific guidelines	(Han et al., 2023; Bui et al., 2023)

References

Abuhasira, R., Ron, A., Sikorin, I., Novack, V., 2019. Medical cannabis for older patients—treatment protocol and initial results. J. Clin. Med. 2077-0383. 8 (11), 1819. doi:10.3390/jcm8111819.

Ahmed, A.I.A., van den Elsen, G.A.H., Colbers, A., van der Marck, M.A., Burger, D.M., Feuth, T.B., Rikkert, M.G.M.O., Kramers, C., 2014. Safety and pharmacokinetics of oral delta-9-tetrahydrocannabinol in healthy older subjects: a randomized controlled trial. Eur. Neuropsychopharmacol. 1873-7862. 24 (9), 1475–1482. Available from: www.elsevier.com/locate/euroneuro. doi:10.1016/j.euroneuro.2014.06.007.

Bar-Sela, G., Avisar, A., Batash, R., Schaffer, M., Goldberg, Y., Ben-Aharon, I., Haim, N., 2022. Cannabis in palliative oncology: a review of the evidence for its use in cancer pain, nausea, appetite, and sleep. Curr. Oncol. Rep. 24 (2), 183–192.

Beal, J.E., Olson, R., Laubenstein, L., Morales, J.O., Bellman, P., Yangco, B., Lefkowitz, L., Plasse, T.F., Shepard, K.V., 1995. Dronabinol as a treatment for anorexia associated with weight loss in patients with AIDS. J. Pain Symptom Manage. 0885-3924. 10 (2), 89–97. doi:10.1016/0885-3924(94)00117-4.

Beedham, W., Sbai, M., Allison, I., Coary, R., Shipway, D., 2020. Cannabinoids in the older person: a literature review. Geriatrics. 2308-3417. 5 (1), 2. doi:10.3390/geriatrics5010002.

Bhaskar, A., Bell, A., Boivin, M., Briques, W., Brown, M., Clarke, H., Cyr, C., Eisenberg, E., de Oliveira Silva, R.F., Frohlich, E., Georgius, P., Hogg, M., Horsted, T.I., MacCallum, C.A., Müller-Vahl, K.R., O'Connell, C., Sealey, R., Seibolt, M., Sihota, A., Smith, B.K., Sulak, D., Vigano, A., Moulin, D.E., 2021. Consensus recommendations on dosing and administration of medical cannabis to treat chronic pain: results of a modified Delphi process. J. Cannabis Res. 2522-5782. 3 (1). Available from: https://jcannabisresearch.biomedcentral.com/. doi:10.1186/s42238-021-00073-1.

Bui, K., Worster, B., Ryan, J., Pant, S., Kelly, E.L., 2023. Physicians' perspectives on their training for and role within Pennsylvania's Medical Cannabis Program. J. Am. Board Fam. Med. 1558-7118. 36 (4), 670–681. Available from: http://www.jabfm.org/lookup/doi/10.3122/jabfm.2022.220304R1. doi:10.3122/jabfm.2022.220304R1.

Chesney, E., Oliver, D., Green, A., Sovi, S., Wilson, J., Englund, A., Freeman, T.P., McGuire, P., 2020. Adverse effects of cannabidiol: a systematic review and meta-analysis of randomized clinical trials. Neuropsychopharmacology. 1740-634X. 45 (11), 1799–1806. Available from: http://www.nature.com/npp/index.html. doi:10.1038/s41386-020-0667-2.

Ginsburg, B.C., Hensler, J.G., 2022. Age-related changes in CB1 receptor expression and function and the behavioral effects of cannabinoid receptor ligands. Pharmacol. Biochem. Behav. 1873-5177. 213. Available from: www.elsevier.com/locate/pharmbiochembeh. doi:10.1016/j.pbb.2022.173339.

Grayson, L., Vines, B., Nichol, K., Szaflarski, J.P., 2018. An interaction between warfarin and cannabidiol, a case report. Epilepsy Behav. Case Rep. 2213-3232. 9, 10–11. Available from: http://www.journals.elsevier.com/epilepsy-and-behavior-case-reports/. doi:10.1016/j.ebcr.2017.10.001.

Grotenhermen, F., 2003. Pharmacokinetics and pharmacodynamics of cannabinoids. Clin. Pharmacokinet. 0312-5963. 42 (4), 327–360. doi:10.2165/00003088-200342040-00003.

Han, B.H., Brennan, J.J., Orozco, M.A., Moore, A.A., Castillo, E.M., 2023. Trends in emergency department visits associated with cannabis use among older adults in California, 2005–2019. J. Am. Geriatr. Soc. 1532-5415. 71 (4), 1267–1274. Available from: http://onlinelibrary.wiley.com/journal/10.1111/(ISSN)1532-5415. doi:10.1111/jgs.18180.

Han, B.H., Palamar, J.J., 2020. Trends in cannabis use among older adults in the United States, 2015–2018. JAMA Intern. Med. 180 (4), 609–611. doi:10.1001/jamainternmed.2019.7517.

Kepner, W.E., Han, B.H., Nguyen, D., Han, S.S., Lopez, F.A., Palamar, J.J., 2023. Past-month binge drinking and cannabis use among middle-aged and older adults in the United States, 2015–2019. Alcohol. 1873-6823. 107, 32–37. Available from: www.elsevier.com/locate/alcohol. doi:10.1016/j.alcohol.2022.07.006.

Koppel, B.S., Brust, J.C.M., Fife, T., Bronstein, J., Youssof, S., Gronseth, G., Gloss, D., 2014. Systematic review: efficacy and safety of medical marijuana in selected neurologic disorders. Neurology. 1526-632X. 82 (17), 1556–1563. Available from: http://www.neurology.org. doi:10.1212/WNL.0000000000000363.

Lo, L.A., Christiansen, A., Eadie, L., Strickland, J.C., Kim, D.D., Boivin, M., Barr, A.M., MacCallum, C.A., 2023. Cannabidiol-associated hepatotoxicity: a systematic review and meta-analysis. J. Intern. Med. 1365-2796. 293 (6), 724–752. Available from: http://onlinelibrary.wiley.com/journal/10.1111/(ISSN)1365-2796. doi:10.1111/joim.13627.

MacCallum, C.A., Russo, E.B., 2018. Practical considerations in medical cannabis administration and dosing. Eur. J. Intern. Med. 1879-0828. 49, 12–19. Available from: www.elsevier.com/locate/ejim. doi:10.1016/j.ejim.2018.01.004.

Marconi, A., Di Forti, M., Lewis, C.M., Murray, R.M., Vassos, E., 2016. Meta-analysis of the association between the level of cannabis use and risk of psychosis. Schizophr. Bull. 1745-1701. 42 (5), 1262–1269. Available from: http://schizophreniabulletin.oxfordjournals.org/. doi:10.1093/schbul/sbw003.

Minerbi, A., Häuser, W., Fitzcharles, M.A., 2019. Medical cannabis for older patients. Drugs Aging. 1179-1969. 36 (1), 39–51. Available from: http://rd.springer.com/journal/40266. doi:10.1007/s40266-018-0616-5.

National Academies of Sciences, Engineering, and Medicine, 2017. The Health Effects of Cannabis and Cannabinoids: The Current State of Evidence and Recommendations for Research. The National Academies Press, Washington, DC.

National Survey on Drug Use and Health (NSDUH), 2023. Key Substance Use and Mental Health Indicators in the United States: Results from the 2023 National Survey on Drug

Use and Health. Substance Abuse and Mental Health Services Administration, Rockville, MD.

Nidadavolu, P., Bilkei-Gorzo, A., Effah, F., Leidmaa, E., Schürmann, B., Berger, M., Bindila, L., Schmid, M., Lutz, B., Zimmer, A., Bailey, A., 2022. Dynamic changes in the endocannabinoid system during the aging process: focus on the middle-age crisis. Int. J. Mol. Sci. 1422-0067. 23 (18), 10254. doi:10.3390/ijms231810254.

Scott, E.P., Brennan, E., Benitez, A., 2019. A systematic review of the neurocognitive effects of cannabis use in older adults. Curr. Addict. Rep. 2196-2952. 6 (4), 443–455. Available from: https://link.springer.com/journal/40429. doi:10.1007/s40429-019-00285-9.

Shelef, A., Barak, Y., Berger, U., Paleacu, D., Tadger, S., Plopsky, I., Baruch, Y., 2016. Safety and efficacy of medical cannabis oil for behavioral and psychological symptoms of dementia: an-open label, add-on, pilot study. J. Alzheimers Dis. 1875-8908. 51 (1), 15–19. Available from: www.iospress.nl/journal/journal-of-alzheimers-disease/. doi:10.3233/JAD-150915.

Sznitman, S.R., Vulfsons, S., Meiri, D., Weinstein, G., 2020. Medical cannabis and insomnia in older adults with chronic pain: a cross-sectional study. BMJ Support. Palliat. Care. 2045-4368. 10 (4), 415–420. Available from http://www.spcare.bmj.com/. doi:10.1136/bmjspcare-2019-001938.

Tramèr, M.R., Carroll, D., Campbell, F.A., Reynolds, D.J.M., Moore, R.A., McQuay, H.J., 2001. Cannabinoids for control of chemotherapy induced nausea and vomiting: quantitative systematic review. Br. Med. J. 0959-8146. 323 (7303), 16–21.

Van Den Elsen, G.A.H., Ahmed, A.I.A., Verkes, R.J., Kramers, C., Feuth, T., Rosenberg, P.B., Marck, M.A.V.D., Rikkert, M.G.M.O., 2015. Tetrahydrocannabinol for neuropsychiatric symptoms in dementia: a randomized controlled trial. Neurology. 1526-632X. 84 (23), 2338–2346. Available from: http://www.neurology.org. doi:10.1212/WNL.0000000000001675.

Velayudhan, L., Pisani, S., Dugonjic, M., McGoohan, K., Bhattacharyya, S., 2024. Adverse events caused by cannabinoids in middle aged and older adults for all indications: a meta-analysis of incidence rate difference. Age Ageing. 0002-0729. 53 (11). doi:10.1093/ageing/afae261.

Walsh, Z., Gonzalez, R., Crosby, K., Thiessen, M.S., Carroll, C., Bonn-Miller, M.O., 2017. Medical cannabis and mental health: a guided systematic review. Clin. Psychol. Rev. 1873-7811. 51, 15–29. Available from: www.elsevier.com/locate/clinpsychrev. doi:10.1016/j.cpr.2016.10.002.

Ware, M.A., Wang, T., Shapiro, S., Collet, J.P., 2015. Cannabis for the management of pain: assessment of safety study (COMPASS). J. Pain. 1528-8447. 16 (12), 1233–1242. Available from: http://www.elsevier.com/inca/publications/store/6/2/3/1/4/7/index.htt. doi:10.1016/j.jpain.2015.07.014.

Whiting, P.F., Wolff, R.F., Deshpande, S., Di Nisio, M., Duffy, S., Hernandez, A.V., Keurentjes, J.C., Lang, S., Misso, K., Ryder, S., Schmidlkofer, S., Westwood, M., Kleijnen, J., 2015. Cannabinoids for medical use: a systematic review and meta-analysis. JAMA. 1538-3598. 313 (24), 2456–2473. Available from: http://jama.jamanetwork.com/article.aspx?articleid=2338251. doi:10.1001/jama.2015.6358.

CHAPTER

Medicolegal considerations of medical marijuana

19

Hemant Kalia[a] and Alaa Abd-Elsayed[b]

[a]*C.R.I.S.P. Center for Research & Innovation in Spine & Pain, Rochester, NY, United States*
[b]*Departments of Anesthesiology and Chronic Pain Physician, University of Wisconsin-Madison, Madison, WI, United States*

Introduction

The landscape of medical marijuana regulation presents one of the most complex medicolegal challenges in contemporary healthcare. As of 2024, the fundamental tension between federal prohibition and state-level authorization continues to create a patchwork of conflicting laws that healthcare providers must navigate carefully. This chapter examines the historical development of cannabis regulation, existing federal and state legal frameworks, and the critical medicolegal considerations that physicians and healthcare institutions must address when recommending or discussing medical marijuana with patients.

Historical development and current federal framework

The Controlled Substances Act and federal prohibition

The modern federal approach to cannabis regulation began with the Controlled Substances Act (CSA) of 1970, which consolidated all prior federal laws related to potentially abused substances (FDA, n.d.; Drug Enforcement Administration, n.d.). Under the CSA, cannabis was classified as a Schedule I substance, defined as having a high potential for abuse, no currently accepted medical use in the United States, and a lack of accepted safety for use under medical supervision. This classification places cannabis alongside substances such as heroin, lysergic acid diethylamide, and methylenedioxymethamphetamine (FDA, n.d.).

The Schedule I classification has profound implications for medical practice and research. It prevents physicians from formally prescribing cannabis, limits research opportunities to federally approved programs, and creates barriers to clinical trials that would typically inform evidence-based medicine. Substances classified as Schedule I can only be legally possessed or dispensed as part of federally approved research grants, with all involved parties required to obtain Schedule I drug-specific registration (Gregorio, 2014).

Cannabinoids for Chronic Pain. https://doi.org/10.1016/B978-0-443-36773-1.00018-0

The 2018 Farm Bill and hemp differentiation

A significant development occurred with the passage of the Agricultural Improvement Act of 2018, commonly known as the Farm Bill, which created an important legal distinction within the cannabis plant family. The Farm Bill legalized hemp at the federal level, defining it specifically as the *Cannabis sativa* L. plant with a delta-9-tetrahydrocannabinol (THC) concentration of 0.3% or less on a dry weight basis. This legislation effectively removed hemp from the CSA, making hemp-derived products, including cannabidiol (CBD), legal under federal law (Hudak, 2018).

However, the Farm Bill explicitly preserved the Food and Drug Administration's (FDA) authority to regulate products containing cannabis or cannabis-derived compounds under existing federal food, drug, and cosmetic laws (FDA, n.d.). The FDA has since approved one cannabis-derived medication: Epidiolex, which contains purified CBD for treating seizures associated with specific rare epilepsy syndromes (FDA, n.d.). Despite this regulatory framework, the CBD market remains largely unregulated, leading to concerns about product quality and safety (FDA, n.d.).

Drug Enforcement Administration's rescheduling efforts

The most significant development in federal cannabis policy began in October 2022, when President Joe Biden asked the Department of Health and Human Services (HHS) and the Drug Enforcement Administration (DEA) to review marijuana's scheduling under the CSA. Following this directive, HHS conducted a comprehensive scientific and medical evaluation and recommended, in August 2023, that marijuana be transferred from Schedule I to Schedule III.

In May 2024, the DEA published a Notice of Proposed Rulemaking, proposing to reclassify cannabis from Schedule I to Schedule III. This proposed rescheduling represents the agency's most significant policy change in over 50 years. Schedule III substances are defined as having a moderate-to-low potential for physical and psychological dependence and the accepted medical use.

If finalized, rescheduling to Schedule III would allow businesses to claim federal tax deductions currently prohibited under 26 U.S.C. § 280E, potentially making research easier to conduct. It could eventually allow pharmacies to dispense marijuana-based medications, pending FDA approval (Stains et al., 2025).

State-level medical cannabis programs: A complex patchwork

Existing state authorization status

The state-level medical cannabis landscape has evolved dramatically since California became the first state to effectively legalize medical cannabis in 1996 with the passage of Proposition 215. As of 2024, medical cannabis use is legal in 40 states, four out of five permanently inhabited US territories, and the District of Columbia.

An additional nine states have more restrictive laws limiting THC content to allow access to products rich in CBD (Hudak, 2018).

This expansion reflects growing public acceptance of medical cannabis. However, the regulatory frameworks vary significantly among jurisdictions, creating a complex legal environment that healthcare providers must navigate.

Variation in state regulatory approaches

State medical cannabis programs demonstrate remarkable variation in several key areas. **Qualifying conditions** differ substantially between jurisdictions, ranging from specific lists of approved conditions to broader provisions allowing physician discretion. For example, some states limit medical cannabis to patients with cancer, epilepsy, and other severe conditions, while others include chronic pain, anxiety, and posttraumatic stress disorder (Stains et al., 2025; Josefson, 2002).

Product regulations also vary significantly. States differ in approved forms of medical cannabis, with some allowing only oils and tinctures, while others permit smoking or vaporization. Potency limits, testing requirements, and packaging standards are inconsistent across jurisdictions.

Physician requirements present another area of variation. Most states require healthcare providers to complete specific educational courses before recommending medical cannabis, but the duration and content of these requirements differ. Some states mandate ongoing education or limit the number of patients a physician may certify (Crowley et al., 2024).

Registration and oversight systems

States employ various oversight mechanisms for medical cannabis programs. Patient registration systems are standard, requiring patients to obtain identification cards after receiving physician recommendations. These systems often include background checks, fees (ranging from nominal amounts to several hundred dollars), and renewal requirements.

Healthcare provider registration varies similarly. Some states maintain public directories of registered physicians, while others protect physicians' privacy. Disciplinary mechanisms exist in all programs, with state medical boards retaining authority to investigate complaints related to medical cannabis recommendations.

Critical medicolegal considerations for healthcare providers

The prescription-recommendation distinction

One of the most fundamental medicolegal concepts in medical cannabis practice is the distinction between **prescribing** and **recommending**. Under federal law, physicians cannot prescribe cannabis because it remains a Schedule I controlled substance. However, the landmark 2002 Ninth Circuit Court of Appeals decision in *Conant v.*

Walters established that physicians have a First Amendment right to discuss medical cannabis with patients and make recommendations based on sincere medical judgment.

This distinction has practical implications for clinical practice. When recommending medical cannabis, physicians are not ordering specific dosages or formulations as they would with FDA-approved medications. Instead, they are certifying that a patient has a qualifying condition and that the potential benefits of cannabis use outweigh the risks. This gatekeeping function requires careful clinical judgment but does not involve the same level of prescriptive control typical in pharmaceutical medicine.

Federal legal protections for healthcare providers

Healthcare providers receive several layers of federal protection when appropriately recommending medical cannabis. The Medical Marijuana and Cannabidiol Research Expansion Act, signed into law in December 2022, explicitly permits state-licensed physicians to discuss the currently known potential harms and benefits of marijuana and marijuana derivatives with patients or their legal guardians. This federal legislation reinforces the First Amendment protections established in *Conant v. Walters*.

However, these protections are contingent upon physicians acting within the bounds of **sincere medical judgment**. Physicians can face criminal liability if recommendations are not based on legitimate medical evaluation or if they fail to comply with state law requirements. Through prosecutorial discretion, the federal government has indicated that it will not pursue actions against appropriately certifying physicians in states with legal medical cannabis programs.

Professional liability and malpractice considerations

Medical malpractice liability in the context of medical cannabis recommendations presents unique challenges. While no reported malpractice cases related to cannabis recommendations have established precedent, legal experts emphasize that physicians remain subject to professional liability standards. A successful malpractice claim would likely need to demonstrate that the physician failed to meet the standard of care in making the recommendation.

Key risk mitigation strategies include thorough documentation of the clinical rationale for recommendations, comprehensive patient education about risks and benefits, and appropriate follow-up care. Physicians should document the failure of conventional treatments, the review of patient medical history, the discussion of contraindications, and the ongoing monitoring of patient response. Given the lack of an established standard of care for medical cannabis, many experts recommend following models used for controlled substance prescribing, particularly opioid prescribing protocols.

Insurance coverage presents a significant concern for physicians recommending medical cannabis. Many traditional medical malpractice insurance policies exclude coverage for activities involving non-FDA-approved substances or federal law violations. Physicians should explicitly discuss medical cannabis activities with their malpractice carriers and obtain written confirmation of coverage. Specialized cannabis malpractice insurance products have emerged to address this coverage gap.

Hospital and institutional policies

Healthcare institutions face challenges regarding medical cannabis recommendations. Hospitals and health systems participating in Medicare and Medicaid programs must certify compliance with all federal laws, creating potential conflicts with state medical cannabis activities. Many institutions prohibit on-site medical cannabis consultations and may restrict affiliated physicians from recommending cannabis, even in off-site practices.

Risk managers should establish protocols addressing physician privileges, documentation requirements, and institutional liability. These protocols should consider state-specific laws, accreditation requirements, and federal compliance obligations. Some institutions adopt policies requiring physicians to provide medical cannabis services only through separate, unaffiliated practices.

Patient safety and clinical guidelines

The evidence base for medical cannabis efficacy remains limited for many conditions commonly cited by patients. According to a comprehensive review by the American Society of Pain and Neuroscience, cancer-related pain has the strongest evidence base for medical cannabis treatment. However, this evidence yields only a Grade D recommendation due to limitations in supporting clinical trials (FDA, n.d.). Neuropathic pain has Grade C evidence, while other common indications, such as chronic musculoskeletal pain, have even weaker evidentiary support (Strand et al., 2023).

Healthcare providers must carefully weigh limited efficacy evidence against potential risks. Common adverse effects include dizziness, fatigue, and cognitive disturbances. Cannabis use may interact with other medications, particularly those with narrow therapeutic margins. Providers should be especially cautious with vulnerable populations, including pregnant patients, adolescents, and individuals with psychiatric conditions or substance use disorders.

Professional board regulation and discipline

State medical boards retain authority to regulate physicians who recommend medical cannabis, and disciplinary actions provide insight into regulatory expectations. Common violations leading to discipline include:

- **Inadequate patient evaluation**: Failing to establish proper physician-patient relationships or conducting insufficient examinations before making recommendations.

- **Inappropriate certification patterns**: Writing excessive numbers of recommendations without adequate clinical justification.
- **Financial conflicts of interest**: Maintaining improper financial relationships with dispensaries or cultivation facilities.
- **Documentation deficiencies**: Failing to maintain adequate medical records or comply with state reporting requirements.

The available enforcement data suggest serious disciplinary actions remain rare when physicians follow state requirements and maintain appropriate clinical standards. However, regulatory scrutiny is increasing as programs mature and patient numbers grow.

Clinical practice recommendations

Establishing appropriate clinical protocols

Healthcare providers considering medical cannabis recommendations should develop systematic clinical protocols addressing patient evaluation, risk assessment, and ongoing care (Crowley et al., 2024). The **initial evaluation** should include a comprehensive medical history, physical examination, a review of previous treatments, and a contraindication assessment. Providers should document the failure or inadequacy of conventional therapies and establish the medical rationale for considering cannabis.

Patient education must address potential benefits and risks, including impairment effects that may impact driving or work performance. Providers should discuss proper storage (significant for preventing pediatric exposures), possible drug interactions, and signs of problematic use. Written educational materials and signed acknowledgment forms can help document this critical communication.

Ongoing monitoring and follow-up

Medical cannabis recommendations require ongoing clinical oversight, similar to other therapeutic interventions. Regular follow-up appointments should assess treatment effectiveness, adverse effects, and any changes in clinical status. Providers should monitor for signs of cannabis use disorder, particularly in patients with personal or family histories of substance abuse.

Documentation requirements vary by state but generally include initial certification details, patient response monitoring, and any modifications to recommendations. Some states require periodic recertification, creating natural opportunities for clinical reassessment.

Integration with existing medical care

Medical cannabis should be integrated into comprehensive treatment plans rather than considered as an isolated therapy. Coordination with other healthcare providers is essential, particularly given that primary care physicians may not be comfortable with medical cannabis, but specialists increasingly report using it in their practices. Clear communication with other healthcare team members helps ensure comprehensive, coordinated care.

Regulatory compliance and risk management

State-specific regulatory adherence

Healthcare providers must thoroughly understand their state's requirements, as legal frameworks vary significantly between jurisdictions. Key areas requiring attention include:

- **Educational requirements**: Most states mandate specific training before physicians can recommend medical cannabis.
- **Registration obligations**: Provider enrollment in state programs, including background checks and fees.
- **Qualifying conditions**: Understanding which conditions allow for legal recommendations in each jurisdiction.
- **Documentation standards**: Meeting state-specific record-keeping and reporting requirements
- **Patient limits**: Some states restrict the number of patients physicians may certify.

Interstate practice considerations

For healthcare providers practicing across state lines, medical cannabis recommendations present additional complexity. Telemedicine regulations vary significantly, and some states restrict out-of-state physicians from providing cannabis recommendations. Providers must ensure compliance with regulations in each jurisdiction where they practice or offer telehealth services.

Emerging legal trends

The medical cannabis legal landscape continues to evolve rapidly. The existing trends include expanded qualifying condition lists, reduced regulatory barriers, and increased integration with mainstream medical practice. However, providers should also be aware of potential regulatory tightening, as some jurisdictions have implemented stricter oversight in response to program growth.

Federal rescheduling, if completed, would likely trigger additional regulatory changes at both federal and state levels. Healthcare providers should stay informed about these developments through professional organizations, legal counsel, and state regulatory agencies.

Conclusion and future directions

The medicolegal landscape for medical marijuana continues to evolve as state and federal policies adapt to changing scientific understanding and public opinion. Healthcare providers face the challenge of navigating complex, sometimes contradictory legal frameworks while providing appropriate patient care based on limited but growing evidence.

Key principles for safe, legal practice include a thorough understanding of applicable state laws, appropriate clinical protocols, comprehensive documentation, adequate insurance coverage, and ongoing education about legal and clinical developments. The distinction between prescription and recommendation remains fundamental, with First Amendment protections providing essential safeguards for appropriate physician-patient communication.

As the evidence base expands and regulatory frameworks mature, medical cannabis may become more integrated into mainstream medical practice. However, the current environment requires careful attention to legal compliance, professional standards, and patient safety. Healthcare providers should approach medical cannabis with the same clinical rigor applied to other therapeutic interventions, while remaining sensitive to the unique legal and social considerations that distinguish this area of practice.

The ongoing federal rescheduling process represents a potential watershed moment that could significantly alter the medical cannabis landscape. Regardless of its outcome, healthcare providers must remain committed to evidence-based practice, regulatory compliance, and patient welfare as this dynamic field develops.

Healthcare institutions and individual providers are strongly encouraged to seek legal counsel familiar with medical cannabis law in their jurisdictions, maintain regular communication with professional liability insurers, and participate in continuing education to stay current with rapidly evolving legal and clinical standards. Only through such diligent attention to both clinical excellence and legal compliance can the medical community appropriately serve patients seeking medical cannabis treatment while protecting professional interests and advancing the field responsibly.

References

Crowley, R., Cline, K., Hilden, D., Beachy, M., Hilden, D., Beachy, M., Carney, J.K., Foster, H., Hoeg, A., Hollon, M., Jumper, C., Mascarenhas, S.J., Sagar, A., Silberger, J.R., Slocum, J., Tan, M., Yousef, E.A., 2024. Regulatory framework for cannabis: a position paper from the American College of Physicians. Ann. Intern. Med. 0003-4819. 177 (8), 1104–1105. doi:10.7326/m24-0638.

Drug Enforcement Administration, n.d. The Controlled Substances Act. Available at: https://www.dea.gov/drug-information/csa.

FDA, n.d. Regulation of Cannabis and Cannabis-Derived Products, Including Cannabidiol (CBD). Available from: https://www.fda.gov/news-events/public-health-focus/fda-regulation-cannabis-and-cannabis-derived-products-including-cannabidiol-cbd.

Gregorio, J., 2014. Physicians, medical marijuana, and the law. Virt. Mentor. 1937-7010. 16 (9), 732–738. Available from: https://journalofethics.ama-assn.org/sites/journalofethics.ama-assn.org/files/2018-05/stas1-1409.pdf. doi:10.1001/virtualmentor.2014.16.9.hlaw1-1409.

Hudak, J., 2018. The Farm Bill, Hemp Legalization, and the Status of CBD: An Explainer, Brookings.

Josefson, D., 2002. US government cannot revoke licences of doctors who recommend marijuana. BMJ. 1468-5833. 325 (7372), 1058. doi:10.1136/bmj.325.7372.1058/e.

Stains, E.L., Kennalley, A.L., Tian, M., Boehnke, K.F., Kraus, C.K., Piper, B.J., 2025. Medical cannabis in the United States: comparing 2017 and 2024 state qualifying conditions to the 2017 National Academies of Sciences Report. Mayo Clin. Proc. Innov. Qual. Outcomes 9 (2), 100590. PMID: 40066150; PMCID: PMC11891685. doi:10.1016/j.mayocpiqo.2025.100590.

Strand, N., D'souza, R.S., Karri, J., Kalia, H., Weisbein, J., Kassa, B.J., Hussain, N., Chitneni, A., Budwany, R.R., Hagedorn, J., Pope, J.E., Deer, T.R., Sayed, D., Abd-Elsayed, A., 2023. Medical cannabis: a review from the American Society of Pain and Neuroscience. J. Pain Res. 1178-7090. 16, 4217–4228. Available from:: https://www.dovepress.com/getfile.php?fileID=95077. doi:10.2147/JPR.S425862.

Index

Note: Page numbers followed by "*f*" indicate figures, "*t*" indicate tables.

D

E

F

G

H

I

L

M

N

Printed and bound by CPI Group (UK) Ltd, Croydon, CR0 4YY
14/09/2026
02216408-0011